11–19 PROGRESSION

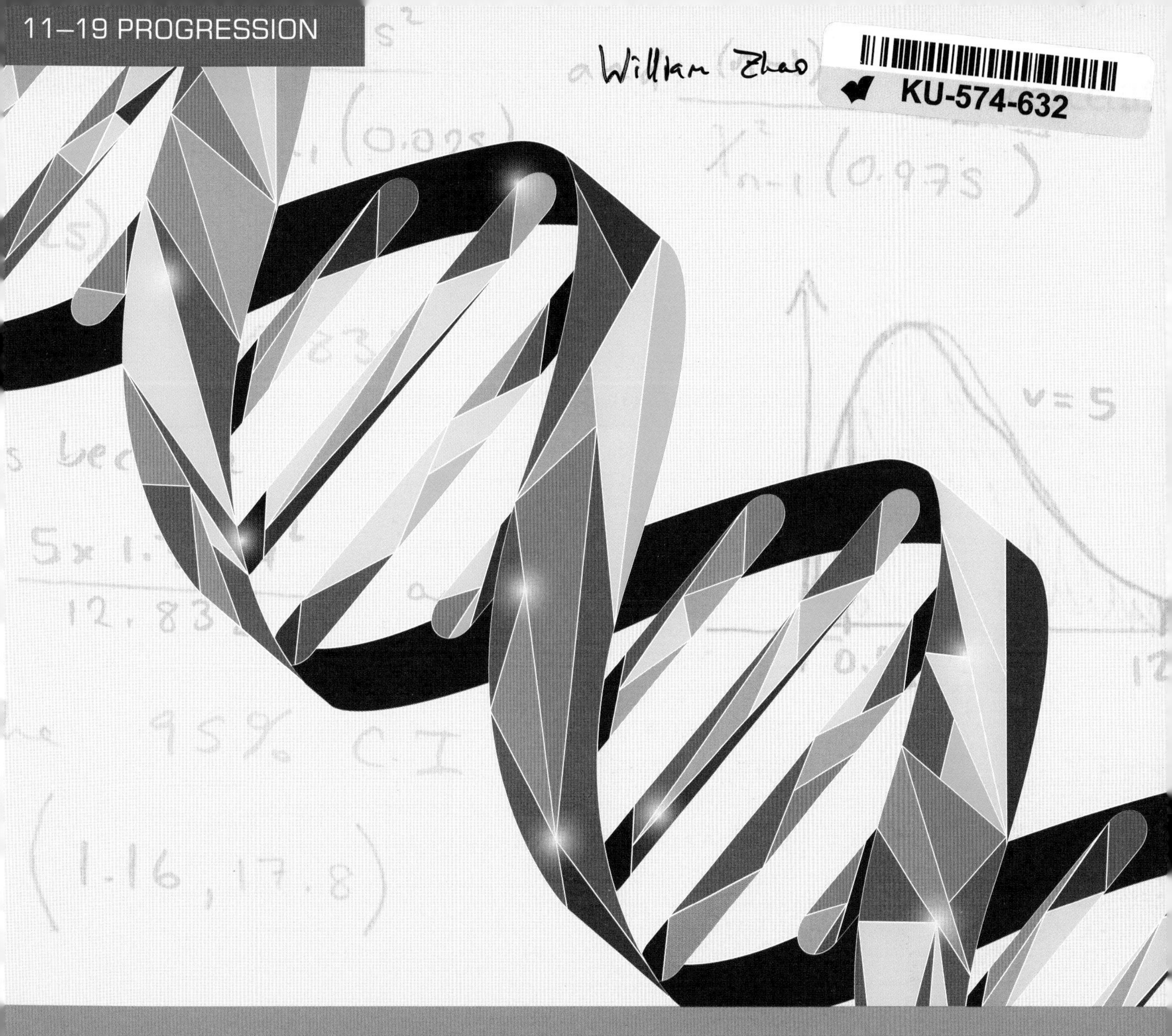

Edexcel AS and A level Further Mathematics

Further Statistics 1

FS1

Series Editor: Harry Smith

Authors: Greg Attwood, Tom Begley, Ian Bettison, Alan Clegg, Gill Dyer, Jane Dyer, John Kinoulty, Guilherme Frederico Lima, Harry Smith

Published by Pearson Education Limited, 80 Strand, London WC2R 0RL.

www.pearsonschoolsandfecolleges.co.uk

Copies of official specifications for all Pearson qualifications may be found on the website: qualifications.pearson.com

Edited by Tech-Set Ltd, Gateshead
Typeset by Tech-Set Ltd, Gateshead

Cover illustration Marcus@kja-artists

First published 2017

20 19 18 17
10 9 8 7 6 5 4 3 2 1

British Library Cataloguing in Publication Data
A catalogue record for this book is available from the British Library

ISBN 978 1 292 18337 4

Printed in the UK by Bell and Bain Ltd, Glasgow

Acknowledgements
The authors and publisher would like to thank the following for their kind permission to reproduce their photographs:

(Key: b-bottom; c-centre; l-left; r-right; t-top)

123RF: Artem Kevorkov 019, Somsak Sudthangtum 146, 173r; **Alamy Stock Photo:** Cosmo Condina Stock Market 001, Carolyn Jenkins 043, RichardBakerNews 076; **Shutterstock:** Harvepino 058, 173l, Leigh Prather 091, 173cl, Monkey Business Images128, 173cr

A note from the publisher
In order to ensure that this resource offers high-quality support for the associated Pearson qualification, it has been through a review process by the awarding body. This process confirms that this resource fully covers the teaching and learning content of the specification or part of a specification at which it is aimed. It also confirms that it demonstrates an appropriate balance between the development of subject skills, knowledge and understanding, in addition to preparation for assessment.

Endorsement does not cover any guidance on assessment activities or processes (e.g. practice questions or advice on how to answer assessment questions), included in the resource nor does it prescribe any particular approach to the teaching or delivery of a related course.

While the publishers have made every attempt to ensure that advice on the qualification and its assessment is accurate, the official specification and associated assessment guidance materials are the only authoritative source of information and should always be referred to for definitive guidance.

Pearson examiners have not contributed to any sections in this resource relevant to examination papers for which they have responsibility.

Examiners will not use endorsed resources as a source of material for any assessment set by Pearson.

Endorsement of a resource does not mean that the resource is required to achieve this Pearson qualification, nor does it mean that it is the only suitable material available to support the qualification, and any resource lists produced by the awarding body shall include this and other appropriate resources.

Pearson has robust editorial processes, including answer and fact checks, to ensure the accuracy of the content in this publication, and every effort is made to ensure this publication is free of errors. We are, however, only human, and occasionally errors do occur. Pearson is not liable for any misunderstandings that arise as a result of errors in this publication, but it is our priority to ensure that the content is accurate. If you spot an error, please do contact us at resourcescorrections@pearson.com so we can make sure it is corrected.

Contents

● = A level only

Overarching themes iv
Extra online content vi

1 Discrete random variables 1
1.1 Expected value of a discrete random variable 2
1.2 Variance of a discrete random variable 5
1.3 Expected value and variance of a function of X 7
1.4 Solving problems involving random variables 11
Mixed exercise 1 14

2 Poisson distributions 19
2.1 The Poisson distribution 20
2.2 Modelling with the Poisson distribution 23
2.3 Adding Poisson distributions 27
2.4 Mean and variance of a Poisson distribution 30
2.5 Mean and variance of the binomial distribution 32
2.6 Using the Poisson distribution to approximate the binomial distribution 34
Mixed exercise 2 38

● **3 Geometric and negative binomial distributions** 43
● **3.1** The geometric distribution 44
● **3.2** Mean and variance of a geometric distribution 47
● **3.3** The negative binomial distribution 49
● **3.4** Mean and variance of the negative binomial distribution 52
Mixed exercise 3 55

4 Hypothesis testing 58
4.1 Testing for the mean of a Poisson distribution 59
4.2 Finding critical regions for a Poisson distribution 62
● **4.3** Hypothesis testing for the parameter p of a geometric distribution 66
● **4.4** Finding critical regions for a geometric distribution 69
Mixed exercise 4 72

● **5 Central limit theorem** 76
● **5.1** The central limit theorem 77
● **5.2** Applying the central limit theorem to other distributions 80
Mixed exercise 5 82

Review exercise 1 85

6 Chi-squared tests 91
6.1 Goodness of fit 92
6.2 Degrees of freedom and the chi-squared family of distributions 96
6.3 Testing a hypothesis 99
6.4 Testing the goodness of fit with discrete data 103
6.5 Using contingency tables 113
● **6.6** Apply goodness-of-fit tests to geometric distributions 119
Mixed exercise 6 122

● **7 Probability generating functions** 128
● **7.1** Probability generating functions 129
● **7.2** Probability generating functions of standard distributions 132
● **7.3** Mean and variance of a distribution 135
● **7.4** Sums of independent random variables 139
Mixed exercise 7 143

● **8 Quality of tests** 146
● **8.1** Type I and Type II errors 147
● **8.2** Finding Type I and Type II errors using the normal distribution 153
● **8.3** Calculate the size and power of a test 157
● **8.4** The power function 162
Mixed exercise 8 167

Review exercise 2 173

Exam-style practice: AS 180

● **Exam-style practice: A level** 182

Appendix 185

Answers 193

Index 213

Overarching themes

The following three overarching themes have been fully integrated throughout the Pearson Edexcel AS and A level Mathematics series, so they can be applied alongside your learning and practice.

1. Mathematical argument, language and proof

- Rigorous and consistent approach throughout
- Notation boxes explain key mathematical language and symbols
- Dedicated sections on mathematical proof explain key principles and strategies
- Opportunities to critique arguments and justify methods

2. Mathematical problem solving

- Hundreds of problem-solving questions, fully integrated into the main exercises
- Problem-solving boxes provide tips and strategies
- Structured and unstructured questions to build confidence
- Challenge boxes provide extra stretch

The Mathematical Problem-solving cycle

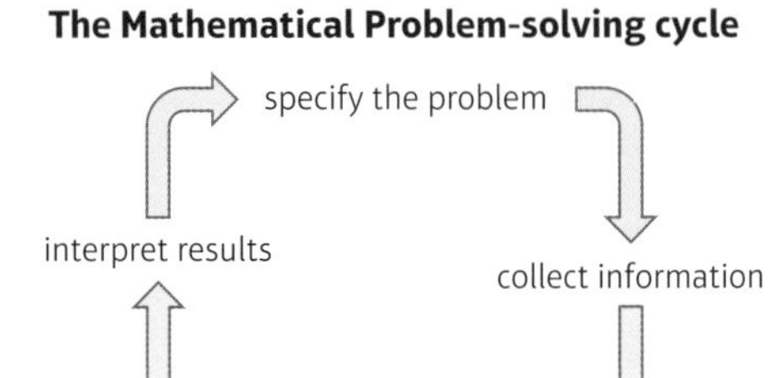

3. Mathematical modelling

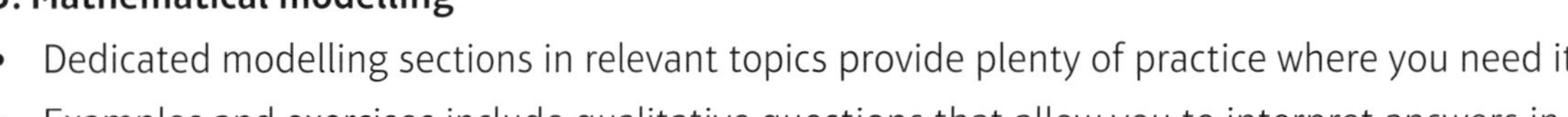

- Dedicated modelling sections in relevant topics provide plenty of practice where you need it
- Examples and exercises include qualitative questions that allow you to interpret answers in the context of the model
- Dedicated chapter in Statistics & Mechanics Year 1/AS explains the principles of modelling in mechanics

Finding your way around the book

Access an online digital edition using the code at the front of the book.

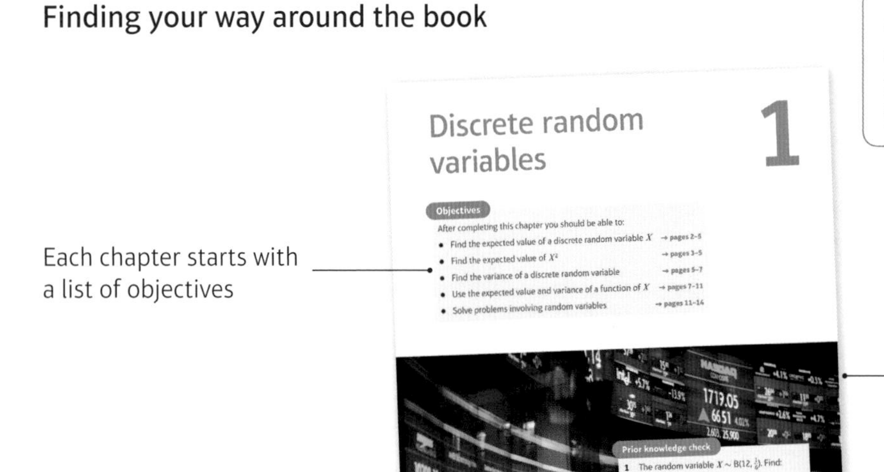

Each chapter starts with a list of objectives

The real world applications of the maths you are about to learn are highlighted at the start of the chapter with links to relevant questions in the chapter

The *Prior knowledge check* helps make sure you are ready to start the chapter

A level content is clearly flagged

Exercises are packed with exam-style questions to ensure you are ready for the exams

Challenge boxes give you a chance to tackle some more difficult questions

Each section begins with explanation and key learning points

Each chapter ends with a *Mixed exercise* and a *Summary of key points*

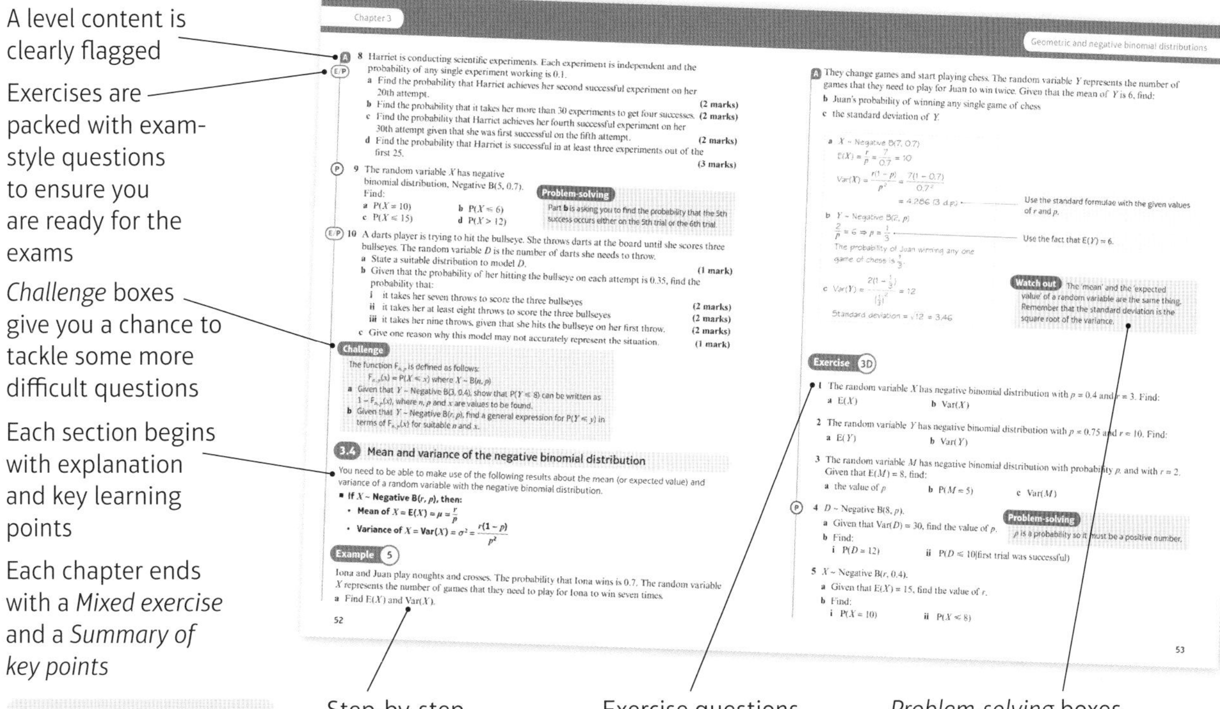
Chapter 3

Geometric and negative binomial distributions

8 Harriet is conducting scientific experiments. Each experiment is independent and the probability of any single experiment working is 0.1.
a Find the probability that Harriet achieves her second successful experiment on her 20th attempt. (2 marks)
b Find the probability that it takes her more than 30 experiments to get four successes. (2 marks)
c Find the probability that Harriet achieves her fourth successful experiment on her 30th attempt given that she was first successful on the fifth attempt. (2 marks)
d Find the probability that Harriet is successful in at least three experiments out of the first 25. (3 marks)

9 The random variable X has negative binomial distribution, Negative B(5, 0.7).
Find:
a $P(X = 10)$ b $P(X \leqslant 6)$
c $P(X \leqslant 15)$ d $P(X > 12)$

Problem-solving
Part **b** is asking you to find the probability that the 5th success occurs either on the 5th trial or the 6th trial.

10 A darts player is trying to hit the bullseye. She throws darts at the board until she scores three bullseyes. The random variable D is the number of darts she needs to throw.
a State a suitable distribution to model D. (1 mark)
b Given that the probability of her hitting the bullseye on each attempt is 0.35, find the probability that:
i it takes her seven throws to score the three bullseyes (2 marks)
ii it takes her at least eight throws to score the three bullseyes (2 marks)
iii it takes her nine throws, given that she hits the bullseye on her first throw. (2 marks)
c Give one reason why this model may not accurately represent the situation. (1 mark)

Challenge
The function $F_{n,p}$ is defined as follows:
$F_{n,p}(x) = P(X \leqslant x)$ where $X \sim B(n, p)$
a Given that $Y \sim$ Negative B(3, 0.4), show that $P(Y \leqslant 8)$ can be written as $1 - F_{n,p}(x)$, where n, p and x are values to be found.
b Given that $Y \sim$ Negative B(r, p), find a general expression for $P(Y \leqslant y)$ in terms of $F_{n,p}(x)$ for suitable n and x.

3.4 Mean and variance of the negative binomial distribution

You need to be able to make use of the following results about the mean (or expected value) and variance of a random variable with the negative binomial distribution.
■ If $X \sim$ Negative B(r, p), then:
• Mean of $X = E(X) = \mu = \frac{r}{p}$
• Variance of $X = Var(X) = \sigma^2 = \frac{r(1-p)}{p^2}$

Example 5

Iona and Juan play noughts and crosses. The probability that Iona wins is 0.7. The random variable X represents the number of games that they need to play for Iona to win seven times.
a Find E(X) and Var(X).

52

They change games and start playing chess. The random variable Y represents the number of games that they need to play for Juan to win twice. Given that the mean of Y is 6, find:
b Juan's probability of winning any single game of chess
c the standard deviation of Y.

a $X \sim$ Negative B(7, 0.7)
$E(X) = \frac{r}{p} = \frac{7}{0.7} = 10$
$Var(X) = \frac{r(1-p)}{p^2} = \frac{7(1-0.7)}{0.7^2}$
$= 4.286$ (3 d.p.)
Use the standard formulae with the given values of r and p.

b $Y \sim$ Negative B(2, p)
$\frac{2}{p} = 6 \Rightarrow p = \frac{1}{3}$
The probability of Juan winning any one game of chess is $\frac{1}{3}$.
Use the fact that E(Y) = 6.

c $Var(Y) = \frac{2(1-\frac{1}{3})}{(\frac{1}{3})^2} = 12$
Standard deviation $= \sqrt{12} = 3.46$

Watch out The mean and the expected value of a random variable are the same thing. Remember that the standard deviation is the square root of the variance.

Exercise 3D

1 The random variable X has negative binomial distribution with $p = 0.4$ and $r = 3$. Find:
a E(X) b Var(X)

2 The random variable Y has negative binomial distribution with $p = 0.75$ and $r = 10$. Find:
a E(Y) b Var(Y)

3 The random variable M has negative binomial distribution with probability p, and with $r = 2$. Given that E(M) = 8, find:
a the value of p b $P(M = 5)$ c Var(M)

4 $D \sim$ Negative B(8, p).
a Given that Var(D) = 30, find the value of p.
b Find:
i $P(D = 12)$ ii $P(D \leqslant 10$|first trial was successful)

Problem-solving
p is a probability so it must be a positive number.

5 $X \sim$ Negative B(r, 0.4).
a Given that E(X) = 15, find the value of r.
b Find:
i $P(X = 10)$ ii $P(X \leqslant 8)$

53

Step-by-step worked examples focus on the key types of questions you'll need to tackle

Exercise questions are carefully graded so they increase in difficulty and gradually bring you up to exam standard

Problem-solving boxes provide hints, tips and strategies, and *Watch out* boxes highlight areas where students often lose marks in their exams

Exam-style questions are flagged with Ⓔ

Problem-solving questions are flagged with Ⓟ

Every few chapters a *Review exercise* helps you consolidate your learning with lots of exam-style questions

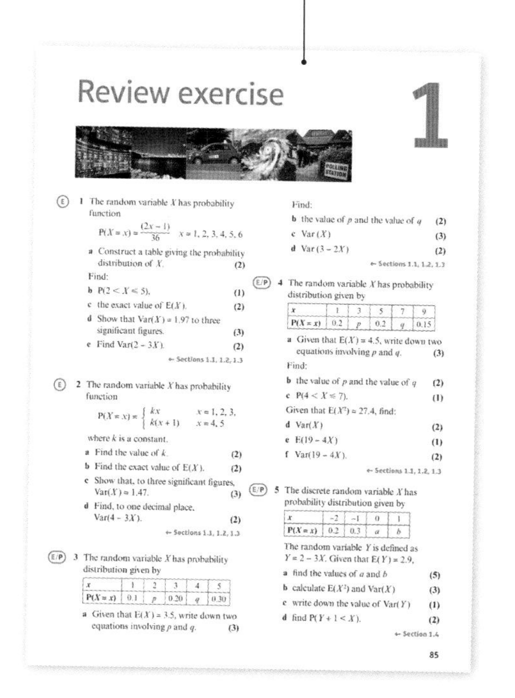
Review exercise 1

1 The random variable X has probability function
$P(X = x) = \frac{(2x-1)}{36}$ $x = 1, 2, 3, 4, 5, 6$
a Construct a table giving the probability distribution of X. (2)
Find:
b $P(2 < X \leqslant 5)$, (1)
c the exact value of E(X). (2)
d Show that Var(X) = 1.97 to three significant figures. (3)
e Find Var(2 − 3X). (2)
← Sections 1.1, 1.2, 1.3

2 The random variable X has probability function
$P(X = x) = \begin{cases} kx & x = 1, 2, 3, \\ k(x+1) & x = 4, 5 \end{cases}$
where k is a constant.
a Find the value of k. (2)
b Find the exact value of E(X). (2)
c Show that, to three significant figures, Var(X) = 1.47. (3)
d Find, to one decimal place, Var(4 − 3X). (2)
← Sections 1.1, 1.2, 1.3

3 The random variable X has probability distribution given by

x	1	2	3	4	5
$P(X = x)$	0.1	p	0.20	q	0.30

a Given that E(X) = 3.5, write down two equations involving p and q. (3)

Find:
b the value of p and the value of q (2)
c Var(X) (3)
d Var(3 − 2X) (2)
← Sections 1.1, 1.2, 1.3

4 The random variable X has probability distribution given by

x	1	3	5	7	9
$P(X = x)$	0.2	p	0.2	q	0.15

a Given that E(X) = 4.5, write down two equations involving p and q. (3)
Find:
b the value of p and the value of q (2)
c $P(4 < X \leqslant 7)$. (1)
Given that E(X^2) = 27.4, find:
d Var(X) (2)
e E(19 − 4X) (1)
f Var(19 − 4X) (2)
← Sections 1.1, 1.2, 1.3

5 The discrete random variable X has probability distribution given by

x	−2	−1	0	1
$P(X = x)$	0.2	0.3	a	b

The random variable Y is defined as $Y = 2 - 3X$. Given that E(Y) = 2.9,
a find the values of a and b (5)
b calculate E(X^2) and Var(X) (3)
c write down the value of Var(Y) (1)
d find $P(Y + 1 < X)$. (2)
← Section 1.4

85

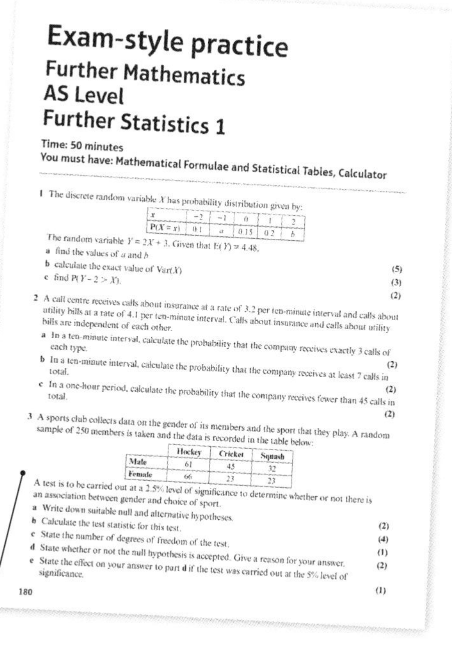
Exam-style practice
Further Mathematics
AS Level
Further Statistics 1

Time: 50 minutes
You must have: Mathematical Formulae and Statistical Tables, Calculator

1 The discrete random variable X has probability distribution given by:

x	−2	−1	0	1	2
$P(X = x)$	0.1	a	0.15	0.2	b

The random variable $Y = 2X + 3$. Given that E(Y) = 4.48,
a find the values of a and b (5)
b calculate the exact value of Var(X) (3)
c find $P(Y - 2 > X)$. (2)

2 A call centre receives calls about insurance at a rate of 3.2 per ten-minute interval and calls about utility bills at a rate of 4.1 per ten-minute interval. Calls about insurance and calls about utility bills are independent of each other.
a In a ten-minute interval, calculate the probability that the company receives exactly 3 calls of each type. (2)
b In a ten-minute interval, calculate the probability that the company receives at least 7 calls in total. (2)
c In a one-hour period, calculate the probability that the company receives fewer than 45 calls in total. (2)

3 A sports club collects data on the gender of its members and the sport that they play. A random sample of 250 members is taken and the data is recorded in the table below:

	Hockey	Cricket	Squash
Male	61	45	32
Female	66	23	23

A test is to be carried out at a 2.5% level of significance to determine whether or not there is an association between gender and choice of sport.
a Write down suitable null and alternative hypotheses. (2)
b Calculate the test statistic for this test. (4)
c State the number of degrees of freedom of the test. (1)
d State whether or not the null hypothesis is accepted. Give a reason for your answer. (2)
e State the effect on your answer to part **d** if the test was carried out at the 5% level of significance. (1)

180

AS and A level practice papers at the back of the book help you prepare for the real thing.

Extra online content

Whenever you see an *Online* box, it means that there is extra online content available to support you.

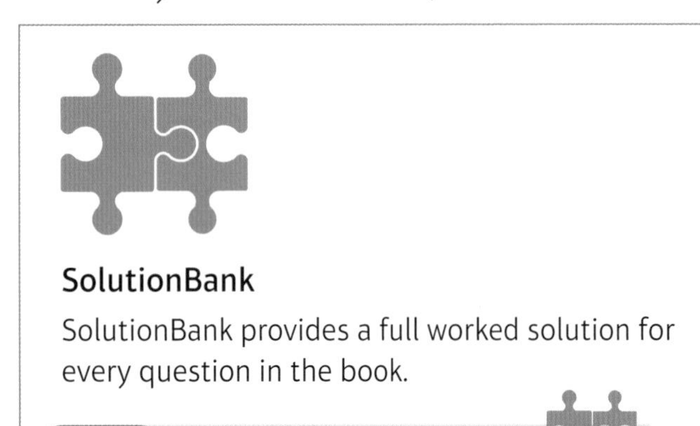

SolutionBank

SolutionBank provides a full worked solution for every question in the book.

Online Full worked solutions are available in SolutionBank.

Download all the solutions as a PDF or quickly find the solution you need online

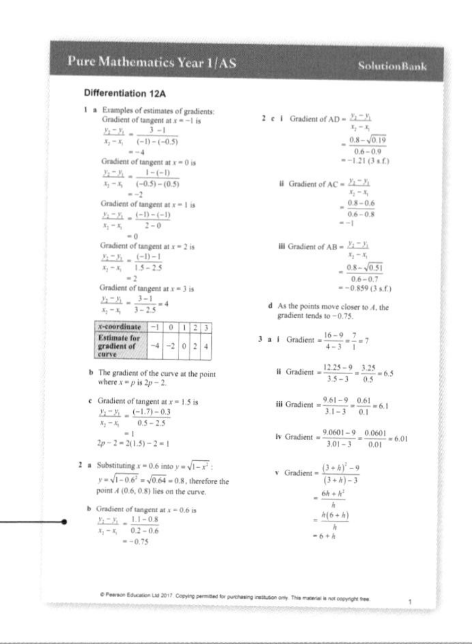

Pure Mathematics Year 1/AS — SolutionBank

Differentiation 12A

1 a Examples of estimates of gradients:

Gradient of tangent at $x = -1$ is

$\frac{y_2 - y_1}{x_2 - x_1} = \frac{3 - 1}{(-1) - (-0.5)} = -4$

Gradient of tangent at $x = 0$ is

$\frac{y_2 - y_1}{x_2 - x_1} = \frac{1 - (-1)}{(-0.5) - (0.5)} = -2$

Gradient of tangent at $x = 1$ is

$\frac{y_2 - y_1}{x_2 - x_1} = \frac{(-1) - (-1)}{2 - 0} = 0$

Gradient of tangent at $x = 2$ is

$\frac{y_2 - y_1}{x_2 - x_1} = \frac{(-1) - 1}{1.5 - 2.5} = 2$

Gradient of tangent at $x = 3$ is

$\frac{y_2 - y_1}{x_2 - x_1} = \frac{3 - 1}{3 - 2.5} = 4$

x-coordinate	−1	0	1	2	3
Estimate for gradient of curve	−4	−2	0	2	4

b The gradient of the curve at the point where $x = p$ is $2p - 2$.

c Gradient of tangent at $x = 1.5$ is

$\frac{y_2 - y_1}{x_2 - x_1} = \frac{(-1.7) - 0.3}{0.5 - 2.5} = 1$

$2p - 2 = 2(1.5) - 2 = 1$

2 a Substituting $x = 0.6$ into $y = \sqrt{1 - x^2}$:

$y = \sqrt{1 - 0.6^2} = \sqrt{0.64} = 0.8$, therefore the point A (0.6, 0.8) lies on the curve.

b Gradient of tangent at $x = 0.6$ is

$\frac{y_2 - y_1}{x_2 - x_1} = \frac{1.1 - 0.8}{0.2 - 0.6} = -0.75$

2 c i Gradient of AD $= \frac{y_2 - y_1}{x_2 - x_1} = \frac{0.8 - \sqrt{0.19}}{0.6 - 0.9} = -1.21$ (3 s.f.)

ii Gradient of AC $= \frac{y_2 - y_1}{x_2 - x_1} = \frac{0.8 - 0.6}{0.6 - 0.8} = -1$

iii Gradient of AB $= \frac{y_2 - y_1}{x_2 - x_1} = \frac{0.8 - \sqrt{0.51}}{0.6 - 0.7} = -0.859$ (3 s.f.)

d As the points move closer to A, the gradient tends to −0.75.

3 a i Gradient $= \frac{16 - 9}{4 - 3} = \frac{7}{1} = 7$

ii Gradient $= \frac{12.25 - 9}{3.5 - 3} = \frac{3.25}{0.5} = 6.5$

iii Gradient $= \frac{9.61 - 9}{3.1 - 3} = \frac{0.61}{0.1} = 6.1$

iv Gradient $= \frac{9.0601 - 9}{3.01 - 3} = \frac{0.0601}{0.01} = 6.01$

v Gradient $= \frac{(3 + h)^2 - 9}{(3 + h) - 3} = \frac{6h + h^2}{h} = \frac{h(6 + h)}{h} = 6 + h$

© Pearson Education Ltd 2017. Copying permitted for purchasing institution only. This material is not copyright free. 1

Use of technology

Explore topics in more detail, visualise problems and consolidate your understanding using pre-made GeoGebra activities.

Online Find the point of intersection graphically using technology.

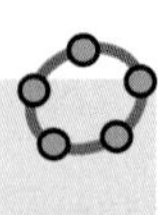

GeoGebra

GeoGebra-powered interactives

Interact with the maths you are learning using GeoGebra's easy-to-use tools

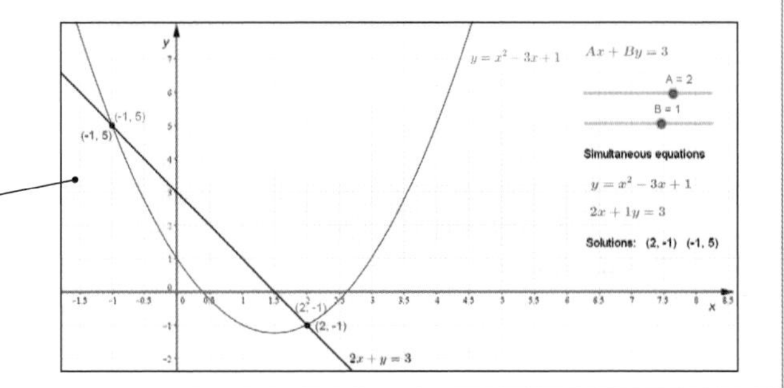

Access all the extra online content for free at:

www.pearsonschools.co.uk/fs1maths

You can also access the extra online content by scanning this QR code:

Discrete random variables

1

Objectives

After completing this chapter you should be able to:

- Find the expected value of a discrete random variable X → pages 2–5
- Find the expected value of X^2 → pages 3–5
- Find the variance of a discrete random variable → pages 5–7
- Use the expected value and variance of a function of X → pages 7–11
- Solve problems involving random variables → pages 11–14

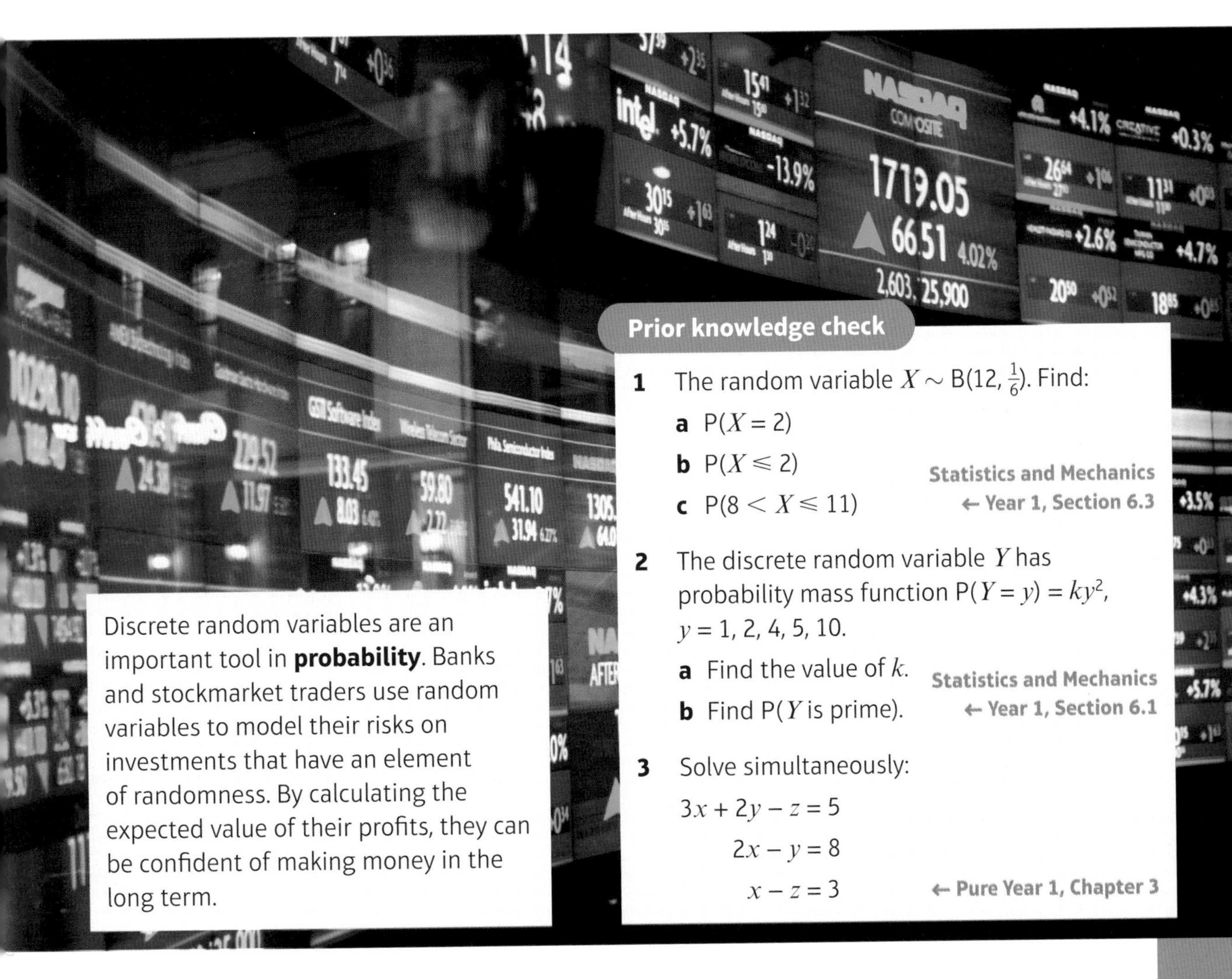

Discrete random variables are an important tool in **probability**. Banks and stockmarket traders use random variables to model their risks on investments that have an element of randomness. By calculating the expected value of their profits, they can be confident of making money in the long term.

Prior knowledge check

1. The random variable $X \sim \text{B}(12, \frac{1}{6})$. Find:
   a $\text{P}(X = 2)$
   b $\text{P}(X \leqslant 2)$
   c $\text{P}(8 < X \leqslant 11)$
   Statistics and Mechanics ← Year 1, Section 6.3

2. The discrete random variable Y has probability mass function $\text{P}(Y = y) = ky^2$, $y = 1, 2, 4, 5, 10$.
   a Find the value of k.
   b Find $\text{P}(Y \text{ is prime})$.
   Statistics and Mechanics ← Year 1, Section 6.1

3. Solve simultaneously:
   $$3x + 2y - z = 5$$
   $$2x - y = 8$$
   $$x - z = 3$$
   ← Pure Year 1, Chapter 3

1.1 Expected value of a discrete random variable

Recall that a **random variable** is a variable whose value depends on a random event. The random variable is **discrete** if it can only take certain numerical values.

Links The probabilities of any discrete random variable add up to 1. For a discrete random variable, X, you write $\sum P(X = x) = 1$.
← Statistics and Mechanics Year 1, Chapter 6

If you take a set of observations from a discrete random variable, you can find the mean of those observations. As the number of observations increases, this value will get closer and closer to the **expected value** of the discrete random variable.

Watch out The expected value is a theoretical quantity, and gives information about the probability distribution of a random variable.

- **The expected value of the discrete random variable X is denoted E(X) and defined as E(X) = $\sum x$P($X = x$).**

Notation The expected value is sometimes referred to as the **mean**, and is denoted by μ.

Example 1

A fair six-sided dice is rolled. The number on the uppermost face is modelled by the random variable X.

a Write down the probability distribution of X.

b Use the probability distribution of X to calculate E(X).

a

x	1	2	3	4	5	6
P($x = x$)	$\frac{1}{6}$	$\frac{1}{6}$	$\frac{1}{6}$	$\frac{1}{6}$	$\frac{1}{6}$	$\frac{1}{6}$

Since the dice is fair, each side is equally likely to end facing up, so the probability of any face ending up as the uppermost is $\frac{1}{6}$

b The expected value of X is:

$$\text{E}(X) = \sum x\text{P}(X = x) = \frac{1}{6} + \frac{2}{6} + \ldots + \frac{6}{6}$$

$$= \frac{21}{6} = \frac{7}{2} = 3.5$$

Substitute values from the probability distribution into the formula then simplify.

If you know the probability distribution of X then you can calculate the expected value. Notice that in Example 1 the expected value is 3.5, but P($X = 3.5$) = 0. The expected value of a random variable does not have to be a value that the random variable can actually take. Instead this tells us that in the long run, we would expect the average of all rolls to get close to 3.5.

Example 2

The random variable X has a probability distribution as shown in the table.

x	1	2	3	4	5
p(x)	0.1	p	0.3	q	0.2

a Given that E(X) = 3, write down two equations involving p and q.

b Find the value of p and the value of q.

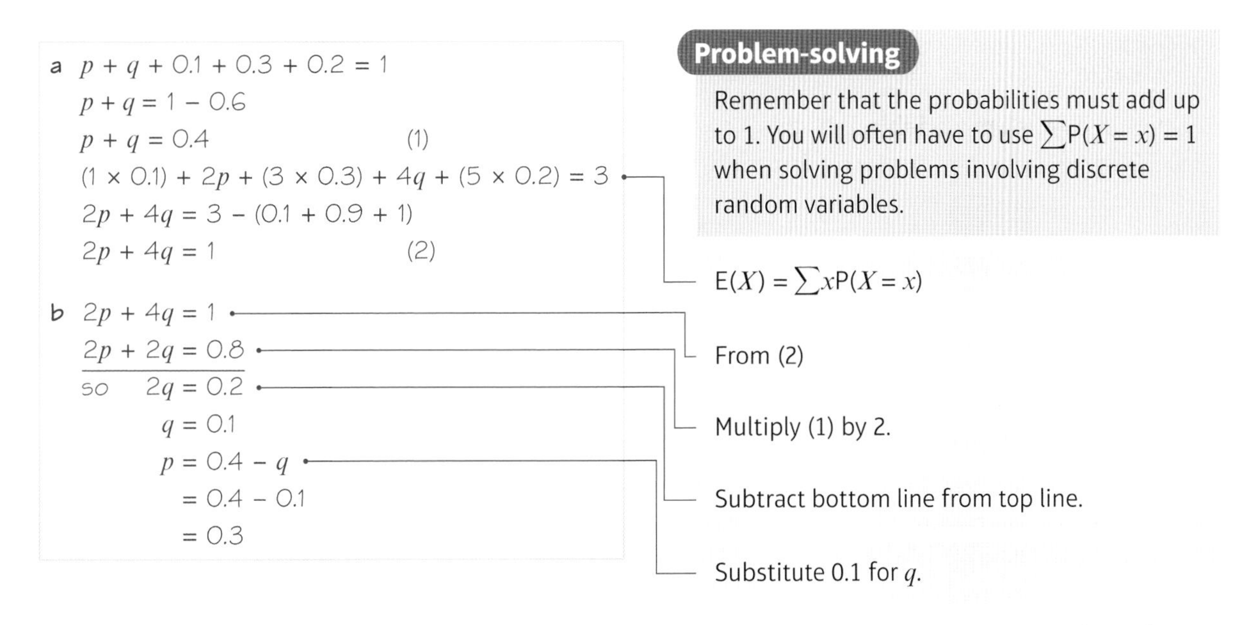

a $p + q + 0.1 + 0.3 + 0.2 = 1$

$p + q = 1 - 0.6$

$p + q = 0.4 \quad (1)$

$(1 \times 0.1) + 2p + (3 \times 0.3) + 4q + (5 \times 0.2) = 3$ — $E(X) = \sum xP(X = x)$

$2p + 4q = 3 - (0.1 + 0.9 + 1)$

$2p + 4q = 1 \quad (2)$

b $2p + 4q = 1$ — From (2)

$2p + 2q = 0.8$ — Multiply (1) by 2.

so $2q = 0.2$ — Subtract bottom line from top line.

$q = 0.1$

$p = 0.4 - q$ — Substitute 0.1 for q.

$= 0.4 - 0.1$

$= 0.3$

Problem-solving

Remember that the probabilities must add up to 1. You will often have to use $\sum P(X = x) = 1$ when solving problems involving discrete random variables.

If X is a discrete random variable, then X^2 is also a discrete random variable. You can use this rule to determine the expected value of X^2.

- $\mathbf{E(X^2) = \sum x^2P(X = x)}$

Links Any function of a random variable is also a random variable. → Section 1.3

Example 3

A discrete random variable X has the following probability distribution

x	1	2	3	4
$P(X = x)$	$\frac{12}{25}$	$\frac{6}{25}$	$\frac{4}{25}$	$\frac{3}{25}$

a Write down the probability distribution for X^2.

b Find $E(X^2)$.

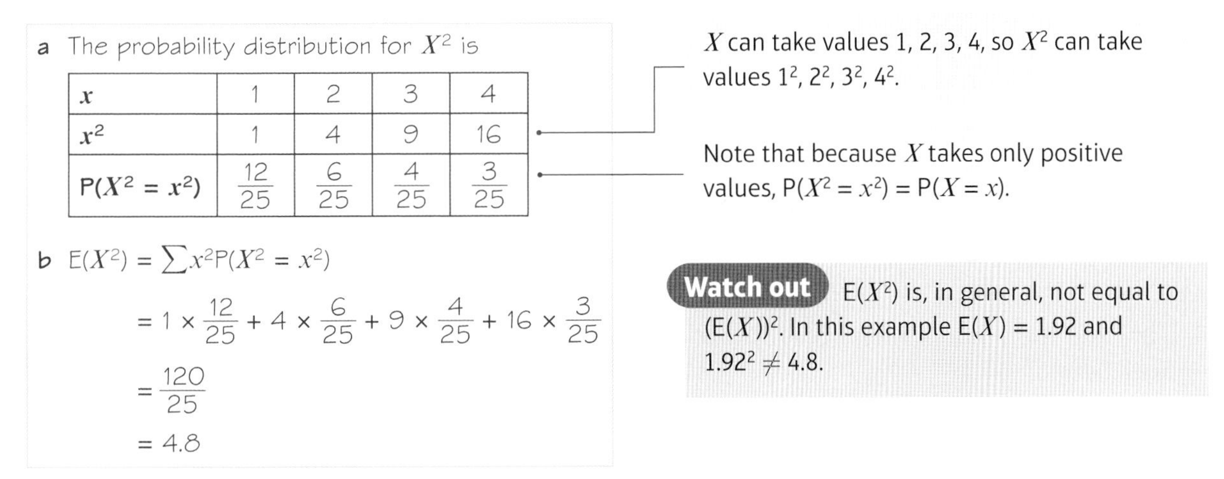

a The probability distribution for X^2 is

x	1	2	3	4
x^2	1	4	9	16
$P(X^2 = x^2)$	$\frac{12}{25}$	$\frac{6}{25}$	$\frac{4}{25}$	$\frac{3}{25}$

X can take values 1, 2, 3, 4, so X^2 can take values $1^2, 2^2, 3^2, 4^2$.

Note that because X takes only positive values, $P(X^2 = x^2) = P(X = x)$.

b $E(X^2) = \sum x^2P(X^2 = x^2)$

$= 1 \times \frac{12}{25} + 4 \times \frac{6}{25} + 9 \times \frac{4}{25} + 16 \times \frac{3}{25}$

$= \frac{120}{25}$

$= 4.8$

Watch out $E(X^2)$ is, in general, not equal to $(E(X))^2$. In this example $E(X) = 1.92$ and $1.92^2 \neq 4.8$.

Exercise 1A

1 For each of the following probability distributions write out the distribution of X^2 and calculate both $E(X)$ and $E(X^2)$.

a

x	2	4	6	8
$P(X = x)$	0.3	0.3	0.2	0.2

b

x	−2	−1	1	2
$P(X = x)$	0.1	0.4	0.1	0.4

Watch out Note that, for example, $P(X^2 = 4) = P(X = 2) + P(X = -2)$.

2 The score on a biased dice is modelled by a random variable X with probability distribution

x	1	2	3	4	5	6
$P(X = x)$	0.1	0.1	0.1	0.2	0.4	0.1

Find $E(X)$ and $E(X^2)$.

3 The random variable X has a probability function

$$P(X = x) = \frac{1}{x} \qquad x = 2, 3, 6$$

a Construct tables giving the probability distributions of X and X^2.

b Work out $E(X)$ and $E(X^2)$.

c State whether or not $(E(X))^2 = E(X^2)$.

4 The random variable X has a probability function given by

$$P(X = x) = \begin{cases} 2^{-x} & x = 1, 2, 3, 4 \\ 2^{-4} & x = 5 \end{cases}$$

a Construct a table giving the probability distribution of X.

b Calculate $E(X)$ and $E(X^2)$.

c State whether or not $(E(X))^2 = E(X^2)$.

E/P **5** The random variable X has the following probability distribution:

x	1	2	3	4	5
$P(X = x)$	0.1	a	b	0.2	0.1

Given that $E(X) = 2.9$, find the value of a and the value of b. **(5 marks)**

E/P **6** The random variable X has the following probability distribution:

x	−2	−1	1	2
$P(X = x)$	0.1	a	b	c

Given that $E(X) = 0.3$ and $E(X^2) = 1.9$, find a, b and c. **(7 marks)**

Problem-solving

You can use the given information to write down simultaneous equations for a, b and c which can be solved using the matrix inverse operation on your calculator.
← Core Pure Book 1, Section 6.6

(E/P) **7** The discrete random variable X has probability function

$$P(X = x) = \begin{cases} a(1 - x) & x = -2, -1, 0 \\ b & x = 5 \end{cases}$$

Given that $E(X) = 1.2$, find the value of a and the value of b. **(6 marks)**

(E/P) **8** A biased six-sided dice has a $\frac{1}{8}$ chance of landing on any of the numbers 1, 2, 3 or 4. The probabilities of landing on 5 or 6 are unknown. The outcome is modelled as a random variable, X. Given that $E(X) = 4.1$,

a find the probability distribution of X. **(5 marks)**

The dice is rolled 10 times.

b Calculate the probability that the dice lands on 6 at least 3 times. **(3 marks)**

(E) **9** Jorge has designed a game for his school fete. Students can pay £1 to roll a fair six-sided dice. If they score a 6 they win a prize of £5. If they score a 4 or a 5, they win a smaller prize of £P.

By modelling the amount paid out in prize money as a discrete random variable, determine the maximum value of P in order for Jorge to not make a loss on his game. **(5 marks)**

Hint The expected profit from the game is the cost of playing the game minus the expected value of the amount paid out in prize money.

Challenge

Three fair six-sided dice are rolled. The discrete random variable X is defined as the largest value of the three values shown. Find $E(X)$.

1.2 Variance of a discrete random variable

If you take a set of observations from a discrete random variable, you can find the variance of those observations. As the number of observations increases, this value will get closer and closer to the **variance** of the discrete random variable.

Notation The variance is sometimes denoted by σ^2, where σ is the standard deviation.

- **The variance of X is usually written as Var(X) and is defined as Var(X) = E((X – E(X))²)**

The random variable $(X - E(X))^2$ is the squared deviation from the expected value of X. It is large when X takes values that are very different to $E(X)$.

Online Explore probability distributions of a discrete random variable and compare the theoretical distribution with obseerved results generated from that discrete random variable using GeoGebra.

- **Sometimes it is easier to calculate the variance using the formula $\text{Var}(X) = E(X^2) - (E(X))^2$**

From the definition you can see that $\text{Var}(X) \geqslant 0$ for any random variable X. The larger Var(X) the more variable X is. In other words, the more likely it is to take values very different to its expected value.

Example 4

A fair six-sided dice is rolled. The number on the uppermost face is modelled by the random variable X.

Find $\mathrm{Var}(X)$.

Method 1

We have that $\mathrm{E}(X) = 3.5$.

This was calculated in the first example of the previous section.

The distributions of X, X^2 and $(X - \mathrm{E}(X))^2$ are given by

x	1	2	3	4	5	6
x^2	1	4	9	16	25	36
$(x - \mathrm{E}(X))^2$	6.25	2.25	0.25	0.25	2.25	6.25
$\mathrm{P}(X = x)$	$\frac{1}{6}$	$\frac{1}{6}$	$\frac{1}{6}$	$\frac{1}{6}$	$\frac{1}{6}$	$\frac{1}{6}$

So the variance is

$$\mathrm{Var}(X) = \sum(x - \mathrm{E}(X))^2\mathrm{P}(X = x)$$

$$= 6.25 \times \frac{2}{6} + 2.25 \times \frac{2}{6} + 0.25 \times \frac{2}{6}$$

Substitute values into the formula for variance.

$$= (6.25 + 2.25 + 0.25) \times \frac{1}{3} = \frac{35}{12}$$

Method 2

The expected value of X^2 is

$$\mathrm{E}(X^2) = \sum x^2\mathrm{P}(X = x) = \frac{1}{6}(1 + 4 + \ldots + 36) = \frac{91}{6}$$

So using the alternative formula

$$\mathrm{Var}(X) = \mathrm{E}(X^2) - (\mathrm{E}(X))^2 = \frac{91}{6} - \frac{49}{4} = \frac{35}{12}$$

It is usually quicker to use this method to find the variance of a random variable.

Exercise 1B

1 The random variable X has a probability distribution given by

x	−1	0	1	2	3
$\mathrm{P}(X = x)$	$\frac{1}{5}$	$\frac{1}{5}$	$\frac{1}{5}$	$\frac{1}{5}$	$\frac{1}{5}$

a Find $\mathrm{E}(X)$.
b Find $\mathrm{Var}(X)$.

2 Find the expected value and variance of the random variable X with probability distributions given by:

a

x	1	2	3
$\mathrm{P}(X = x)$	$\frac{1}{3}$	$\frac{1}{2}$	$\frac{1}{6}$

b

x	−1	0	1
$\mathrm{P}(X = x)$	$\frac{1}{4}$	$\frac{1}{2}$	$\frac{1}{4}$

c

x	−2	−1	1	2
$\mathrm{P}(X = x)$	$\frac{1}{3}$	$\frac{1}{3}$	$\frac{1}{6}$	$\frac{1}{6}$

3 Given that Y is the score when a single, unbiased, eight-sided dice is rolled, find $\mathrm{E}(Y)$ and $\mathrm{Var}(Y)$.

(P) **4** Two fair, cubical dice are rolled and S is the sum of their scores. Find:

a the distribution of S **b** $E(S)$

c $Var(S)$ **d** the standard deviation, σ

Hint The standard deviation of a random variable is the square root of its variance.

(P) **5** Two fair, tetrahedral (four-sided) dice are rolled and D is the difference between their scores. Find:

a the distribution of D

b $E(D)$

c $Var(D)$

(E) **6** A fair coin is spun repeatedly until a head appears or three spins have been made. The random variable T represents the number of spins of the coin.

a Show that the probability distribution of T is

t	1	2	3
$P(T = t)$	$\frac{1}{2}$	$\frac{1}{4}$	$\frac{1}{4}$

(3 marks)

b Find the expected value and variance of T. **(6 marks)**

(E/P) **7** The random variable X has a probability distribution given by

x	1	2	3
$P(X = x)$	a	b	a

where a and b are constants.

a Write down an expression for $E(X)$ in terms of a and b. **(2 marks)**

b Given that $Var(X) = 0.75$, find the values of a and b. **(5 marks)**

1.3 Expected value and variance of a function of X

If X is a discrete random variable, and g is a function, then $g(X)$ is also a discrete random variable. You can calculate the expected value of $g(X)$ using the formula:

- $\mathbf{E(g(X)) = \sum g(x)P(X = x)}$

This is a more general version of the formula for $E(X^2)$. For simple functions, such as addition and multiplication by a constant, you can learn the following rules:

- **If X is a random variable and a and b are constants, then $E(aX + b) = aE(X) + b$**
- **If X and Y are random variables, then $E(X + Y) = E(X) + E(Y)$**

You can use a similar rule to simplify variance calculations for some functions of random variables:

- **If X is a random variable and a and b are constants then $Var(aX + b) = a^2Var(X)$**

Example 5

A discrete random variable X has the probability distribution

x	1	2	3	4
$P(X = x)$	$\frac{12}{25}$	$\frac{6}{25}$	$\frac{4}{25}$	$\frac{3}{25}$

a Write down the probability distribution for Y where $Y = 2X + 1$.

b Find $E(Y)$.

c Compute $E(X)$ and verify that $E(Y) = 2E(X) + 1$.

a The probability distribution for Y is

x	1	2	3	4
y	3	5	7	9
$P(Y = y)$	$\frac{12}{25}$	$\frac{6}{25}$	$\frac{4}{25}$	$\frac{3}{25}$

When $x = 1, y = 2 \times 1 + 1 = 3$
$x = 2, y = 2 \times 2 + 1 = 5$
etc.

Notice how the probabilities relating to X are still being used, for example, $P(X = 3) = P(Y = 7)$.

b $E(Y) = \sum yP(Y = y)$

$= 3 \times \frac{12}{25} + 5 \times \frac{6}{25} + 7 \times \frac{4}{25} + 9 \times \frac{3}{25}$

$= \frac{121}{25}$

$= 4.84$

c $E(X) = \sum xP(X = x) = 1 \times \frac{12}{25} + 2 \times \frac{6}{25}$

$+ 3 \times \frac{4}{25} + 4 \times \frac{3}{25} = \frac{48}{25} = 1.92$

Therefore $2E(X) + 1 = 2 \times 1.92 + 1 = 4.84$

Therefore $E(Y) = 2E(X) + 1$

If you know or are given $E(X)$ you can use the formula to find $E(Y)$ quickly.

Example 6

A random variable X has $E(X) = 4$ and $Var(X) = 3$. Find:

a $E(3X)$ **b** $E(X - 2)$

c $Var(3X)$ **d** $Var(X - 2)$

e $E(X^2)$

a $E(3X) = 3E(X) = 3 \times 4 = 12$

b $E(X - 2) = E(X) - 2 = 4 - 2 = 2$

c $Var(3X) = 3^2 Var(X) = 9 \times 3 = 27$

d $Var(X - 2) = Var(X) = 3$

e $E(X^2) = Var(X) + (E(X))^2 = 3 + 4^2 = 19$

Rearrange $Var(X) = E(X^2) - (E(X))^2$.

Example 7

Two fair 10p coins are spun. The random variable X pence represents the total value of the coins that land heads up.

a Find $E(X)$ and $Var(X)$.

The random variables S and T are defined as follows:

$$S = X - 10 \text{ and } T = \tfrac{1}{2}X - 5$$

b Show that $E(S) = E(T)$.

c Find $Var(S)$ and $Var(T)$.

A large number of observations of S and T are taken.

d Comment on any likely differences or similarities.

a The probability distribution of X is

x	0	10	20
$P(X = x)$	$\frac{1}{4}$	$\frac{1}{2}$	$\frac{1}{4}$

$E(X) = 10$ by inspection

$Var(X) = E(X^2) - (E(X))^2$

$Var(X) = 0^2 \times \frac{1}{4} + 10^2 \times \frac{1}{2} + 20^2 \times \frac{1}{4} - 10^2 = 50$

b $E(S) = E(X - 10) = E(X) - 10 = 10 - 10 = 0$

Use $E(aX + b) = aE(X) + b$

$E(T) = E\left(\frac{1}{2}X - 5\right) = \frac{1}{2}E(X) - 5 = \frac{1}{2} \times 10 - 5 = 0$

c $Var(S) = Var(X) = 50$

Subtracting a constant doesn't change the variance, so $Var(S) = Var(X)$.

$Var(T) = \left(\frac{1}{2}\right)^2 Var(X) = \frac{50}{4} = 12.5$

d The means of both set of observations should be close to zero. The observed values of S will be more spread out than the observed values of T.

You could also say that the sum of the observed values of each random variable will be close to 0.

Example 8

The random variable X has the following probability distribution:

x	0°	30°	60°	90°
$P(X = x)$	0.4	0.2	0.1	0.3

Calculate $E(\sin X)$.

The distribution of $\sin X$ is

$\sin x$	0	$\frac{1}{2}$	$\frac{\sqrt{3}}{2}$	1
$P(X = x)$	0.4	0.2	0.1	0.3

$$E(\sin X) = \sum \sin x \; P(X = x)$$

$$= 0 \times 0.4 + \frac{1}{2} \times 0.2 + \frac{\sqrt{3}}{2} \times 0.1 + 1 \times 0.3$$

Use the general formula for $E(g(X))$.

$$= \frac{8 + \sqrt{3}}{20} \approx 0.487$$

Exercise 1C

1 The random variable X has a probability distribution given by

x	1	2	3	4
$P(X = x)$	0.1	0.3	0.2	0.4

a Write down the probability distribution for Y where $Y = 2X - 3$.

b Find $E(Y)$.

c Calculate $E(X)$ and verify that $E(2X - 3) = 2E(X) - 3$.

2 The random variable X has a probability distribution given by

x	−2	−1	0	1	2
$P(X = x)$	0.1	0.1	0.2	0.4	0.2

a Write down the probability distribution for Y where $Y = X^3$.

b Calculate $E(Y)$.

3 The random variable X has $E(X) = 1$ and $\text{Var}(X) = 2$. Find:

a $E(8X)$ **b** $E(X + 3)$ **c** $\text{Var}(X + 3)$

d $\text{Var}(3X)$ **e** $\text{Var}(1 - 2X)$ **f** $E(X^2)$

4 The random variable X has $E(X) = 3$ and $E(X^2) = 10$. Find:

a $E(2X)$ **b** $E(3 - 4X)$ **c** $E(X^2 - 4X)$

d $\text{Var}(X)$ **e** $\text{Var}(3X + 2)$

5 The random variable X has a mean μ and standard deviation σ.

Find, in terms of μ and σ:

a $E(4X)$ **b** $E(2X + 2)$ **c** $E(2X - 2)$

d $\text{Var}(2X + 2)$ **e** $\text{Var}(2X - 2)$

6 In a board game, players roll a fair, six-sided dice each time they make it around the board. The score on the dice is modelled as a discrete random variable X.

a Write down $E(X)$.

They are paid £200 plus £100 times the score on the dice. The amount paid to each player is modelled as a discrete random variable Y.

b Write Y in terms of X.

c Find the expected pay-out each time a player makes it around the board.

(P) **7** John runs a pizza parlour that sells pizza in three sizes: small (20 cm diameter), medium (30 cm diameter) and large (40 cm diameter). Each pizza base is 1 cm thick. John has worked out that on average, customers order a small, medium or large pizza with probabilities $\frac{3}{10}$, $\frac{9}{20}$ and $\frac{5}{20}$ respectively. Calculate the expected amount of pizza dough needed per customer.

(E/P) **8** Two tetrahedral dice are rolled. The random variable X represents the result of subtracting the smaller score from the larger.

a Find $E(X)$ and $Var(X)$. **(7 marks)**

The random variables Y and Z are defined as $Y = 2^X$ and $Z = \frac{4X + 1}{2}$

b Show that $E(Y) = E(Z)$. **(3 marks)**

c Find $Var(Z)$. **(2 marks)**

Challenge

Show that $E((X - E(X))^2) = E(X^2) - (E(X))^2$.

Hint You can assume that $E(X + Y) = E(X) + E(Y)$.

1.4 Solving problems involving random variables

Suppose you have two random variables X and $Y = g(X)$. If g is one-to-one, and you know the mean and variance of Y, then it is possible to deduce the mean and variance of X.

Example 9

X is a discrete random variable. The discrete random variable Y is defined as $Y = \frac{X - 150}{50}$

Given that $E(Y) = 5.1$ and $Var(Y) = 2.5$, find:

a $E(X)$

b $Var(X)$.

a $Y = \dfrac{X - 150}{50}$

$X = 50Y + 150$ — Rearrange to get an expression for X in terms of Y.

$E(X) = E(50Y + 150)$
$= 50E(Y) + 150$
$= 255 + 150$
$= 405$

b $\text{Var}(X) = \text{Var}(50Y + 150)$ — Use your expression for X in terms of Y. Remember that the '+150' does not affect the variance, and that you have to multiply Var(Y) by 50^2 to get Var(X).

$= 50^2\text{Var}(Y)$
$= 50^2 \times 2.5$
$= 6250$

Example 10

The discrete random variable X has a probability distribution given by

x	−2	−1	0	1	2
P($X = x$)	0.3	a	0.25	b	c

The discrete random variable Y is defined as $Y = 3X - 1$.

Given that $E(Y) = -2.5$ and $\text{Var}(Y) = 13.95$, find:

a $E(X)$ and $E(X^2)$

b the values of a, b and c

c $P(X > Y)$

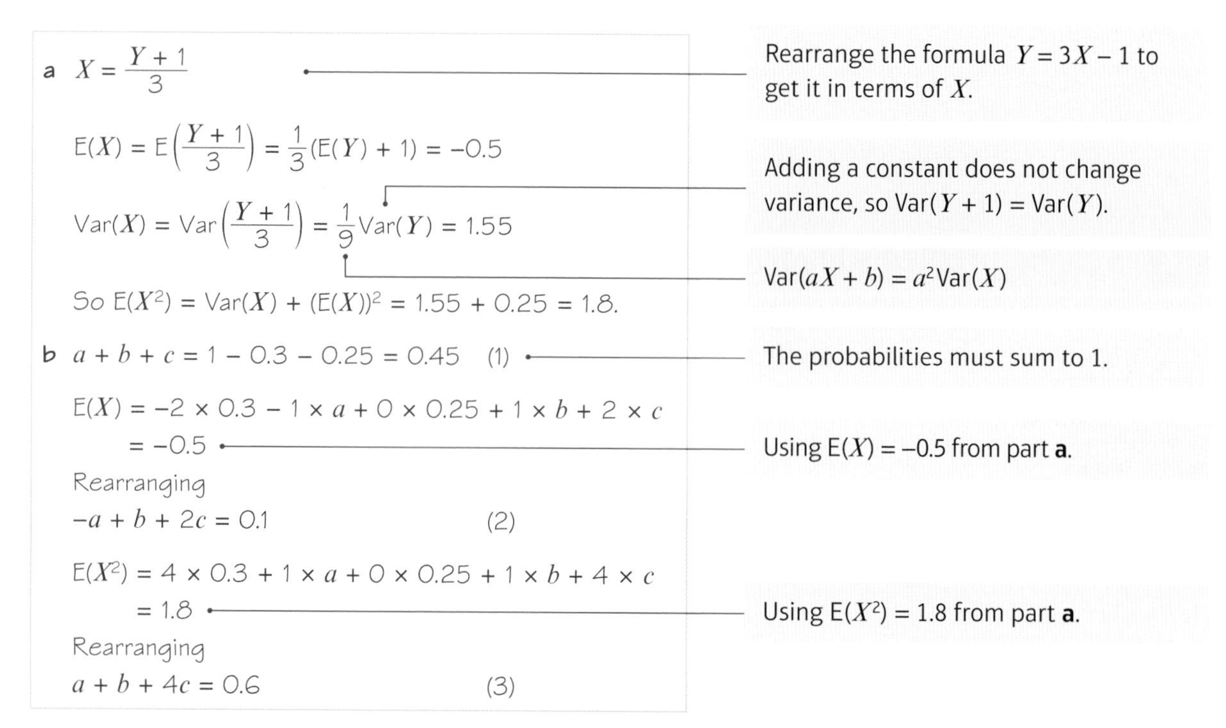

a $X = \dfrac{Y + 1}{3}$ — Rearrange the formula $Y = 3X - 1$ to get it in terms of X.

$E(X) = E\left(\dfrac{Y+1}{3}\right) = \dfrac{1}{3}(E(Y) + 1) = -0.5$

$\text{Var}(X) = \text{Var}\left(\dfrac{Y+1}{3}\right) = \dfrac{1}{9}\text{Var}(Y) = 1.55$

Adding a constant does not change variance, so $\text{Var}(Y + 1) = \text{Var}(Y)$.

$\text{Var}(aX + b) = a^2\text{Var}(X)$

So $E(X^2) = \text{Var}(X) + (E(X))^2 = 1.55 + 0.25 = 1.8$.

b $a + b + c = 1 - 0.3 - 0.25 = 0.45$ (1) — The probabilities must sum to 1.

$E(X) = -2 \times 0.3 - 1 \times a + 0 \times 0.25 + 1 \times b + 2 \times c$
$= -0.5$ — Using $E(X) = -0.5$ from part **a**.

Rearranging

$-a + b + 2c = 0.1$ (2)

$E(X^2) = 4 \times 0.3 + 1 \times a + 0 \times 0.25 + 1 \times b + 4 \times c$
$= 1.8$ — Using $E(X^2) = 1.8$ from part **a**.

Rearranging

$a + b + 4c = 0.6$ (3)

Writing equations (1), (2) and (3) as a single matrix equation:

$$\begin{pmatrix} 1 & 1 & 1 \\ -1 & 1 & 2 \\ 1 & 1 & 4 \end{pmatrix}\begin{pmatrix} a \\ b \\ c \end{pmatrix} = \begin{pmatrix} 0.45 \\ 0.1 \\ 0.6 \end{pmatrix}$$

So, by inverting the matrix we find

$$\begin{pmatrix} a \\ b \\ c \end{pmatrix} = \frac{1}{6}\begin{pmatrix} 2 & -3 & 1 \\ 6 & 3 & -3 \\ -2 & 0 & 2 \end{pmatrix}\begin{pmatrix} 0.45 \\ 0.1 \\ 0.6 \end{pmatrix} = \begin{pmatrix} 0.2 \\ 0.2 \\ 0.05 \end{pmatrix}$$

So $a = 0.2$, $b = 0.2$ and $c = 0.05$.

Enter the 3 × 3 matrix representing the three simultaneous equations into your calculator. Find its inverse and multiply to find the values of a, b and c.

c $P(X > Y) = P(X > 3X - 1) = P(X < \frac{1}{2})$

So $P(X > Y) = 0.3 + 0.2 + 0.25 = 0.75$

Use the expression for Y to write everything in terms of X only.

Add the probabilities of all the values of x that satisfy the inequality.

Exercise 1D

1 X is a discrete random variable. The random variable Y is defined by $Y = 4X - 6$. Given that $E(Y) = 2$ and $Var(Y) = 32$, find:

a $E(X)$

b $Var(X)$

c the standard deviation of X.

2 X is a discrete random variable. The random variable Y is defined by $Y = \frac{4 - 3X}{2}$

Given that $E(Y) = -1$ and $Var(Y) = 9$, find:

a $E(X)$

b $Var(X)$

c $E(X^2)$

(P) **3** The discrete random variable X has a probability distribution given by

x	1	2	3	4
$P(X = x)$	0.3	a	b	0.2

The random variable Y is defined by $Y = 2X + 3$. Given that $E(Y) = 8$, find the values of a and b.

(E/P) **4** The discrete random variable X has a probability distribution given by

x	90°	180°	270°
$P(X = x)$	a	b	0.3

The random variable Y is defined as $Y = \sin X°$.

a Find the range of possible values of $E(Y)$. **(5 marks)**

b Given that $E(Y) = 0.2$, write down the values of a and b. **(2 marks)**

(P) **5** The discrete random variable X has a probability distribution given by

x	-2	-1	0	1	2
$\mathrm{P}(X = x)$	a	b	c	b	a

The random variable Y is defined $Y = (X + 1)^2$.

a Given that $\mathrm{E}(Y) = 2.4$ and $\mathrm{P}(Y > 2) = 0.4$, show that:

$$2a + 2b + c = 1$$
$$10a + 4b + c = 2.4$$
$$a + b = 0.4$$

b Hence find the values of a, b, and c.

c Find $\mathrm{P}(2X + 3 \leqslant Y)$.

(E/P) **6** The discrete random variable X has a probability distribution given by

$$\mathrm{P}(X = x) = \begin{cases} a & x = 1, 2, 3 \\ b & x = 4, 5 \\ c & x = 6 \end{cases}$$

Suppose that Y is defined by $Y = 1 - 2X$.

a Given that $\mathrm{E}(Y) = -5.6$ and $\mathrm{P}(Y \leqslant -5) = 0.6$, write down the value of $\mathrm{E}(X)$. **(1 mark)**

b Show that:

$$3a + 2b + c = 1$$
$$2a + 3b + 2c = 1.1$$
$$a + 2b + c = 0.6$$

(4 marks)

c Solve the system to find values for a, b, c. **(2 marks)**

d Find $\mathrm{P}(X > 5 + Y)$. **(2 marks)**

Mixed exercise 1

1 The random variable X has the probability function

$$\mathrm{P}(X = x) = \frac{x}{21} \qquad x = 1, 2, 3, 4, 5, 6$$

a Construct a table giving the probability distribution of X.

Find:

b $\mathrm{P}(2 < X \leqslant 5)$ **c** $\mathrm{E}(X)$ **d** $\mathrm{Var}(X)$

e $\mathrm{Var}(3 - 2X)$ **f** $\mathrm{E}(X^3)$

2 The discrete random variable X has the probability distribution given in the table below.

x	-2	-1	0	1	2	3
$\mathrm{P}(X = x)$	0.1	0.2	0.3	r	0.1	0.1

Find:

a r **b** $\mathrm{P}(-1 \leqslant X < 2)$ **c** $\mathrm{E}(2X + 3)$ **d** $\mathrm{Var}(2X + 3)$

3 A discrete random variable X has the probability distribution shown in the table below.

x	0	1	2
P($X = x$)	$\frac{1}{5}$	b	$\frac{1}{5} + b$

a Find the value of b.

b Show that E(X) = 1.3.

c Find the exact value of Var(X).

d Find the exact value of P($X \leqslant 1.5$).

(E) **4** The discrete random variable X has a probability function

$$P(X = x) = \begin{cases} k(1 - x) & x = 0, 1 \\ k(x - 1) & x = 2, 3 \\ 0 & \text{otherwise} \end{cases}$$

where k is a constant.

a Show that $k = \frac{1}{4}$ **(2 marks)**

b Find E(X) and show that E(X^2) = 5.5. **(4 marks)**

c Find Var($2X - 2$). **(4 marks)**

5 A discrete random variable X has the probability distribution

x	0	1	2	3
P($X = x$)	$\frac{1}{4}$	$\frac{1}{2}$	$\frac{1}{8}$	$\frac{1}{8}$

Find:

a P($1 < X \leqslant 2$)

b E(X)

c E($3X - 1$)

d Var(X)

e E(log($X + 1$))

6 A discrete random variable X has the probability distribution

x	1	2	3	4
P($X = x$)	0.4	0.2	0.1	0.3

Find:

a P($3 < X^2 < 10$)

b E(X)

c Var(X)

d $E\left(\frac{3 - X}{2}\right)$

e $E(\sqrt{X})$

f $E(2^{-x})$

7 A discrete random variable is such that each of its values is assumed to be equally likely.

a Write the name of the distribution.

b Give an example of such a distribution.

A random variable X has discrete uniform distribution and can take values 0, 1, 2, 3 and 4.

Find:

c E(X)

d Var(X)

(P) **8** The random variable X has the probability distribution

x	1	2	3	4	5
$P(X = x)$	0.1	p	q	0.3	0.1

a Given that $E(X) = 3.1$, write down two equations involving p and q.

Find:

b the value of p and the value of q

c $Var(X)$

d $Var(2X - 3)$

(E) **9** The random variable X has the probability function

$$P(X = x) = \begin{cases} kx & x = 1, 2 \\ k(x - 2) & x = 3, 4, 5 \end{cases}$$

where k is a constant.

a Find the value of k. **(2 marks)**

b Find the exact value of $E(X)$. **(1 mark)**

c Show that, to three significant figures, $Var(X) = 2.02$. **(2 marks)**

d Find, to one decimal place, $Var(3 - 2X)$. **(1 mark)**

(E) **10** The random variable X has the discrete uniform distribution

$$P(X = x) = \tfrac{1}{6} \qquad x = 1, 2, 3, 4, 5, 6$$

a Write down $E(X)$ and show that $Var(X) = \frac{35}{12}$ **(4 marks)**

b Find $E(2X - 1)$. **(2 marks)**

c Find $Var(3 - 2X)$. **(2 marks)**

d Find $E(2^X)$. **(3 marks)**

11 The random variable X has the probability function

$$P(X = x) = \frac{3x - 1}{26} \qquad x = 1, 2, 3, 4$$

a Construct a table giving the probability distribution of X.

Find:

b $P(2 < X \leqslant 4)$

c the exact value of $E(X)$.

d Show that $Var(X) = 0.92$ to two significant figures.

e Find $Var(1 - 3X)$.

12 The random variable Y has mean 2 and variance 9.

Find:

a $E(3Y + 1)$

b $E(2 - 3Y)$

c $Var(3Y + 1)$

d $Var(2 - 3Y)$

e $E(Y^2)$

f $E((Y - 1)(Y + 1))$

13 The random variable T has a mean of 20 and a standard deviation of 5.
The random variable S is defined as $S = 3T + 4$.
Find $E(S)$ and $Var(S)$.

14 A fair spinner is made from the disc in the diagram and the random variable X represents the number it lands on after being spun.

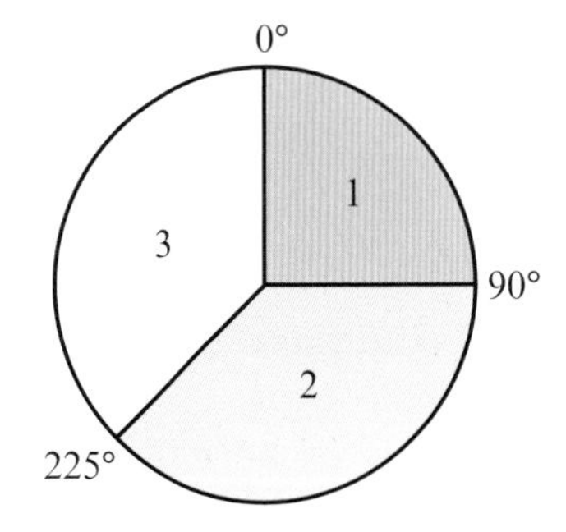

a Write down the distribution of X.
b Work out $E(X)$.
c Find $Var(X)$.
d Find $E(2X + 1)$.
e Find $Var(3X - 1)$.

15 The discrete variable X has the probability distribution

x	−1	0	1	2
$P(X = x)$	0.2	0.5	0.2	0.1

Find:

a $E(X)$ **b** $Var(X)$ **c** $E(\frac{1}{3}X + 1)$ **d** $Var(\frac{1}{3}X + 1)$

E/P **16** The discrete random variable X has a probability distribution given by

x	−1	0	1	2
$P(X = x)$	0.1	0.3	a	b

The random variable Y is defined $Y = 3X - 1$. Given that $E(Y) = 1.1$,

a find the values of a and b. **(5 marks)**
b Calculate $E(X^2)$ and $Var(X)$ using the values of a and b that you found in part **a**. **(3 marks)**
c Write down the value of $Var(Y)$. **(1 mark)**
d Find $P(Y + 2 > X)$. **(2 marks)**

E/P **17** The discrete random variable X has a probability distribution given by

x	−2	0	2	3	4
$P(X = x)$	a	b	a	b	c

The random variable Y is defined as $Y = \frac{2 - 3X}{5}$

You are given that $E(Y) = -0.98$ and $P(Y \geqslant -1) = 0.4$.

a Write down three simultaneous equations in a, b and c. **(4 marks)**
b Solve this system to find the values of a, b and c. **(3 marks)**
c Find $P(-2X > 10Y)$. **(2 marks)**

Challenge

A Let n be a positive integer and suppose that X is a discrete random variable with $P(X = i) = \frac{1}{n}$ for $i = 1, \ldots, n$.

Show that $E(X) = \frac{n+1}{2}$ and $Var(X) = \frac{(n+1)(n-1)}{12}$

Hint You can make use of the following results:

$\sum_{i=1}^{n} i = \frac{n(n+1)}{2}$

$\sum_{i=1}^{n} i^2 = \frac{n(n+1)(2n+1)}{6}$

← Core Pure Book 1, Chapter 3

Summary of key points

1. The expected value of the discrete random variable X is denoted $\mathrm{E}(X)$ and defined as
   $\mathbf{E}(X) = \sum x\mathbf{P}(X = x)$

2. The expected value of X^2 is $\mathbf{E}(X^2) = \sum x^2\mathbf{P}(X = x)$

3. The variance of X is usually written as $\mathrm{Var}(X)$ and is defined as
   $\mathbf{Var}(X) = \mathbf{E}((X - \mathbf{E}(X))^2)$

4. Sometimes, it is easier to calculate the variance using the formula
   $\mathbf{Var}(X) = \mathbf{E}(X^2) - (\mathbf{E}(X))^2$

5. If X is a discrete random variable, and g is a function, then $\mathrm{g}(X)$ is also a discrete random variable.

   You can calculate the expected value of $\mathrm{g}(X)$ using the formula:

   $\mathbf{E}(\mathbf{g}(X)) = \sum \mathbf{g}(x)\,\mathbf{P}(X = x)$

6. If X is a random variable and a and b are constants, then $\mathbf{E}(aX + b) = a\mathbf{E}(X) + b$.

7. If X and Y are random variables, then $\mathbf{E}(X + Y) = \mathbf{E}(X) + \mathbf{E}(Y)$

8. If X is a random variable and a and b are constants then $\mathbf{Var}(aX + b) = a^2\mathbf{Var}(X)$

Poisson distributions

2

Objectives

After completing this chapter you should be able to:

- Use the Poisson distribution to model real-world situations → pages 20–27
- Use the additive property of the Poisson distribution → pages 27–29
- Understand and use the mean and variance of the Poisson distribution → pages 30–31
- Understand and use the mean and variance of the binomial distribution → pages 32–34
- Use the Poisson distribution as an approximation to the binomial distribution → pages 34–38

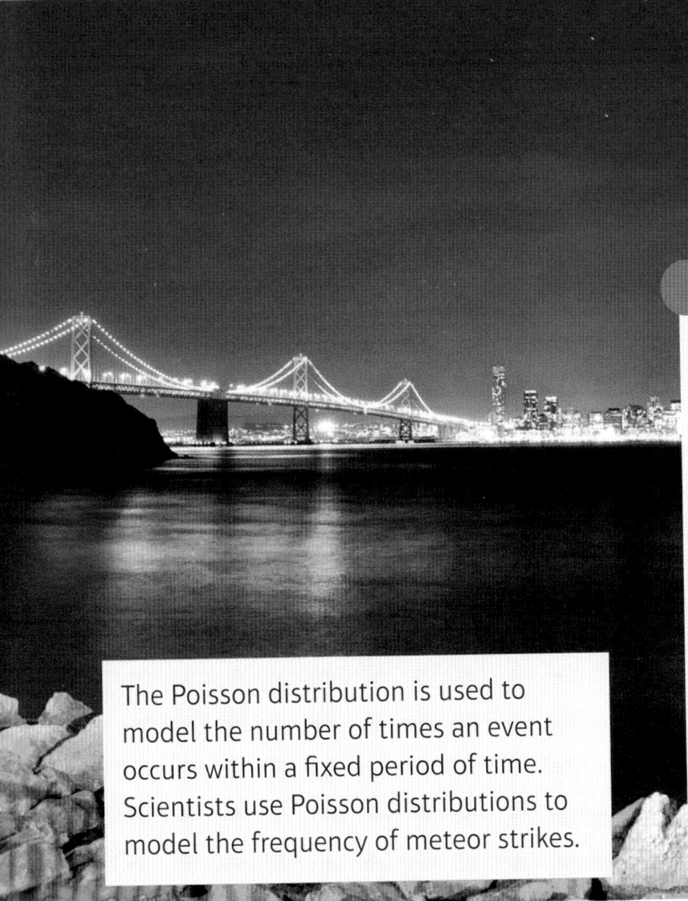

The Poisson distribution is used to model the number of times an event occurs within a fixed period of time. Scientists use Poisson distributions to model the frequency of meteor strikes.

Prior knowledge check

1 The random variable $X \sim B(35, 0.4)$. Find:

a $P(X = 20)$ **b** $P(X < 6)$

c $P(15 \leqslant X < 20)$

← Statistics and Mechanics Year 1, Chapter 6

2 A biased dice is modelled by a random variable X with the following probability distribution.

x	1	2	3	4	5	6
$P(X = x)$	0.2	0.1	0.1	0.2	0.1	0.3

Find:

a $E(X)$ **b** $E(X^2)$

c $Var(X)$

← Sections 1.1, 1.2

2.1 The Poisson distribution

The exponential function e^x can be defined as an infinite series expansion:

$$e^x = 1 + \frac{x^1}{1!} + \frac{x^2}{2!} + \frac{x^3}{3!} + \ldots + \frac{x^r}{r!} + \ldots$$

Links This is the Maclaurin expansion of e^x.
← **Core Pure 2, Chapter 2**

This definition can be used to generate a probability distribution with parameter λ, where $\lambda > 0$.

$$e^\lambda = \lambda^0 + \frac{\lambda^1}{1!} + \frac{\lambda^2}{2!} + \frac{\lambda^3}{3!} + \ldots + \frac{\lambda^r}{r!} + \ldots$$

Dividing both sides by e^λ gives

$$1 = \lambda^0 e^{-\lambda} + \frac{\lambda^1 e^{-\lambda}}{1!} + \frac{\lambda^2 e^{-\lambda}}{2!} + \frac{\lambda^3 e^{-\lambda}}{3!} + \ldots + \frac{\lambda^r e^{-\lambda}}{r!} + \ldots$$

Notice that the sum of the probabilities of the infinite series on the right-hand side equals 1 and so you could use these values as probabilities to define a probability distribution.

Links The sum of the probabilities in any probability distribution must equal 1.
← **Statistics and Mechanics Year 1, Section 6.1**

If we let X be a discrete random variable such that X takes the values 0, 1, 2, 3, … then the probability distribution for X could be:

X	0	1	2	3	…	r	…
P($X = x$)	$e^{-\lambda}$	$\frac{e^{-\lambda}\lambda^1}{1!}$	$\frac{e^{-\lambda}\lambda^2}{2!}$	$\frac{e^{-\lambda}\lambda^3}{3!}$		$\frac{e^{-\lambda}\lambda^r}{r!}$	

This distribution is called the Poisson distribution.

- **If $X \sim \text{Po}(\lambda)$, then the Poisson distribution is given by**

$$\mathbf{P}(X = x) = \frac{e^{-\lambda}\lambda^x}{x!} \qquad x = \mathbf{0, 1, 2, 3, \ldots}$$

Watch out This is an infinite probability distribution. $P(X = x) > 0$ for any positive integer x, although as x gets large, the probabilities get very small.

Notation You say that X has a Poisson distribution with parameter λ.

Example 1

The random variable $X \sim \text{Po}(2.1)$. Find:

a $P(X = 3)$

b $P(X \geqslant 1)$

c $P(1 < X \leqslant 4)$

a $P(X = 3) = \frac{e^{-2.1} \times 2.1^3}{3!}$
$= 0.189011\ldots$
$= 0.1890$ (4 d.p.)

Use the formula $P(X = x) = \frac{e^{-\lambda}\lambda^x}{x!}$ with $x = 3$ and $\lambda = 2.1$.

You can work this out using the Poisson probability distribution function on your calculator.

b $P(X \geqslant 1) = 1 - P(X = 0)$
$= 1 - e^{-2.1}$
$= 1 - 0.1224\ldots$
$= 0.8775$ (4 d.p.)

X can only take positive integer values.

Round probabilities to 4 decimal places.

c $P(1 < X \leqslant 4)$
$= P(X = 2) + P(X = 3) + P(X = 4)$
$= \frac{e^{-2.1} \times 2.1^2}{2!} + \frac{e^{-2.1} \times 2.1^3}{3!} + \frac{e^{-2.1} \times 2.1^4}{4!}$
$= 0.2700... + 0.1890... + 0.0992...$
$= 0.5583$ (4 d.p.)

Add together all the possible probabilities. The positive integers that satisfy the inequality are 2, 3 and 4.

Online Explore the Poisson distribution using GeoGebra.

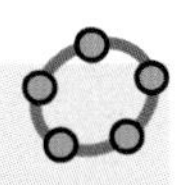

Exercise 2A

1 The discrete random variable $X \sim \text{Po}(2.5)$. Find:
 a $P(X = 3)$ b $P(X > 1)$ c $P(1 < X \leqslant 3)$

2 The discrete random variable $X \sim \text{Po}(3.1)$. Find:
 a $P(X = 4)$ b $P(X \geqslant 2)$ c $P(1 \leqslant X \leqslant 4)$

3 The discrete random variable $X \sim \text{Po}(4.2)$. Find:
 a $P(X = 2)$ b $P(X \leqslant 3)$ c $P(3 \leqslant X \leqslant 5)$

4 The discrete random variable $X \sim \text{Po}(0.84)$. Find:
 a $P(X = 1)$ b $P(X \geqslant 1)$ c $P(1 < X \leqslant 3)$

(P) 5 The discrete random variable $X \sim \text{Po}(\lambda)$. Given that $P(X = 2) = P(X = 3)$, find λ.

(P) 6 The discrete random variable $X \sim \text{Po}(\lambda)$. Given that $P(X = 4) = 3 \times P(X = 2)$, find λ.

Calculations involving the Poisson distribution can often be simplified by using the cumulative distribution tables given on page 191. These tables will be given in the *Mathematical Formulae and Statistical Tables* booklet in your exam. These will tell you $P(X \leqslant x)$ for values of λ between 0 and 10, in steps of 0.5, and for values of x from 0 to 22.

You can also use the Poisson cumulative distribution function on your calculator to find $P(X \leqslant x)$ for other values of λ and x.

Example 2

The random variable $X \sim \text{Po}(5)$. Find, using tables:
a $P(X \leqslant 3)$ **b** $P(X \geqslant 2)$ **c** $P(1 \leqslant X \leqslant 4)$

a $P(X \leqslant 3) = 0.2650$

$\lambda =$	0.5	4.5	5.0
$x = 0$	0.6065	0.0111	0.0067
1	0.9098	0.0611	0.0404
2	0.9856	0.1736	0.1247
3	0.9982	0.3423	0.2650
4	0.9998	0.5321	0.4405

b $P(X \geqslant 2) = 1 - P(X \leqslant 1) = 1 - 0.0404$
$= 0.9596$

Be careful. The inequality is $\geqslant$ so you need to work out $1 - P(X \leqslant 1)$.

c $P(1 \leqslant X \leqslant 4) = P(X \leqslant 4) - P(X \leqslant 0)$
$= 0.4405 - 0.0067$
$= 0.4338$

Example 3

The random variable $X \sim Po(7.5)$. Find the values of a, b and c such that:

a $P(X \leqslant a) = 0.2414$ **b** $P(X < b) = 0.5246$ **c** $P(X \geqslant c) = 0.3380$

a $P(X \leqslant a) = 0.2414$
so $a = 5$

Use tables with $\lambda = 7.5$.
$P(X \leqslant 5) = 0.2414$

b $P(X < b) = P(X \leqslant b - 1) = 0.5246$
so $b - 1 = 7$
$b = 8$

Use tables with $\lambda = 7.5$.
$P(X \leqslant 7) = 0.5246$

c $P(X \geqslant c) = 1 - P(X \leqslant c - 1) = 0.3380$
so $P(X \leqslant c - 1) = 1 - 0.3380$
$= 0.6620$
so $c - 1 = 8$
$c = 9$

Use tables with $\lambda = 7.5$.
$P(X \leqslant 8) = 0.6620$

Exercise 2B

Use the Poisson cumulative distribution tables on page 191 to answer these questions.

1 The discrete random variable $X \sim Po(5.5)$. Find:
a $P(X \leqslant 3)$ **b** $P(X \geqslant 6)$ **c** $P(3 \leqslant X \leqslant 7)$

2 The discrete random variable $X \sim Po(10)$. Find:
a $P(X \geqslant 8)$ **b** $P(7 \leqslant X \leqslant 12)$ **c** $P(4 < X < 9)$

3 The discrete random variable $X \sim Po(3.5)$. Find:
a $P(X \geqslant 2)$ **b** $P(3 \leqslant X \leqslant 6)$ **c** $P(2 < X \leqslant 5)$

4 The discrete random variable $X \sim Po(4.5)$. Find:
a $P(X \geqslant 5)$ **b** $P(3 < X \leqslant 5)$ **c** $P(1 \leqslant X < 7)$

5 The discrete random variable $X \sim Po(8)$. Find the values of a, b, c and d such that:
a $P(X \leqslant a) = 0.3134$ **b** $P(X \leqslant b) = 0.7166$ **c** $P(X < c) = 0.0996$ **d** $P(X > d) = 0.8088$

6 The discrete random variable $X \sim Po(3.5)$. Find the values of a, b, c and d such that:
a $P(X \leqslant a) = 0.8576$ **b** $P(X > b) = 0.6792$ **c** $P(X \leqslant c) \geqslant 0.95$ **d** $P(X > d) \leqslant 0.005$

2.2 Modelling with the Poisson distribution

You need to be able to recognise situations that can be modelled with a Poisson distribution. The Poisson distribution is used to model the number of times, X, that a particular event occurs within a given interval of time or space.

■ **In order for the Poisson distribution to be a good model, the events must occur:**

- **independently**
- **singly, in space or time**
- **at a constant average rate so that the mean number in an interval is proportional to the length of the interval**

The parameter, λ, in the Poisson distribution is the average number of times that the event will occur in a single interval.

Examples of where a Poisson distribution might be appropriate are:

- the number of radioactive particles being emitted by a certain source in a 5-minute period
- the number of telephone calls to a switchboard in a 10-minute interval
- the number of spelling mistakes on a page of a newspaper
- the number of cars passing the front of a school in a 3-minute interval
- the number of raisins in a fruit scone.

Example 4

An internet service provider has a large number of users regularly connecting to the internet. On average, 4 users every hour fail to connect to the internet at their first attempt.

a Give two reasons why a Poisson distribution might be a suitable model for the number of failed connections every hour.

b Find the probability that in a randomly chosen hour:

i 2 users fail to connect at their first attempt

ii more than 6 users fail to connect at their first attempt.

c Find the probability that in a randomly chosen 90-minute period:

i 5 users fail to connect at their first attempt

ii fewer than 7 users fail to connect at their first attempt.

a Failed connections occur singly and at a constant rate of 4 users per hour.

b X = the number of failed connections in one hour

$X \sim \text{Po}(4)$

i $P(X = 2) = 0.1465$

ii $P(X > 6) = 1 - P(X \leqslant 6)$
$= 1 - 0.88932...$
$= 0.1107$ (4 d.p.)

Define your random variable, and write down the model you are using.

Use the tables, or your calculator, with $\lambda = 4$ to find $P(X \leqslant 6)$.

c Y = the number of failed connections in 90 minutes

$Y \sim \text{Po}(6)$

i $P(Y = 5) = 0.1606$ (4 d.p.)

ii $P(Y < 7) = P(Y \leqslant 6)$
$= 0.6063$ (4 d.p.)

Problem-solving

Because the failures occur at a **constant average rate** the value of the parameter λ will be $\frac{90}{60} \times 4 = 6$ for a 90-minute period.

Exercise 2C

1 The maintenance department of a school receives requests for replacement light bulbs at a rate of 3 per week.
The number of requests, X, in a given week is modelled as $X \sim \text{Po}(3)$.

a Find the probability that, in a randomly chosen week, the number of requests for replacement light bulbs is:
i exactly 4
ii more than 5.

b Find the probability that, in a randomly chosen fortnight, the number of requests for replacement light bulbs is:
i exactly 6
ii no more than 4.

Hint The number of requests, Y, in a given fortnight can be modelled as $Y \sim \text{Po}(6)$.

2 A botanist suggests that the number of weeds growing in a field can be modelled by a Poisson distribution.

a Write down two conditions that must apply for this model to be applicable.

Assuming this model and that weeds occur at a rate of 1.3 per m^2, find:

b the probability that, in a randomly chosen plot of size $4\,m^2$, there will be fewer than 3 weeds

c the probability that, in a randomly chosen plot of $5\,m^2$, there will be more than 8 weeds.

Problem-solving

The number of weeds, X, in a plot of $4\,m^2$ can be modelled as $X \sim \text{Po}(4 \times 1.3)$, i.e $X \sim \text{Po}(5.2)$.

3 An electronics company manufactures a component for use in computer hardware. At the end of the manufacturing process, each component is checked to see if it is faulty. Faulty components are detected at a rate of 2.5 per hour.

a Suggest a suitable model for the number of faulty components detected per hour.

b Describe, in the context of this question, two assumptions you have made in part **a** for this model to be suitable.

c Find the probability of 2 faulty components being detected in a 1-hour period.

d Find the probability of at least 6 faulty components being detected in a 3-hour period.

e Find the probability of at least 7 faulty components being detected in a 4-hour period.

4 A call-centre agent handles telephone calls at a rate of 15 per hour.

a Find the probability that, in any randomly selected 20-minute interval, the agent handles:
i exactly 4 calls **ii** more than 8 calls.

b Find the probability that, in a randomly selected 30-minute interval, the agent handles:
i at least 6 calls **ii** no more than 10 calls.

5 The average number of cars crossing over a bridge is 180 per hour. Assuming a Poisson distribution, find the probability that:

a more than 5 cars will cross in any given minute

b fewer than 4 cars will cross in any 2-minute period.

6 A café serves breakfast every morning. Customers arrive for breakfast at random at an average rate of 1 every 4 minutes.
Find the probability that on a Friday morning between 10 am and 10:20 am:

a fewer than 3 customers arrive for breakfast

b more than 10 customers arrive for breakfast.

E/P **7** An estate agent has been selling houses at a rate of 1.8 per week.

a Find the probability that in a particular week she sells:
i no houses **ii** 3 houses **iii** at least 3 houses. **(6 marks)**

The estate agent meets her weekly target if she sells at least 3 houses in one week.

b Find the probability that over a period of 4 consecutive weeks she meets her weekly target exactly once. **(3 marks)**

Problem-solving

Use a binomial model for part **b**.

← Statistics and Mechanics Year 1, Chapter 6

E **8** Patients arrive at a hospital accident and emergency department at random at a rate of 5 per hour.

a Find the probability that, during any 30-minute period, the number of patients arriving at the hospital accident and emergency department is:
i exactly 4 **ii** at least 3. **(5 marks)**

A patient arrives at 11:00 am.

b Find the probability that the next patient arrives before 11:15 am. **(3 marks)**

(E) **9** The lift in a block of flats breaks down at random at a mean rate of three times per four-week period.

a Find the probability that the lift breaks down:
i at least once in one week
ii exactly twice in one week. **(5 marks)**

In one particular week, the lift broke down twice.

b Write down the probability that the lift will break down at some point in the next week. Give a reason for your answer. **(2 marks)**

(E/P) **10** Flaws occur at random in a particular type of material at a mean rate of 1.5 per 50 m.

a Find the probability that, in a randomly chosen 50 m length of this material, there will be exactly 3 flaws. **(2 marks)**

This material is sold in rolls of length 200 m.

b Find the probability that a single roll has fewer than 4 flaws. **(3 marks)**

Priya buys 5 rolls of this material.

c Find the probability that at least two of these rolls will have fewer than 4 flaws. **(5 marks)**

(E/P) **11** A company produces chocolate chip biscuits. The number of chocolate chips per biscuit has a Poisson distribution with mean 5.

a Find the probability that one of these biscuits, selected at random, contains fewer than 3 chocolate chips. **(2 marks)**

A packet contains 6 of these biscuits, selected at random.

b Find the probability that exactly half of the biscuits in the packet contain fewer than 3 chocolate chips. **(4 marks)**

(E/P) **12** A company has minibuses that can only be hired for a week at a time. All hiring starts on a Sunday. During the summer, the mean number of requests for minibuses each Sunday is 5.

a Calculate the probability that fewer than 4 requests for minibuses are made on a particular Sunday in summer. **(2 marks)**

The company wants to be at least 99% sure they can fulfil all requests on any particular Sunday.

b Calculate the number of minibuses the company must have in order to satisfy this condition. **(3 marks)**

(E/P) **13** On a typical summer's day, a boat company hires out rowing boats at a rate of 9 per hour.

a Find the probability of hiring out at least 6 boats in a randomly selected 30-minute period. **(2 marks)**

The company has 8 boats and decides to hire them out for 20-minute periods.

b Show that the probability of running out of boats is less than 1%. **(3 marks)**

c Find the number of boats that the company should have in order to be 99% sure of meeting all demands if the hire period is extended to 30 minutes. **(3 marks)**

(E/P) **14** Breakdowns on a particular machine occur at a rate of 1.5 per week.

a Find the probability that no more than 2 breakdowns occur in a randomly chosen week. **(2 marks)**

b Find the probability of at least 5 breakdowns in a randomly chosen two-week period. **(3 marks)**

A maintenance firm offers a contract for repairing breakdowns over a six-week period.
The firm will give a full refund if there are more than n breakdowns in a six-week period.
The firm wants the probability of having to pay a refund to be 5% or less.

c Find the smallest possible value of n. **(3 marks)**

2.3 Adding Poisson distributions

If two Poisson variables X and Y are independent, then the variable $Z = X + Y$ also has a Poisson distribution.

- **If $X \sim \text{Po}(\lambda)$ and $Y \sim \text{Po}(\mu)$, then $X + Y \sim \text{Po}(\lambda + \mu)$**

Watch out For $X + Y$ to be meaningful in this context, the random variables X and Y must both model events occurring within the same interval of time or space.

Example 5

If $X \sim \text{Po}(3.6)$ and $Y \sim \text{Po}(4.4)$, find:

a $P(X + Y = 7)$ **b** $P(X + Y \leqslant 5)$

$X + Y \sim \text{Po}(3.6 + 4.4)$ — Add the parameters.

$X + Y \sim \text{Po}(8)$

a $P(X + Y = 7) = \dfrac{e^{-8} \times 8^7}{7!} = 0.1396$ (4 d.p.)

b $P(X + Y \leqslant 5) = 0.1912$ (4 d.p.)

Use the tables, or your calculator, with $\lambda = 8$.

Example 6

The number of cars passing an observation point in a 5-minute interval is modelled by a Poisson distribution with mean 2. The number of other vehicles passing the observation point in a 15-minute interval is modelled by a Poisson distribution with mean 3.

Find the probability that:

a exactly 5 vehicles, of any type, pass the observation point in a 10-minute interval

b more than 8 vehicles, of any type, pass the observation point in a 15-minute interval.

a X_1 = number of cars passing in a 10-minute interval

Y_1 = number of other vehicles passing in a 10-minute interval

$X_1 \sim \text{Po}(4)$, $Y_1 \sim \text{Po}(2)$

$X_1 + Y_1 \sim \text{Po}(4 + 2)$

$X_1 + Y_1 \sim \text{Po}(6)$

$P(X_1 + Y_1 = 5) = 0.1606$ (4 d.p.)

Problem-solving

You need to model the number of cars passing in a 10-minute interval, and the number of other vehicles passing in a 10-minute interval. The time intervals must be the same before you can add the parameters.

b X_2 = number of cars passing in a 15-minute interval

Y_2 = number of other vehicles passing in a 15-minute interval

$X_2 \sim \text{Po}(6)$, $Y_2 \sim \text{Po}(3)$

$X_2 + Y_2 \sim \text{Po}(6 + 3)$

$X_2 + Y_2 \sim \text{Po}(9)$

$\text{P}(X_2 + Y_2 > 8) = 1 - \text{P}(X_2 + Y_2 \leqslant 8)$

$= 1 - 0.45565\ldots$

$= 0.5443$ (4 d.p.)

Define new random variables for the number of cars, and other types of vehicle, passing in a 15-minute interval.

This can be calculated using the tables, or your calculator, with $\lambda = 9$.

Exercise 2D

1 X and Y are independent random variables such that $X \sim \text{Po}(3.3)$ and $Y \sim \text{Po}(2.7)$. Find:

a $\text{P}(X + Y = 5)$ **b** $\text{P}(X + Y \leqslant 7)$ **c** $\text{P}(X + Y > 4)$

2 A and B are independent random variables such that $A \sim \text{Po}(3.25)$ and $B \sim \text{Po}(4.25)$. Find:

a $\text{P}(A + B = 7)$ **b** $\text{P}(A + B \leqslant 5)$ **c** $\text{P}(A + B > 9)$

(P) **3** X and Y are independent random variables such that $X \sim \text{Po}(2.5)$ and $Y \sim \text{Po}(3.5)$. Find:

a $\text{P}(X = 2 \text{ and } Y = 2)$ **b** P(both X and Y are greater than 2)

c $\text{P}(X + Y = 5)$ **d** $\text{P}(X + Y \leqslant 4)$

(P) **4** The number of emissions per minute from two different sources of radioactivity are modelled as independent Poisson random variables X and Y, with parameters of 3 and 5 respectively. Calculate the probability that, in a given one-minute period,

a the number of emissions from each source is at least 3

b the total number of emissions from the two sources is no more than 6.

(P) **5** During a weekday at a certain point of a road, cars pass by at a rate of 24 per minute, while lorries pass by at a rate of 8 per minute.

a Find the probability that, in any 15-second period,

i at least 4 of each type of vehicle passes by

ii the total number of cars and lorries that pass by is no more than 9.

b Write down one assumption that you have made in your calculations.

(E) **6** A taxi company supplies two particular organisations independently.
Company A orders taxis at a rate of 1.25 cars per day.
Company B orders taxis at a rate of 0.75 cars per day.

a On a given day, find the probability that 2 cars are ordered by company A. **(2 marks)**

b On a given day, find the probability that the total number of cars ordered by both companies is 2. **(2 marks)**

c In a given 5-day working week, find the probability that the total number of cars ordered by both companies is less than 10. **(2 marks)**

E/P **7** A restaurant has two coffee machines, C and D. Machine C breaks down at a rate of 0.1 times per week while, independently, machine D breaks down at a rate of 0.05 times per week. Find the probability that, in a 12-week period,

a machine C breaks down at least once **(2 marks)**

b each machine breaks down at least once **(3 marks)**

c there will be a total of 3 breakdowns. **(2 marks)**

E **8** A secretary receives internal calls at a rate of 1 every 5 minutes and external calls at a rate of 2 every 5 minutes.
Calculate the probability that the total number of calls is:

a 3 in a 4-minute period **(2 marks)**

b at least 2 in a 2-minute period **(2 marks)**

c no more than 5 in a 10-minute period. **(2 marks)**

E/P **9** An office is situated on 3 floors of a building. On each floor it has a photocopier. The ground-floor photocopier breaks down at a rate of 0.4 times per week, the first-floor photocopier breaks down at a rate of 0.2 times per week and the second-floor photocopier breaks down at a rate of 0.8 times per week. Find the probability, in a given week, that:

a each photocopier will break down exactly once **(3 marks)**

b at least one photocopier breaks down **(3 marks)**

c there will be a total of 2 breakdowns. **(2 marks)**

E/P **10** During the working day the emails arriving to the account of a company director are classified into three types: personal, business and advertising. Personal emails arrive at a mean rate of 1.8 per hour, business emails arrive at a mean rate of 3.7 per hour and advertising emails arrive at a mean rate of 1.5 per hour. Find the probability that she receives:

a at least one of each type of email during a 30-minute period of the working day **(3 marks)**

b more than 50 emails in an 8-hour working day. **(3 marks)**

c Find the probability that she receives more than 50 emails on exactly two days out of a 5-day working week.

Use a binomial model for part **c**.
← **Statistics and Mechanics Year 1, Chapter 6**

(3 marks)

Challenge

$X \sim \text{Po}(\lambda)$ and $Y \sim \text{Po}(\mu)$. The random variable $Q = X + Y$.

a Prove that $\text{P}(Q = 0) = \text{e}^{-(\lambda + \mu)}$

b Prove that $\text{P}(Q = 1) = (\lambda + \mu)\text{e}^{-(\lambda + \mu)}$

2.4 Mean and variance of a Poisson distribution

It can be shown that if the random variable, X, has a Poisson distribution with parameter, λ, then the mean and variance of X, are both equal to λ.

- **If $X \sim \text{Po}(\lambda)$**
  - **Mean of $X = \text{E}(X) = \lambda$**
  - **Variance of $X = \text{Var}(X) = \sigma^2 = \lambda$**

The fact that the mean is equal to the variance is an important property of a Poisson distribution. The presence or absence of this property can be a useful indicator of whether or not a Poisson distribution is a suitable model for a particular situation.

Example 7

A botanist counts the number of daisies, x, in each of 80 randomly selected squares within a field. The results are summarised below.

$$\Sigma x = 295,\ \Sigma x^2 = 1386$$

a Calculate the mean and the variance of the number of daisies per square for the 80 squares. Give your answers to 2 decimal places.

b Explain how the answers from part **a** support the choice of a Poisson distribution as a model.

c Using a suitable value for λ, estimate the probability that exactly 3 daisies will be found in a randomly selected square.

a Mean $= \bar{x} = \frac{\Sigma x}{80} = \frac{295}{80} = 3.69$ (2 d.p.)

Variance $= \sigma^2 = \frac{\Sigma x^2}{80} - \bar{x}^2$

$= \frac{1386}{80} - \left(\frac{295}{80}\right)^2$

$= 3.73$ (2 d.p.)

b Both the mean and the variance are 3.7 correct to one decimal place. The fact that the mean is close to the variance supports the choice of a Poisson distribution as a model.

c Using $\lambda = 3.7$

Use $\lambda = 3.7$, which is the mean and variance from part **b**.

X = the number of daisies per square

$X \sim \text{Po}(3.7)$

$\text{P}(X = 3) = 0.2087$ (4 d.p.)

This can be calculated using the tables, or your calculator, with $\lambda = 3.7$.

Exercise 2E

1 A student is investigating the numbers of cherries in a fruit scone. A random sample of 100 fruit scones is taken and the results can be summarised as:

$\Sigma x = 143,\ \Sigma x^2 = 347$

a Calculate the mean and the variance of the data.

b Explain why the results in part **a** suggest that a Poisson distribution may be a suitable model for the number of cherries in a fruit scone.

c Using a suitable value for λ, estimate the probability that exactly 3 cherries will be found in a randomly selected fruit scone.

2 The number of cars passing a checkpoint during 200 periods of 5 minutes is recorded.

Number of cars	0	1	2	3	4	5	6	7	8	$\geqslant 9$
Frequency	7	21	30	41	36	29	21	11	4	0

a Calculate the mean and the variance of the data.

b Explain why the results in part **a** suggest that a Poisson distribution may be a suitable model for the number of cars passing the checkpoint in a 5-minute period.

c Using a suitable value for λ, estimate the probability that no more than 2 cars will pass the checkpoint in a given 5-minute period.

d Compare your answer to part **c** with the relative frequency of obtaining no more than 2 cars from the sample.

(E) **3** Tests for flaws are carried out in a textile factory on a consignment of 120 pieces of cloth. The results of the tests are shown in the table.

Number of flaws	0	1	2	3	4	5	6	7	$\geqslant 8$
Number of pieces	8	19	28	25	19	11	7	3	0

a Calculate the mean and the variance of the data. **(4 marks)**

b Explain why the results in part **a** suggest that a Poisson distribution may be a suitable model for the number of flaws on a piece of cloth. **(1 mark)**

The factory produces 10 000 pieces of cloth each week, and wants to estimate the number that will have 8 or more flaws.

c Explain why an estimate based on the observed relative frequencies would not be useful. **(1 mark)**

d Use a Poisson distribution to estimate the number of pieces of cloth with 8 or more flaws. **(3 marks)**

Challenge

If $X \sim \text{Po}(\lambda)$, then the distribution of X can be written as:

X	0	1	2	3	...	r	...
$\text{P}(X = x)$	$e^{-\lambda}$	$\frac{e^{-\lambda}\lambda^1}{1!}$	$\frac{e^{-\lambda}\lambda^2}{2!}$	$\frac{e^{-\lambda}\lambda^3}{3!}$		$\frac{e^{-\lambda}\lambda^r}{r!}$	

Using this distribution show that $\text{E}(X) = \lambda$ and $\text{Var}(X) = \lambda$.

2.5 Mean and variance of the binomial distribution

You need to know how to calculate the mean and variance of a binomial random variable.

- **If X is a binomial random variable with $X \sim B(n, p)$, then:**
  - **Mean of $X = E(X) = \mu = np$**
  - **Variance of $X = Var(X) = \sigma^2 = np(1 - p)$**

Links The probability mass function for a binomial random variable $X \sim B(n, p)$ is:

$P(X = x) = \binom{n}{x} p^x (1 - p)^{n - x} \quad x = 0, 1, 2, 3, \ldots$

← Statistics and Mechanics Year 1, Chapter 6

Example 8

A fair, five-sided spinner is spun 20 times. The random variable X represents the number of 5s obtained.

a Find the mean and variance of X. **b** Find $P(X < \mu - \sigma)$.

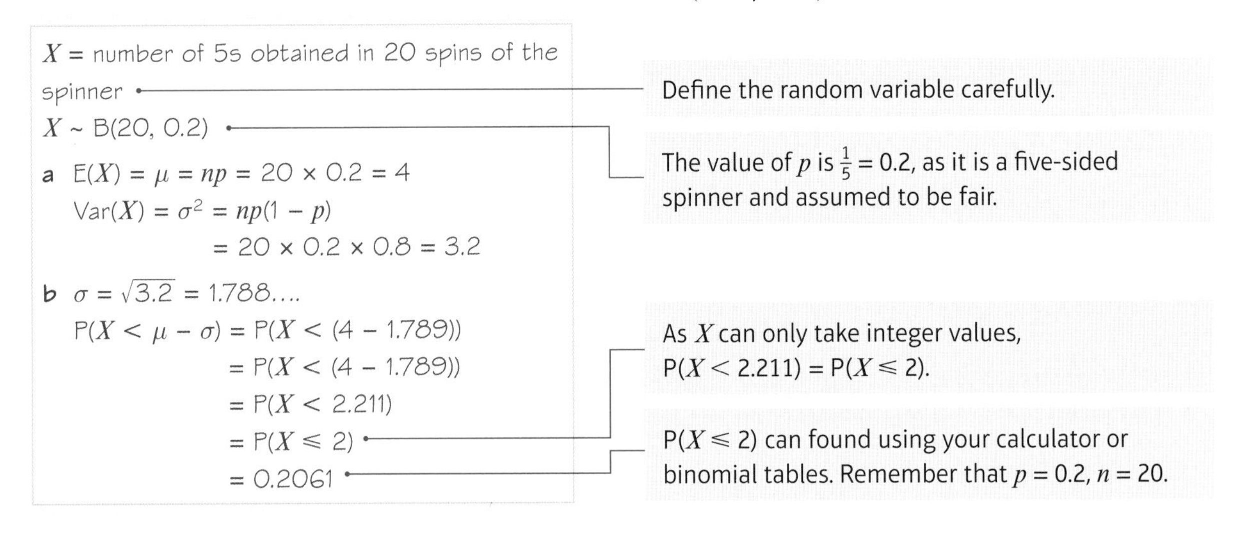

X = number of 5s obtained in 20 spins of the spinner — Define the random variable carefully.

$X \sim B(20, 0.2)$ — The value of p is $\frac{1}{5} = 0.2$, as it is a five-sided spinner and assumed to be fair.

a $E(X) = \mu = np = 20 \times 0.2 = 4$

$Var(X) = \sigma^2 = np(1 - p)$

$= 20 \times 0.2 \times 0.8 = 3.2$

b $\sigma = \sqrt{3.2} = 1.788\ldots$

$P(X < \mu - \sigma) = P(X < (4 - 1.789))$

$= P(X < (4 - 1.789))$

$= P(X < 2.211)$

$= P(X \leqslant 2)$ — As X can only take integer values, $P(X < 2.211) = P(X \leqslant 2)$.

$= 0.2061$ — $P(X \leqslant 2)$ can found using your calculator or binomial tables. Remember that $p = 0.2$, $n = 20$.

Example 9

A company produces a certain type of delicate component. The probability of any one component being defective is p. The probability of obtaining at least one defective component in a sample of 4 is 0.3439.

The company produces 600 components in a day.

Find the mean and variance of the number of defective components produced per day.

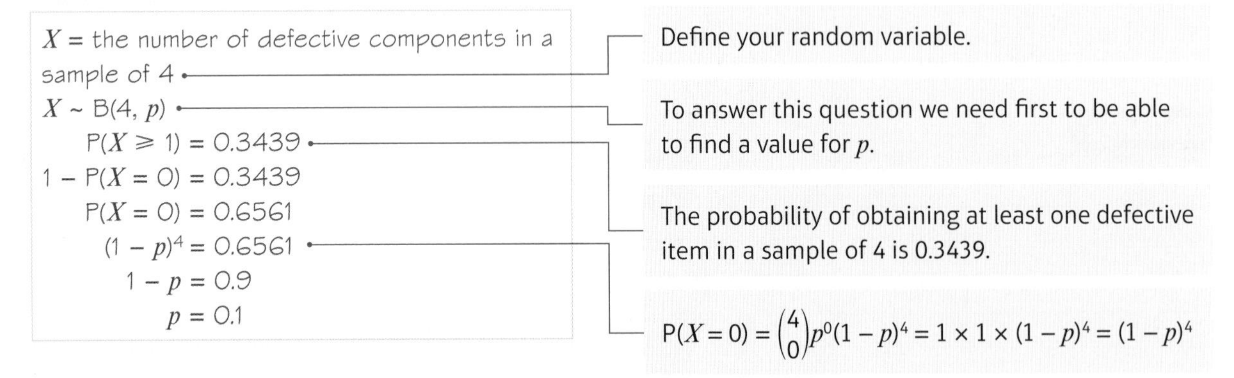

X = the number of defective components in a sample of 4 — Define your random variable.

$X \sim B(4, p)$ — To answer this question we need first to be able to find a value for p.

$P(X \geqslant 1) = 0.3439$ — The probability of obtaining at least one defective item in a sample of 4 is 0.3439.

$1 - P(X = 0) = 0.3439$

$P(X = 0) = 0.6561$

$(1 - p)^4 = 0.6561$ — $P(X = 0) = \binom{4}{0} p^0 (1 - p)^4 = 1 \times 1 \times (1 - p)^4 = (1 - p)^4$

$1 - p = 0.9$

$p = 0.1$

Y = the number of defectives in 600 components

Define a new random variable using the value of p you obtained earlier.

$Y \sim B(600, 0.1)$

Mean = $E(Y) = np = 600 \times 0.1 = 60$

Variance = $\text{Var}(Y) = np(1 - p)$

$= 600 \times 0.1 \times 0.9 = 54$

Exercise 2F

1 X is the random variable such that $X \sim B(12, 0.7)$. Find:

a $E(X)$ **b** $\text{Var}(X)$

2 X is the random variable such that $X \sim B(n, 0.4)$ and $E(X) = 3.2$. Find:

a the value of n **b** $P(X = 5)$ **c** $P(X \leqslant 2)$

3 X is the random variable such that $X \sim B(10, p)$ and $\text{Var}(X) = 2.4$.
Find the two possible values of p.

4 X is the random variable such that $X \sim B(15, p)$ and $\text{Var}(X) = 2.4$.
Find the two possible values of p.

5 X is the random variable such that $X \sim B(n, p)$, $E(X) = 4.8$, $\text{Var}(X) = 2.88$.
Find the values of n and p.

(P) **6** The probability of obtaining a head when a biased coin is spun is p, where $p < \frac{1}{2}$.
An experiment consists of spinning the coin 20 times and recording the number of heads. In a large number of experiments the variance of the number of heads is found to be 4.2.

a Estimate the value of p. **(2 marks)**

b Hence estimate the probability that exactly 7 heads are recorded during a particular experiment. **(2 marks)**

(P) **7** The probability that a canvasser gets a reply when she knocks on the door of a house is 0.65.

a Find the probability that in a street of 10 houses she receives:

i exactly 5 replies

ii at least 5 replies.

b **i** How many houses should she canvas such that the random variable X = 'number of replies' has a mean of 78.

ii What is the variance in this case?

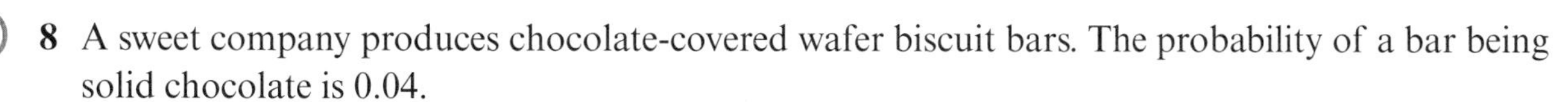

(E/P) **8** A sweet company produces chocolate-covered wafer biscuit bars. The probability of a bar being solid chocolate is 0.04.

a Find the probability that in a box of 48 bars, at least two are solid chocolate. **(3 marks)**

The company produces 120 boxes of biscuits per day.

b Find the mean and variance of the number of boxes of biscuits which contain at least two solid chocolate bars. **(2 marks)**

(E) **9** The random variable X is such that $X \sim B(5, p)$. Given that $P(X \geqslant 1) = 0.83193$, find:

a the value of p **(3 marks)**

b $E(X)$ and $Var(X)$. **(2 marks)**

(P) **10** A biased dice is thrown 5 times and the number of sixes is noted. The experiment is conducted 500 times. The results are shown in the table.

Number of sixes	0	1	2	3	4	5
Frequency	163	208	98	28	3	0

A student wishes to show that the data can be modelled by a binomial distribution.

a Calculate the mean and variance of the number of sixes in 5 throws of the dice.

b Based on the mean of the data, estimate the probability p of getting a six with this dice.

c Using the value found in part **b,** calculate the expected frequencies of 0, 1, 2, 3, 4 and 5 sixes in 500 experiments, using a binomial distribution with parameters $n = 5$ and p. Comment on the student's suggestion.

d How does the variance of the data support the use of a binomial distribution?

Challenge

If $X \sim B(3, p)$, prove that:

a $E(X) = 3p$

b $Var(X) = 3p(1 - p)$

2.6 Using the Poisson distribution to approximate the binomial distribution

Evaluating binomial probabilities when n is large can be quite difficult and in such situations it is sometimes useful to use an approximation.

- **If $X \sim B(n, p)$ and**
  - **n is large**
  - **p is small**

  then X can be approximated by $Po(\lambda)$, where $\lambda = np$.

Links If p is close to 0.5 you can use a normal distribution to model a binomial distribution with large values of n.

← **Statistics and Mechanics Year 2, Chapter 3**

There is no clear rule as to what constitutes 'large n' or 'small p' but usually the value for np will be $\leqslant 10$. Generally, the larger the value of n and the smaller the value of p, the better the approximation will be. In this situation, $(1 - p)$ will be close to 1, so $Var(X) = np(1 - p)$ will be close to the mean of the distribution, $E(X) = np$. This satisfies the condition for a Poisson distribution model that the mean and variance are close.

In general, a question will state whether you need to use a Poisson approximation.

Example 10

The random variable $X \sim B(200, 0.03)$.

a Find $P(X = 4)$.

A Poisson variable $Y \sim Po(\lambda)$ is used to approximate X.

b Write down the value of λ and justify the use of a Poisson approximation in this context.

c Find $P(Y = 4)$ and comment on the accuracy of the approximation.

a $P(X = 4) = \binom{200}{4} 0.03^4 \, 0.97^{196}$
$= 0.1338$ (4 d.p.)

b Under a Poisson approximation,
$Y \sim Po(200 \times 0.03)$, i.e $\lambda = 6$
As n is large and p is small, then X can be approximated by $Po(np)$.

c $P(Y = 4) = \frac{e^{-6} \times 6^4}{4!} = 0.1339$ (4 d.p.)
The answer obtained from the Poisson approximation is close to the value obtained from the underlying binomial distribution, so the approximation is accurate.

Compare your answers for parts **a** and **c**

Example 11

The probability of a component produced by a certain machine being faulty is 0.007. The number of faulty components in a batch of 1000 components is noted.

a Find the probability that exactly 6 components are faulty.

b Use a Poisson approximation to find the probability that more than 7 components are faulty.

c Explain why the approximation in part **b** is valid.

a X = the number of faulty components in a batch of 1000
$X \sim B(1000, 0.007)$
$P(X = 6) = \binom{1000}{6} \times 0.007^6 \times 0.993^{994}$
$= 0.1494$ (4 d.p.)

Define the random variable.

b Under a Poisson approximation,
$X \sim Po(1000 \times 0.007)$ i.e. $X \sim Po(7)$
$P(Y > 7) = 1 - P(X \leqslant 7) = 1 - 0.5987$
$= 0.4013$

This value can be calculated from tables or using your calculator.

c The approximation in part **b** is valid as n is large and p is small.

Exercise 2G

1 The random variable $X \sim B(100, 0.05)$.

a Calculate:

i $P(X = 4)$ **ii** $P(X \leq 2)$

b Use a Poisson approximation to find estimates for the probabilities calculated in part **a**.

2 The random variable $X \sim B(150, 0.04)$.

a Calculate:

i $P(X = 5)$ **ii** $P(X \leq 3)$

b Use a Poisson approximation to find estimates for the probabilities calculated in part **a**.

(P) **3** The random variable $Y \sim B(200, 0.98)$.

a Calculate:

i $P(Y = 197)$ **ii** $P(Y \geq 198)$

b Use a Poisson approximation to find estimates for the probabilities calculated in part **a**.

Problem-solving

Create a variable $X \sim B(200, 0.02)$ which satisfies the conditions for a Poisson approximation. Hence $P(Y = 197)$ becomes $P(X = 3)$.

4 There are 800 pupils in a school. Find the probability that exactly 4 of them have a birthday on 1 April:

a by using a binomial distribution

b by using a Poisson approximation.

c Comment on your answers to parts **a** and **b**.

Hint If X = 'number of pupils out of 800 having birthday on 1 April' then $X \sim B(800, \frac{1}{365})$.

5 In a manufacturing process the proportion of defective items is 3%. For a batch of 100 articles, use a Poisson approximation to find the probability that:

a there are fewer than 4 defective items

b there are exactly 2 defective items.

6 A medical practice screens a random sample of 180 of its patients for a certain condition which is present in 2% of the population. Using a Poisson approximation, find the probability that they find:

a one patient with the condition

b at least two patients with the condition.

7 A researcher has suggested that 1 in 120 people is likely to catch a particular virus. Assuming that a person catching the virus is independent of any other person catching it,

a find the probability that in a random sample of 20 people, exactly one of them catches the virus.

b Using a Poisson approximation, estimate the probability that in a random sample of 900 people fewer than 7 catch the virus.

8 From company records, a manager knows that the probability that a defective article is produced by a particular production line is 0.025.
A random sample of 10 articles is selected from the production line.

a Find the probability that exactly 1 of them is defective.

On another occasion, a random sample of 120 articles is taken.

b Using a Poisson approximation, find the probability that fewer than 4 of them are defective.

9 A manufacturer produces large quantities of pots. 5% of the pots produced are chipped.
A random sample of 10 pots was taken from the production line.

a Define a suitable distribution to model the number of chipped pots in this sample.

b Find the probability that there were exactly 3 chipped pots in the sample.

A new random sample of 140 pots was taken.

c Find the probability that there were between 6 and 9 (inclusive) chipped pots in this sample, using a Poisson approximation.

10 The probability that a tomato plant grows over 2 metres high is 0.08. A random sample of 50 tomato plants is taken and each tomato plant is measured and its height recorded.
Find, using a Poisson approximation, the probability that the number of tomato plants over 2 metres high is between 5 and 8 (inclusive).

(E) **11** Each cell of a certain insect contains 1200 genes. It is known that each gene has a probability 0.005 of being damaged. A cell is chosen at random.

a Suggest a suitable model for the distribution of the number of damaged genes in the cell. **(1 mark)**

b Find the mean and variance of the number of damaged genes in the cell. **(2 marks)**

c Using a Poisson approximation, find the probability that there are at most 4 damaged genes in the cell. **(3 marks)**

(E/P) **12** A machine which manufactures nails is known to produce 2.5% defective nails. The nails are sold in packets of 200.

a Using a Poisson approximation, calculate the probability that a packet contains more than 6 defective nails. **(3 marks)**

A carpenter buys 6 packets of nails.

b Estimate the probability that more than half of these packets contain more than 6 defective nails. **(4 marks)**

(E/P) **13** The probability of an electrical component being defective is 0.0125.
The component is supplied in boxes of 400.

a Using a Poisson approximation, estimate the probability that there are more than 3 defective components in a box. **(3 marks)**

A retailer buys 5 boxes of components.

b Estimate the probability that there are more than 3 defective components in 3 of the boxes. **(3 marks)**

(E/P) **14** It is claimed that 95% of the letters posted 1st class arrive the next day. Based on this claim, calculate, using a Poisson approximation, the probability that in a sample of 180 letters,

a more than 173 arrive the day after posting **(3 marks)**

b fewer than 168 arrive the day after posting. **(3 marks)**

(E/P) **15** A farmer supplies a bakery with eggs. The manager of the bakery claims that the proportion of eggs which are broken on delivery is 1%. The farmer supplies the eggs to the bakery on a daily basis in consignments of 150 eggs.

a Based on the claim of the manager, calculate, using a Poisson approximation, the probability that a consignment contains more than 4 broken eggs. **(3 marks)**

The farmer supplies a consignment to the baker on 5 days every week.

b Calculate the probability that in a particular week one of the consignments contains more than 4 broken eggs. **(4 marks)**

Mixed exercise 2

(E/P) **1** On a stretch of road, accidents occur at a rate of 0.7 per month.
Find the probability of:

a no accidents in the next month **(2 marks)**

b exactly 2 accidents in the next 3-month period **(2 marks)**

c no accidents in exactly 2 of the next 6 months. **(3 marks)**

(E) **2** The random variable X is the number of misprints per chapter in the first edition of a new textbook.

a State two conditions under which a Poisson distribution is a suitable model for X. **(2 marks)**

The number of misprints per chapter has a Poisson distribution with mean 2.25. Find the probability that:

b a randomly chosen chapter has no more than one misprint **(3 marks)**

c the total number of misprints in 2 randomly chosen chapters is more than 6. **(3 marks)**

(E/P) **3** The random variable $Y \sim \text{Po}(\lambda)$.
Find the value of λ such that $\text{P}(Y = 5)$ is 1.25 times the value of $\text{P}(Y = 3)$. **(3 marks)**

(E) **4** A company receives emails at a mean rate of 3 every 5 minutes.

a Give two reasons why a Poisson distribution could be a suitable model for the number of emails received. **(2 marks)**

b Calculate the probability that, in a 10-minute period, the company receives:

i exactly 7 emails **(2 marks)**

ii at least 8 emails. **(2 marks)**

(E) **5** **a** State the conditions under which the Poisson distribution may be used as an approximation to the binomial distribution. **(2 marks)**

Left-handed people make up 8% of a population. A random sample of 50 people is taken from this population. The discrete random variable X represents the number of left-handed people in the sample.

b Calculate $P(X \leqslant 3)$. **(3 marks)**

c Using a Poisson approximation, estimate $P(X \leqslant 3)$. **(3 marks)**

d Calculate the percentage error in using the Poisson approximation. **(2 marks)**

(E/P) **6** The number of telephone calls per hour received by a small business is a random variable with distribution $Po(\lambda)$ where λ is an integer. Natalia records the number of calls, Y, received in an hour.
Given that $P(Y > 10) < 0.1$, find the largest possible value of λ. **(3 marks)**

Hint Use the Poisson distribution tables.

(E) **7** The probability of a plant cutting successfully taking root is 0.075. Find the probability that, in a batch of 20 randomly selected plant cuttings, the number taking root will be:

a **i** exactly 2 **(2 marks)**

ii more than 4. **(2 marks)**

A second random sample of 80 plant cuttings is selected.

b Using a Poisson approximation, estimate the probability of at least 8 plant cuttings taking root. **(3 marks)**

(E/P) **8** An angler is known to catch fish at a mean rate of 2 per hour. The number of fish caught by the angler in an hour follows a Poisson distribution.
The angler takes 5 fishing trips, each lasting 2 hours.
Find the probability that the angler catches at least 5 fish on exactly 3 of these trips. **(5 marks)**

Hint Find $P(X \geqslant 5)$ using a Poisson variable. Then use this value as the parameter p in a binomial model.

(E/P) **9** The number of cherries in a *Megan's* fruit cake follows a Poisson distribution with mean 2.5. A *Megan's* fruit cake is to be selected at random. Find the probability that it contains:

a **i** exactly 4 cherries **(2 marks)**

ii at least 3 cherries. **(2 marks)**

Megan's fruit cakes are sold in packets of 4.

b Calculate the probability that there are more than 12 cherries, in total, in a randomly selected packet of *Megan's* fruit cakes. **(3 marks)**

Eight packets of *Megan's* fruit cakes are selected at random.

c Find the probability that exactly 2 packets contain more than 12 cherries. **(3 marks)**

(E/P) **10** A car salesman sells cars at a mean rate of 6 per week.

a Suggest a suitable model to represent the number of cars sold in a randomly chosen week. Give two reasons to support your model. **(2 marks)**

b Find the probability that in any randomly chosen week the salesman sells exactly 5 cars. **(2 marks)**

c Find the probability that in a period of 4 consecutive weeks there are exactly 2 weeks in which the salesman sells exactly 5 cars. **(3 marks)**

(E/P) **11** Abbie and Ben share a flat. Abbie receives letters at a mean rate of 1.2 letters per day while Ben receives letters at a rate 0.8 letters per day. Assuming their letters are independent, calculate the probability that on a particular day:

a each receives at least 1 letter **(3 marks)**

b they receive a total of 3 letters between them. **(2 marks)**

Given that post is delivered to the flat from Monday to Friday,

c find the probability that in one particular week they receive a total of 3 letters on at least 3 of the days. **(4 marks)**

(E/P) **12** An electrical outlet sells desktop and laptop computers. The desktops are sold at a mean rate of 2.4 per day and the laptops are sold at a mean rate of 1.6 per day. Calculate the probability that on a particular day the outlet sells:

a at least 2 desktops and at least 2 laptops **(3 marks)**

b a combined total of 6 computers. **(2 marks)**

c Calculate the probability that over a two-day period they sell a combined total of no more than 6 computers. **(3 marks)**

(E) **13** An airline knows that overall 4% of passengers do not turn up for flights. The airline has a policy of selling more tickets than there are seats on a flight. For an aircraft with 148 seats, the airline sold 150 tickets for a particular flight.

a Write down a suitable model for the number of passengers who do not turn up for this flight after buying a ticket. **(2 marks)**

By using a Poisson approximation, find the probability that:

b more than 148 passengers turn up for this flight **(2 marks)**

c there is at least one empty seat on this flight. **(3 marks)**

(E) **14** A receptionist routes incoming telephone calls to rooms within a hotel. The probability of the caller being connected to the wrong room is 0.02.

a Find the probability that more than 1 call in 10 consecutive calls is connected to the wrong room. **(3 marks)**

The receptionist receives 500 calls each day for guests in the hotel.

b Find the mean and variance of the number of wrongly connected calls. **(2 marks)**

c Use a Poisson approximation to find the probability that fewer than 8 calls each day are connected to the wrong room. **(3 marks)**

(E) **15** A disease occurs in 2.5% of a population.

a Find the probability of exactly 2 people having the disease in a random sample of 10 people. **(2 marks)**

b Find the mean and variance of the number of people with the disease in a random sample of 120 people. **(2 marks)**

A doctor tests a random sample of 120 patients for the disease. He decides to offer all patients a vaccination to protect them from the disease if more than 6 of the sample have the disease.

c Using a Poisson approximation, find the probability that the doctor will offer all patients a vaccination. **(3 marks)**

(E/P) **16** Accidents occur randomly at a roundabout at a rate of 15 every year.

a Find the probability that there will fewer than 5 accidents at the roundabout in a 6-month period. **(2 marks)**

b Find the probability that there will be at least 1 accident in a single month. **(2 marks)**

c Find the probability that there is at least 1 accident in exactly 4 months of a 6-month period. **(3 marks)**

(E/P) **17** An office photocopier breaks down randomly at a rate of 8 times per year.

a Find the probability that there will be exactly 2 breakdowns in the next month. **(2 marks)**

b Find the probability of at least 2 breakdowns in 3 of the next 4 months. **(3 marks)**

(E) **18** A holiday website receives visits at a rate of 240 per hour.

a State a distribution that is suitable to model the number of visits obtained during a 1-minute interval, and justify your choice of distribution. **(3 marks)**

Find the probability of:

b 8 visits in a given minute **(2 marks)**

c at least 10 visits in 2 minutes. **(2 marks)**

(E) **19** The number of policies sold by a life insurance company employee each week over a 150-week period is recorded.

Number of policies sold	0	1	2	3	4	5	6	7	8
Number of weeks	10	23	35	33	24	14	7	3	1

a Calculate the mean and the variance of the data. **(3 marks)**

b Explain why the results in part **a** suggest that a Poisson distribution may be a suitable model for the number of policies sold in a week. **(1 mark)**

c Use a Poisson distribution to estimate the probability that no more than 2 policies will be sold in a given week. **(3 marks)**

Challenge

A During normal operational hours, planes land at an airport at an average rate of one every four minutes.

Given that exactly 10 planes landed at the airport between 2 pm and 3 pm, find the probability that

a exactly 5 planes landed between 2 pm and 2.30 pm.

b more than 7 planes landed between 2 pm and 2.30 pm.

Summary of key points

1 If $X \sim \text{Po}(\lambda)$, then the Poisson distribution is given by:

$$P(X = x) = \frac{e^{-\lambda}\lambda^x}{x!}, x = 0, 1, 2, 3, \ldots$$

2 In order for the Poisson distribution to be a good model, the events must occur:
- independently
- singly, in space or time
- at a constant average rate in that the mean number in an interval is proportional to the length of an interval

3 If two Poisson variables X and Y are independent, the variable $Z = X + Y$ also has a Poisson distribution.
If $X \sim \text{Po}(\lambda)$ and $Y \sim \text{Po}(\mu)$, then $X + Y \sim \text{Po}(\lambda + \mu)$

4 If X has a Poisson distribution with $X \sim \text{Po}(\lambda)$, then:
- Mean of $X = \text{E}(X) = \lambda$
- Variance of $X = \text{Var}(X) = \sigma^2 = \lambda$

5 If X has a binomial distribution with $X \sim \text{B}(n, p)$, then:
- Mean of $X = \text{E}(X) = \mu = np$
- Variance of $X = \text{Var}(X) = \sigma^2 = np(1 - p)$

6 If X has a binomial distribution with $X \sim \text{B}(n, p)$, and
- n is large
- p is small

then X can be approximated by $\text{Po}(\lambda)$, where $\lambda = np$.

Geometric and negative binomial distributions

3

Objectives

After completing this chapter, you should be able to:

- Understand and use the geometric distribution → pages 44–47
- Calculate and use the mean and variance of the geometric distribution → pages 47–49
- Understand and use the negative binomial distribution → pages 49–52
- Calculate and use the mean and variance of the negative binomial distribution → pages 52–54

The geometric distribution can be used to model the number of times a learner driver needs to take their test before passing. → Exercise 3B, Q3

Prior knowledge check

1 $X \sim \mathrm{B}(20, 0.4)$

a Find:

i $\mathrm{P}(X = 10)$ **ii** $\mathrm{P}(X < 8)$ **iii** $\mathrm{P}(X \geqslant 12)$

b Calculate:

i $\mathrm{E}(X)$ **ii** $\mathrm{Var}(X)$

← Statistics and Mechanics Year 1, Chapter 6

2 A fair dice is rolled repeatedly.
Find the probability that a six first appears on the 3rd roll.

← Statistics and Mechanics Year 1, Chapter 5

3.1 The geometric distribution

If you are carrying out successive, independent trials, each with the same probability of success, you can model the number of trials needed to achieve a single success using the **geometric distribution**. For example, if you are rolling a fair dice repeatedly, the number of rolls needed before a six is rolled can be modelled using the geometric distribution. This might be particularly useful if you are playing a board game where you have to roll a specified number to start the game.

Example 1

Johan is playing a board game where he has to roll a six to start. Find the probability that Johan starts the board game on his fourth roll.

$\text{P(six)} = \frac{1}{6}$

$\text{P(not six)} = \frac{5}{6}$

P(first six occurs on fourth roll)

$= \frac{5}{6} \times \frac{5}{6} \times \frac{5}{6} \times \frac{1}{6}$

$= \frac{125}{1296} = 0.0965$ (4 d.p.)

Johan will have to roll three *not sixes* in a row and then one *six*. You can think of this as three 'failures', with probability $(1 - p)$, and one success with probability p, and you can write it as $(1 - p)^3 \times p$.

Online Explore the geometric distribution using GeoGebra.

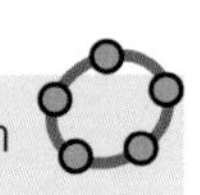

You can define a discrete random variable, X, as the number of rolls needed to obtain a six. X has the following probability distribution:

X	1	2	3	4	...
$\text{P}(X = x)$	$\frac{1}{6}$	$\frac{5}{6} \times \frac{1}{6}$	$\frac{5}{6} \times \frac{5}{6} \times \frac{1}{6}$	$\frac{5}{6} \times \frac{5}{6} \times \frac{5}{6} \times \frac{1}{6}$	...

$\text{P}(X = x) = \left(\frac{5}{6}\right)^{x-1} \times \frac{1}{6}$

Note that X can take any positive integer value. In practice, as x gets large, $\text{P}(X = x)$ gets smaller and smaller, and tends towards 0.

- **For successive independent trials, each with constant probability of success, p, the number of trials needed to get one success, X, has the geometric distribution, with probability function:**

  $\mathbf{P}(X = x) = \mathbf{p}(x) = p(1 - p)^{x-1} \qquad x = 1, 2, 3, \ldots$

Notation You write $X \sim \text{Geo}(p)$.

You can see that the values of $\text{p}(x)$ form a geometric sequence with first term, p, and common ratio, $(1 - p)$. You can derive the **cumulative geometric distribution** by considering the sum of the terms of the geometric series.

$$\sum_{r=1}^{x} \text{p}(r) = \frac{p(1 - (1 - p)^x)}{1 - (1 - p)} = \frac{p(1 - (1 - p)^x)}{p} = 1 - (1 - p)^x$$

Links The sum of the first n terms of a geometric series with first term a and common ratio r is $S_n = \frac{a(1 - r^n)}{1 - r}$ ← **Pure Year 2, Chapter 3**

- **If $X \sim \text{Geo}(p)$, then the cumulative geometric distribution is given by:**

  $\mathbf{P}(X \leqslant x) = 1 - (1 - p)^x \qquad x = 1, 2, 3, \ldots$

A Since $P(X \leqslant x) + P(X > x) = 1$, you can also deduce that $P(X > x) = (1 - p)^x$. This corresponds to the situation where the first x trials result in failure. This result can also be written $P(X \geqslant x) = (1 - p)^{x-1}$

Example 2

The probability that Genevieve passes her driving test on any one attempt is 0.6.

a Find the probability that:

  i she passes on her fifth attempt

  ii she needs five or fewer attempts to pass

  iii she needs more than five attempts to pass.

b State two assumptions you have used in your calculations.

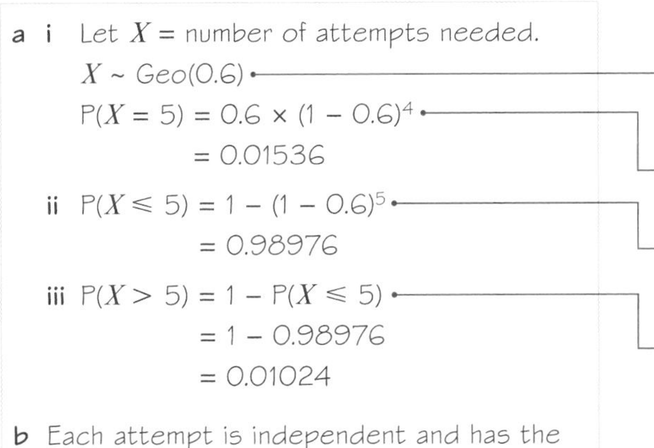

a i Let X = number of attempts needed.

$X \sim \text{Geo}(0.6)$

$P(X = 5) = 0.6 \times (1 - 0.6)^4$

$= 0.01536$

ii $P(X \leqslant 5) = 1 - (1 - 0.6)^5$

$= 0.98976$

iii $P(X > 5) = 1 - P(X \leqslant 5)$

$= 1 - 0.98976$

$= 0.01024$

b Each attempt is independent and has the same probability.

X can be modelled using a geometric distribution with probability 0.6.

Use $P(X = x) = p(1 - p)^{x-1}$.

Use $P(X \leqslant x) = 1 - (1 - p)^x$.

This is the same as the probability of her failing on the first five attempts, or $(1 - 0.6)^5$.

Online Explore the culmulative geometric distribution using GeoGebra.

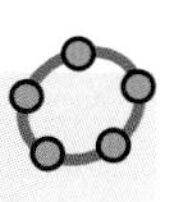

Exercise 3A

1 The random variable $X \sim \text{Geo}(0.15)$. Find:

a $P(X = 10)$ **b** $P(X \leqslant 7)$ **c** $P(3 \leqslant X \leqslant 12)$

2 The random variable $Y \sim \text{Geo}(0.23)$. Find:

a $P(Y = 6)$ **b** $P(Y \geqslant 4)$ **c** $P(2 < Y < 8)$

3 Alice rolls a fair 6-sided dice. She records X, the number of rolls it takes to get a 1. Given that each roll is independent, find:

a $P(X = 4)$ **b** $P(X \leqslant 3)$ **c** $P(X \geqslant 5)$ **d** $P(2 \leqslant X \leqslant 6)$

E **4** Bernhard has to pass an examination to get into law school. He can take the examination as many times as he likes and his probability of passing on any one attempt is 0.3.

a Find the probability that:

  i he passes on his third attempt **(2 marks)**

  ii he takes at least four attempts to pass. **(3 marks)**

b State two assumptions that you have used in your model. **(2 marks)**

A **5** Carolina is playing a board game with a fair four-sided dice. She must roll a 4 to start.

a Given that each roll is independent, find the probability that:

i she starts on her first go

ii she starts on her fifth go

iii she takes no more than four attempts to start.

b State one assumption you have used in your model.

P **6** Donald is taking part in a competition where he has to complete a task within a given time limit. He can have as many attempts as he likes. The probability that he completes the task in the given time on each attempt is 0.45. He uses X to represent the number of attempts he needs and finds that $\text{P}(X = x) = 0.136\,125$. Given that each attempt is independent, calculate:

a the value of x

b the probability that Donald completes the task within 4 attempts.

P **7** $X \sim \text{Geo}(0.032)$

a Given that $\text{P}(X = x) = 0.0203$ (4 d.p.), find the value of x.

Find:

b the largest value of x such that $\text{P}(X \leqslant x) < 0.1$

c the smallest value of x such that $\text{P}(X \geqslant x) < 0.05$.

Hint $\text{P}(X \geqslant x) = (1 - p)^{x-1}$

8 Edith works for a computer company on a telephone help desk. Callers either report a problem with their hardware or their software. The probability that a randomly chosen caller reports a problem with their hardware is 0.1. Given that each call is independent, find the probability that:

a the first caller with a hardware problem is Edith's 7th caller

b the first call from a caller with a hardware problem comes in after the 5th call.

9 Isabelle is surveying people about their eating habits. She asks people who pass her by in the street, at random, whether they like squid pizza. Given that the probability that a randomly chosen person likes squid pizza is 0.05 and that each person is independent, find the probability that:

a the first person to like squid pizza is the 10th person she asks

b she asks at least 15 people before finding someone who likes squid pizza.

E/P **10** Frances, Georgina and Holly agree who is to do the washing up by playing a game. They each roll a fair six-sided dice and record whether the outcome is odd or even. If two of them get the same outcome and the third one gets a different outcome, that person has to wash up. If no decision is reached they play again.

Given that each game is independent, find the probability that:

a a decision is reached on the third game **(4 marks)**

b it takes at least four games to reach a decision. **(2 marks)**

11 Julian works in a village pharmacy and he finds that the long-term average number of customers per hour is 4. Find the probability that:

a at least 5 customers come in during a randomly chosen hour. **(2 marks)**

He records the number of customers during each hour on a particular day.

Find the probability that:

b the first occurrence of 5 or more customers is in the 5th hour **(2 marks)**

c he goes through a whole 8-hour shift with no single hour having 5 or more customers. **(3 marks)**

3.2 Mean and variance of a geometric distribution

You need to be able to make use of the following results about the mean (or expected value) and variance of a geometrically distributed random variable:

- **If $X \sim \text{Geo}(p)$, then:**
  - **Mean of $X = \text{E}(X) = \mu = \frac{1}{p}$**
  - **Variance of $X = \text{Var}(X) = \sigma^2 = \frac{1-p}{p^2}$**

Example 3

Dorothy flips a biased coin until it lands on heads. She records the total number of flips, Y. Given that the mean of Y is 2.5, find:

a the probability of the coin landing on heads on a single flip

b the standard deviation of Y.

a $Y \sim \text{Geo}(p)$

$2.5 = \frac{1}{p}$ — Use $\text{E}(X) = \mu = \frac{1}{p}$

$p = \frac{1}{2.5} = 0.4$

The probability of the coin landing on heads is 0.4.

b $\sigma^2 = \frac{1-p}{p^2}$ — Use $\text{Var}(X) = \sigma^2 = \frac{1-p}{p^2}$

$= \frac{1-0.4}{0.4^2}$

$= 3.75$

$\sigma = \sqrt{3.75} = 1.94$ (3 s.f.)

Watch out The question asks for the **standard deviation**, so find $\text{Var}(Y)$ and then find the square root.

Exercise 3B

1 The random variable $X \sim \text{Geo}(0.2)$. Find:

a $\text{E}(X)$ **b** $\text{Var}(X)$

A

2 Zachariah rolls a fair six-sided dice and records X, the number of rolls it takes for him to get a multiple of 3. Given that each roll is independent, find:

a $E(X)$ **b** $Var(X)$

3 The probability that Yolanda passes her driving test at any one attempt is 0.65. Given that each attempt is independent, find the probability that:

a she passes on the third attempt **b** it takes at least four attempts to pass.

c Find:

i the expected number of times Yolanda will have to take her test

ii the variance.

4 The geometrically distributed random variable $X \sim Geo(p)$, has $E(X) = 4$. Find:

a p **b** $Var(X)$

P

5 Xavier is practising shooting a basketball from the 'free throw' line and records the number of throws X, that he takes to get a basket each time. Given that each throw is independent and that $Var(X) = 20$, find:

a the probability that he hits a basket each throw **b** $E(X)$

P

6 Wilma is a charity collector and goes door-to-door trying to raise money. Given that the probability of her getting a donation at each house is p, that each house call is independent and the variance is 380, find:

a p

b the expected number of house calls Wilma must make before getting a donation.

E

7 Vincent records X, the number of attempts it takes him to parallel-park his car in a particular space.

a State a suitable probability distribution for X. **(1 mark)**

b State two assumptions that must be made for this distribution to be appropriate. **(2 marks)**

Given that $P(X = 2) = 0.16$ and that $p < 0.5$, find:

c p, the probability that Vincent parks correctly on each single attempt **(3 marks)**

d the expected number of attempts Vincent takes **(1 mark)**

e the variance of X. **(1 mark)**

E/P

8 Uma has a bag of marbles, 15% of which are blue. She puts her hand in the bag and pulls out a marble at random. If the marble is not blue, she puts it back in the bag and tries again.

a Calculate:

i the mean

ii the variance of the number of marbles she pulls out, up to and including the first blue one. **(2 marks)**

Calculate the probability that:

b she pulls out 4 marbles **(2 marks)**

c she pulls out at least 8 marbles **(3 marks)**

d she pulls out fewer marbles than expected. **(2 marks)**

9 Tabitha the cat is trying to catch fish out of a garden pond. The probability that she catches a fish at each attempt is 0.12.

a State two assumptions that must be made to model this situation as a geometric distribution. **(2 marks)**

b Find:

i the probability that Tabitha takes two attempts to catch a fish **(2 marks)**

ii the probability that Tabitha takes at least three attempts to catch a fish. **(3 marks)**

c Find the expected number of attempts Tabitha takes to catch a fish and the variance. **(2 marks)**

Once Tabitha has caught one fish, she tries to catch a second one.

d Find the probability that she needs three attempts to catch her first fish, and three further attempts to catch her second fish. **(3 marks)**

e Find the probability that she caught the first fish on the second attempt, and the second fish on the fifth attempt. **(3 marks)**

10 Simeon, a tailor, records the number of faults, X, per metre of cloth.
He finds that there are, on average, 0.8 faults per metre.

a State, giving any assumptions you make, a suitable probability distribution for X. **(2 marks)**

b Find the probability that in a randomly chosen metre of cloth there are more than two faults. **(1 mark)**

Simeon cuts the cloth into one-metre lengths.

c Find the probability that the first time he encounters a metre of cloth with more than two faults is in the 7th metre. **(3 marks)**

d Find the expectation and variance of the number of metres of cloth he cuts before he finds a metre length with more than two faults. **(2 marks)**

If a metre of cloth containing two or more faults occurs before Simeon has cut 3 one-metre lengths, he sends that roll of cloth back to the manufacturers.

e Find the probability that he sends back two consecutive rolls of cloth. **(2 marks)**

3.3 The negative binomial distribution

The binomial distribution, $X \sim \text{B}(n, p)$, models the number of successes in a fixed number of trials, n. The probability of success in each trial, p, is constant and the trials are independent.

Suppose, instead, you want to consider the **number of trials needed** to achieve a fixed number of successes, r. For example, suppose you roll a fair dice repeatedly until you have rolled a total of **three** sixes. You can define a discrete random variable, X, as the number of rolls needed.

Hint If the first three rolls are all sixes, then $X = 3$. X can take any integer value greater than or equal to 3.

To find the probability distribution of X, consider what is required for X to take any particular value. For example, $\text{P}(X = 10)$ is the probability that the third six occurs **on the 10th roll**.
This means that exactly 2 sixes have been rolled in the previous 9 rolls.

P(exactly 2 sixes in first 9 rolls) = $\binom{9}{2}\left(\frac{1}{6}\right)^2\left(\frac{5}{6}\right)^7$

Links The number of sixes rolled in 9 rolls has the binomial distribution $\text{B}(9, \frac{1}{6})$.
← **Statistics and Mechanics Year 1, Chapter 6**

You then need to roll a six on the 10th roll, with probability $\frac{1}{6}$. Since each roll is assumed to be independent:

$$P(X = 10) = P(\text{exactly 2 sixes in first 9 rolls}) \times P(\text{six on the 10th roll})$$
$$= \binom{9}{2}\left(\frac{1}{6}\right)^2\left(\frac{5}{6}\right)^7 \times \frac{1}{6}$$
$$= \binom{9}{2}\left(\frac{1}{6}\right)^3\left(\frac{5}{6}\right)^7$$

You can calculate the probabilities for other values of X in a similar way, giving you the following probability function for X:

$$P(X = x) = \binom{x-1}{2}\left(\frac{1}{6}\right)^3\left(\frac{5}{6}\right)^{x-3}$$

This is an example of a **negative binomial distribution**.

- **For successive trials, each with constant probability of success, p, the number of trials needed to get r successes, X, has the negative binomial distribution, with probability function:**

  $$P(X = x) = p(x) = \binom{x-1}{r-1}p^r(1-p)^{x-r} \qquad x = r, r+1, r+2, \ldots$$

  This is the probability of $r - 1$ successes in $x - 1$ trials multiplied by the probability of success in the xth trial.

Notation There is no standard notation for the negative binomial distribution, but you can write $X \sim \text{NB}(r, p)$ or $X \sim \text{Negative B}(r, p)$.

Example 4

Philomena is practising her piano scales. The probability that she completes a scale correctly on any one attempt is 0.4. She continues practising until she has completed four scales correctly.

a Find the probability that she completes her fourth correct scale on her 12th attempt.

b Find the probability that she completes her fourth correct scale on her 10th attempt, given that her first scale was correct.

Online Explore the negative binominal distribution using GeoGebra.

a X = number of attempts needed to complete four scales without a mistake

$X \sim \text{Negative B}(4, 0.4)$

Model the number of attempts as a negative binomial random variable.

$P(X = 12) = \binom{11}{3} \times 0.4^4 \times 0.6^8$

$= 0.0709$ (4 d.p.)

Use the probability function for a negative binomial random variable.

b Y = number of attempts needed after the first attempt

$Y \sim \text{Negative B}(3, 0.4)$

$P(Y = 9) = \binom{8}{2} \times 0.4^3 \times 0.6^6$

$= 0.0836$ (4 d.p.)

Problem-solving

If the first scale was correct, Philomena now must complete **three** scales correctly in exactly **nine** further attempts. This means she needs to complete 2 correctly in the next 8 attempts, with probability $\binom{8}{2} \times 0.4^2 \times 0.6^6$, and then complete the final scale correctly, with probability 0.4.

Exercise 3C

A

1 Aulden throws a fair four-sided dice. Find the probability that he throws a 4 for the third time on his sixth throw.

2 Billie spins a coin, biased towards heads. If the probability of spinning a head is 0.55, find the probability that Billie spins her fourth head on her seventh spin.

3 Chuck is shooting at a target with a bow and arrow. The probability that he hits the bullseye on any particular shot is 0.15. Find the probability that Chuck hits the bullseye for the second time on his tenth shot.

4 Denise takes part in a multiple-choice quiz where she picks the answers at random. Given that her probability of picking any correct answer is 0.25, find the probability that:

a she picks her first correct answer on her third question

b she picks her fourth correct answer on her seventh question

c she gets exactly two correct answers in the first ten questions

d her third correct answer occurs on or before the tenth question.

Watch out Not all of the parts of this question require you to use the negative binomial distribution.

5 Eliot plays tennis and his probability of winning any particular match is constant such that P(win) = 0.3. Find the probability that:

a he wins his first match on the fourth attempt

b he needs to play more than ten matches to win four times

c he wins his third match on the eighth attempt

d his fifth win occurs on or before his twelfth game.

Hint In part **b** you need to calculate the probability that he wins at most 3 times in his first 10 matches.

E/P

6 Francesca is playing a series of games with her sister. The probability that Francesca wins any particular game is 0.55.

a Find the probability that Francesca wins her fourth game on the sixth attempt. **(2 marks)**

b State two assumptions that have to be made for the model used in part **a** to be valid. **(2 marks)**

c Find the probability that Francesca wins her third game on the fifth attempt, given that she won the first game. **(2 marks)**

d Find the probability that Francesca wins at least seven out of the first ten games. **(3 marks)**

E

7 Gerald is trialling a new drug that in previous trials has cured patients 80% of the time.

a Find the probability that Gerald's drug cures the seventh patient on the tenth trial. **(2 marks)**

b State two assumptions that have to be made for the model used in part **a** to be valid. **(2 marks)**

c Find the probability that Gerald has cured at least seven patients in the first twelve trials. **(3 marks)**

d Find the probability that it takes more than 20 trials to cure 15 patients. **(2 marks)**

8 Harriet is conducting scientific experiments. Each experiment is independent and the probability of any single experiment working is 0.1.

a Find the probability that Harriet achieves her second successful experiment on her 20th attempt. **(2 marks)**

b Find the probability that it takes her more than 30 experiments to get four successes. **(2 marks)**

c Find the probability that Harriet achieves her fourth successful experiment on her 30th attempt given that she was first successful on the fifth attempt. **(2 marks)**

d Find the probability that Harriet is successful in at least three experiments out of the first 25. **(3 marks)**

9 The random variable X has negative binomial distribution, Negative B(5, 0.7). Find:

a $P(X = 10)$ **b** $P(X \leqslant 6)$

c $P(X \leqslant 15)$ **d** $P(X > 12)$

Problem-solving

Part **b** is asking you to find the probability that the 5th success occurs either on the 5th trial or the 6th trial.

E/P **10** A darts player is trying to hit the bullseye. She throws darts at the board until she scores three bullseyes. The random variable D is the number of darts she needs to throw.

a State a suitable distribution to model D. **(1 mark)**

b Given that the probability of her hitting the bullseye on each attempt is 0.35, find the probability that:

i it takes her seven throws to score the three bullseyes **(2 marks)**

ii it takes her at least eight throws to score the three bullseyes **(2 marks)**

iii it takes her nine throws, given that she hits the bullseye on her first throw. **(2 marks)**

c Give one reason why this model may not accurately represent the situation. **(1 mark)**

Challenge

The function $F_{n,p}$ is defined as follows:

$$F_{n,p}(x) = P(X \leqslant x) \text{ where } X \sim B(n, p)$$

a Given that $Y \sim$ Negative B(3, 0.4), show that $P(Y \leqslant 8)$ can be written as $1 - F_{n,p}(x)$, where n, p and x are values to be found.

b Given that $Y \sim$ Negative B(r, p), find a general expression for $P(Y \leqslant y)$ in terms of $F_{n,p}(x)$ for suitable n and x.

3.4 Mean and variance of the negative binomial distribution

You need to be able to make use of the following results about the mean (or expected value) and variance of a random variable with the negative binomial distribution.

- **If $X \sim$ Negative B(r, p), then:**
  - **Mean of X = E(X) = $\mu = \frac{r}{p}$**
  - **Variance of X = Var(X) = $\sigma^2 = \frac{r(1-p)}{p^2}$**

Example 5

Iona and Juan play noughts and crosses. The probability that Iona wins is 0.7. The random variable X represents the number of games that they need to play for Iona to win seven times.

a Find E(X) and Var(X).

They change games and start playing chess. The random variable Y represents the number of games that they need to play for Juan to win twice. Given that the mean of Y is 6, find:

b Juan's probability of winning any single game of chess

c the standard deviation of Y.

a $X \sim$ Negative B(7, 0.7)

$E(X) = \frac{r}{p} = \frac{7}{0.7} = 10$

$\text{Var}(X) = \frac{r(1-p)}{p^2} = \frac{7(1-0.7)}{0.7^2}$

$= 4.286$ (3 d.p.)

Use the standard formulae with the given values of r and p.

b $Y \sim$ Negative B(2, p)

$\frac{2}{p} = 6 \Rightarrow p = \frac{1}{3}$

Use the fact that $E(Y) = 6$.

The probability of Juan winning any one game of chess is $\frac{1}{3}$.

c $\text{Var}(Y) = \frac{2(1-\frac{1}{3})}{\left(\frac{1}{3}\right)^2} = 12$

Standard deviation $= \sqrt{12} = 3.46$

Watch out The 'mean' and the 'expected value' of a random variable are the same thing. Remember that the standard deviation is the square root of the variance.

Exercise 3D

1 The random variable X has negative binomial distribution with $p = 0.4$ and $r = 3$. Find:

a $E(X)$ **b** $\text{Var}(X)$

2 The random variable Y has negative binomial distribution with $p = 0.75$ and $r = 10$. Find:

a $E(Y)$ **b** $\text{Var}(Y)$

3 The random variable M has negative binomial distribution with probability p, and with $r = 2$. Given that $E(M) = 8$, find:

a the value of p **b** $P(M = 5)$ **c** $\text{Var}(M)$

4 $D \sim$ Negative B(8, p).

a Given that $\text{Var}(D) = 30$, find the value of p.

b Find:

i $P(D = 12)$ **ii** $P(D \leqslant 10 | \text{first trial was successful})$

Problem-solving p is a probability so it must be a positive number.

5 $X \sim$ Negative B(r, 0.4).

a Given that $E(X) = 15$, find the value of r.

b Find:

i $P(X = 10)$ **ii** $P(X \leqslant 8)$

A E/P **6** The random variable X has negative binomial distribution with mean 6 and variance 3.
Find:

a the value of p and the value of r

b $P(X = 4)$.

Problem-solving

Write two equations involving r and p and solve them simultaneously.

E **7** Kelly and her classmates are taking part in a competition where students take turns attempting to solve a puzzle. The probability that each student solves the puzzle is 0.7. The random variable X represents the number of students who need to attempt to solve the puzzle before five have solved it.

a State two conditions that are necessary for X to be modelled by a negative binomial distribution. **(2 marks)**

b Using a negative binomial model, find the mean and standard deviation of X. **(3 marks)**

E **8** In each trial of an experiment, four fair coins are spun.

a Find the probability that all four coins show the same result. **(2 marks)**

b Find the probability that all four coins show the same result for the third time on the sixth trial. **(3 marks)**

c Find the expected number of trials needed in order for the coins to show the same result 12 times. **(2 marks)**

E **9** Michelle is playing darts. The probability that she hits 'treble twenty' with any one dart is p. Given that the expected number of throws needed in order for her to hit the 'treble twenty' 3 times is 18.75, find:

a the value of p **(1 mark)**

b the variance. **(2 marks)**

Michelle gets some coaching. Given that her new probability of hitting the 'treble twenty' is 0.24, find:

c the expected number of throws needed to hit the 'treble twenty' 5 times **(1 mark)**

d the probability that it takes her more than the expected number of throws to hit the 'treble twenty' 5 times. **(3 marks)**

E/P **10** Norman picks marbles at random from a bag that contains 100 marbles. He notes the colour and replaces the marble. He repeats the process until he has selected a green marble r times. The random variable X represents the total number of times he selects a marble.

a State a distribution that could be used to model X. **(1 mark)**

b Given that the mean and standard deviation of X are 12 and 6 respectively, calculate:

i the number of green marbles in the bag

ii the value of r. **(4 marks)**

Alison selects marbles from the same bag. She notes the colour but does not replace the marble each time.

c Give a reason why a negative binomial distribution is not a suitable model for this situation. **(1 mark)**

d Find the probability that Alison picks her second green marble on her third pick. **(3 marks)**

Mixed exercise 3

(E) 1 An unbiased eight-sided dice is thrown repeatedly. The first multiple of 3 appears on the rth throw. Calculate the probability that:

a $r = 5$ **(2 marks)**

b the value of r is at least 3. **(3 marks)**

(E) 2 An engineer is checking welds on an oil tanker. The percentage of defective welds is thought to be 10%. If X represents the number of welds checked up to and including the first defective one,

a state the distribution that can be used to model X **(1 mark)**

b find the mean and variance of X **(2 marks)**

c find the probability that the engineer has to check at least 12 welds before finding a defective one. **(2 marks)**

(E) 3 Olivia is playing hoopla and she continues to throw the hoop until she hits the target. The random variable X represents the number of throws she needs.

a State a suitable distribution to model X. **(1 mark)**

Given that the mean of X is 6,

b find the probability that Olivia hits the target first on her fifth attempt **(2 marks)**

c find the variance of X. **(2 marks)**

d State any assumptions you have made in using this model. **(2 marks)**

(E/P) 4 Soujit is designing a game for a charity day. In his game, contestants have to roll a fair ten-sided dice a certain number of times. If the contestant rolls a 10 then they win.

Soujit wants the probability of winning to be less than 0.5.

Find the maximum number of times Soujit should allow contestants to roll the dice. **(4 marks)**

(E/P) 5 A supermarket knows from experience that when they purchase avocados from a particular supplier, any particular one of them has a 0.02 chance of being over-ripe.

A box of 24 avocados will be rejected if more than three are over-ripe.

a Find the probability that a particular box of avocados is rejected. **(2 marks)**

The supermarket is unloading a shipment of boxes.

b Find the probability that the 20th box unloaded is the first to be rejected. **(3 marks)**

(E/P) 6 Pablo is playing a fairground game where his probability of winning a prize is 0.2. He plays the game several times.

a Find the probability that he first wins a prize on his sixth game. **(1 mark)**

b Find the probability that he wins his second prize on his tenth game. **(2 marks)**

c Find the mean and standard deviation of the number of games Pablo needs to play to win his fifth prize. **(3 marks)**

Quinn plays a different game until he has won r prizes. Given that X represents the number of games Quinn plays and that $E(X) = 12$ and $\text{Var}(X) = 16$,

d find the probability of Quinn winning a game. **(3 marks)**

E/P

7 The random variable X is the number of times a biased dice is rolled until 4 sixes have occurred. The variance of X is 15.

a Find the probability of rolling a six. **(3 marks)**

b Find $P(X = 10)$. **(2 marks)**

c Find $P(X > 8)$. **(3 marks)**

d Find $P(X = 9|\text{a six occurs on the first roll})$. **(3 marks)**

E/P

8 Roberta is taking part in a penalty shootout contest. The probability that she scores a goal on any one attempt is 0.65.

a Show that the probability that she first scores a goal on her second attempt is 0.2275. **(2 marks)**

b Find the probability that:

i she scores a goal exactly 5 times during her first 8 attempts **(1 mark)**

ii she scores her fifth goal on her 8th attempt **(2 marks)**

iii she takes more than 9 attempts to score 5 goals **(2 marks)**

iv she scores a goal exactly 4 times in 7 attempts, given that she scores on each of her first two attempts. **(3 marks)**

c Calculate the mean and standard deviation of the number of attempts she needs to score 5 goals. **(3 marks)**

Sukie decides to take part as well. Her probability of scoring a goal on any one attempt is 0.4. Roberta and Sukie take it in turns to shoot at the goal, with Sukie going first. The first girl to score a goal wins.

d Find the probability that Roberta wins on her first attempt. **(2 marks)**

e Find the probability that Sukie wins on her second attempt. **(2 marks)**

f The contest is drawn if neither girl scores a goal with three attempts. Find the probability that the contest is drawn. **(2 marks)**

Challenge

1 In a fairground game, a player throws bean bags at a target. A particular player hits the target with probability p. The random variable X is the number of attempts needed by the player to hit the target twice.

a Write down the distribution of X.

The random variables Y_1 and Y_2 are defined as:

Y_1 = number of attempts needed to hit the target the first time

Y_2 = number of attempts *after the first hit* needed to hit the target the second time

b Write down the distribution of Y_1 and Y_2.

c Write X in terms of Y_1 and Y_2.

d Hence show that $E(X) = \frac{2}{p}$.

2 Use a similar technique to that outlined in question 1 to prove that, for the random variable $X \sim \text{Negative B}(r, p)$,

$E(X) = \frac{r}{p}$ and $\text{Var}(X) = \frac{r(1-p)}{p^2}$

Hint You may assume that, for the random variable $Y \sim \text{Geo}(p)$, $E(Y) = \frac{1}{p}$ and $\text{Var}(Y) = \frac{1-p}{p^2}$

Summary of key points

A

1 For successive independent trials, each with constant probability of success, p, the number of trials needed to get one success, X, has the **geometric distribution**, with probability function:

$\text{P}(X = x) = \text{p}(x) = p(1 - p)^{x-1} \qquad x = 1, 2, 3, \ldots$

2 If $X \sim \text{Geo}(p)$, then the cumulative geometric distribution is given by:

$\text{P}(X \leqslant x) = 1 - (1 - p)^x \qquad x = 1, 2, 3, \ldots$

You can also deduce that

$\text{P}(X > x) = (1 - p)^x$

$\text{P}(X \geqslant x) = (1 - p)^{x-1}$

3 If $X \sim \text{Geo}(p)$, then:

- Mean of $X = \text{E}(X) = \mu = \frac{1}{p}$
- Variance of $X = \text{Var}(X) = \sigma^2 = \frac{1 - p}{p^2}$

4 For successive trials, each with constant probability of success, p, the number of trials needed to get r successes, X, has the **negative binomial distribution**, with probability function:

$\text{P}(X = x) = \text{p}(x) = \binom{x - 1}{r - 1}p^r(1 - p)^{x-r} \qquad x = r, r + 1, r + 2, \ldots$

This is the probability of $r - 1$ successes in $x - 1$ trials multiplied by the probability of success in the xth trial.

5 If $X \sim \text{Negative B}(r, p)$, then:

- Mean of $X = \text{E}(X) = \mu = \frac{r}{p}$
- Variance of $X = \text{Var}(X) = \sigma^2 = \frac{r(1 - p)}{p^2}$

4 Hypothesis testing

Objectives

After completing this chapter you should be able to:

- Use hypothesis tests to test for the mean λ of a Poisson distribution → pages 59–62
- Find critical regions of a Poisson distribution using tables → pages 62–66
- Use hypothesis tests to test for the parameter p in a geometric distribution → pages 66–69
- Find critical regions of a geometric distribution → pages 69–72

The geometric distribution can be used to model the time elapsed between weather events. A hypothesis test for the parameter of a geometric distribution can help determine whether an observed weather event is statistically significant.
→ Mixed exercise, Q11

Prior knowledge check

1 The random variable $X \sim \text{Po}(4)$. Find:

a $\text{P}(X = 5)$ **b** $\text{P}(X \leqslant 3)$

c $\text{P}(6 < X < 11)$ **d** $\text{E}(X)$ ← Chapter 2

2 The probability that a component manufactured by a factory is defective is known to be 0.0037. The random variable X represents the number of components manufactured up to and including the first defective component. Find:

a $\text{P}(X \leqslant 30)$ **b** $\text{P}(X > 50)$

c $\text{E}(X)$ ← Sections 3.1, 3.2

3 A single observation is taken from the random variable $X \sim \text{B}(25, p)$ and is used to test $\text{H}_0: p = 0.2$ against $\text{H}_1: p > 0.2$ at the 10% level of significance.

Find the critical region for this test.
← Statistics and Mechanics Year 1, Chapter 7

4.1 Testing for the mean of a Poisson distribution

A Poisson distribution can be used to model situations when events occur at random, but at a constant average rate.

Links The Poisson distribution $\text{Po}(\lambda)$ has mean λ and variance λ. ← Section 2.4

- **To carry out a hypothesis test for the mean of a Poisson distribution, you form two hypotheses.**
  - **The null hypothesis, H_0: $\lambda = m$ is the value of the mean that you assume to be true.**
  - **The alternative hypothesis, H_1, tells you about the value of the mean if your assumption is shown to be wrong.**

Testing for the mean allows you to answer questions such as:

- Does the servicing of a machine decrease the rate at which it produces defective items?
- Does the introduction of a pelican crossing reduce the number of accidents along a particular stretch of road?

You will need to find values from the cumulative Poisson distribution to carry out hypothesis tests. You can use your calculator, or the table given in the tables on page 191.

Example 1

Accidents used to occur at a certain road junction at the rate of 6 per month. The residents petitioned for traffic lights. In the month after the lights were installed there was only one accident. Test, at the 5% level of significance, whether there is evidence that the lights have reduced the rate of accidents.

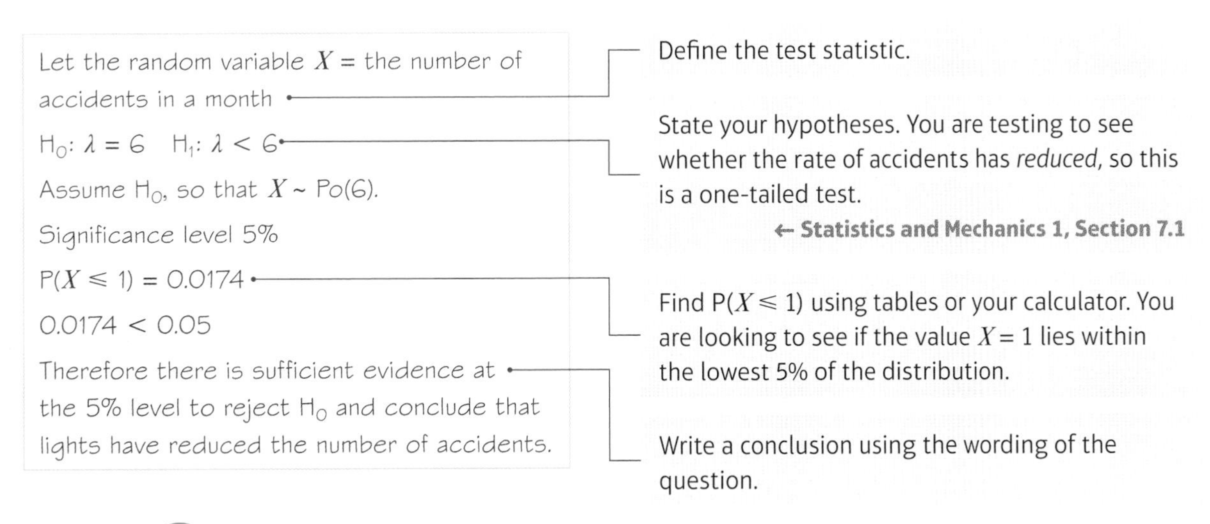

Let the random variable X = the number of accidents in a month

H_0: $\lambda = 6$ $\quad \text{H}_1$: $\lambda < 6$

Assume H_0, so that $X \sim \text{Po}(6)$.

Significance level 5%

$\text{P}(X \leqslant 1) = 0.0174$

$0.0174 < 0.05$

Therefore there is sufficient evidence at the 5% level to reject H_0 and conclude that lights have reduced the number of accidents.

Define the test statistic.

State your hypotheses. You are testing to see whether the rate of accidents has *reduced*, so this is a one-tailed test. ← Statistics and Mechanics 1, Section 7.1

Find $\text{P}(X \leqslant 1)$ using tables or your calculator. You are looking to see if the value $X = 1$ lies within the lowest 5% of the distribution.

Write a conclusion using the wording of the question.

Example 2

Over a long period of time, Fatima found that the bus taking her to school was late at a rate of 2.5 times per month. In the month following the start of the new summer bus schedules, Fatima finds that her bus is late 6 times. Assuming that the number of times the bus is late each month has a Poisson distribution, test at the 2% level of significance, whether or not the new schedules changed the frequency with which the bus is late.

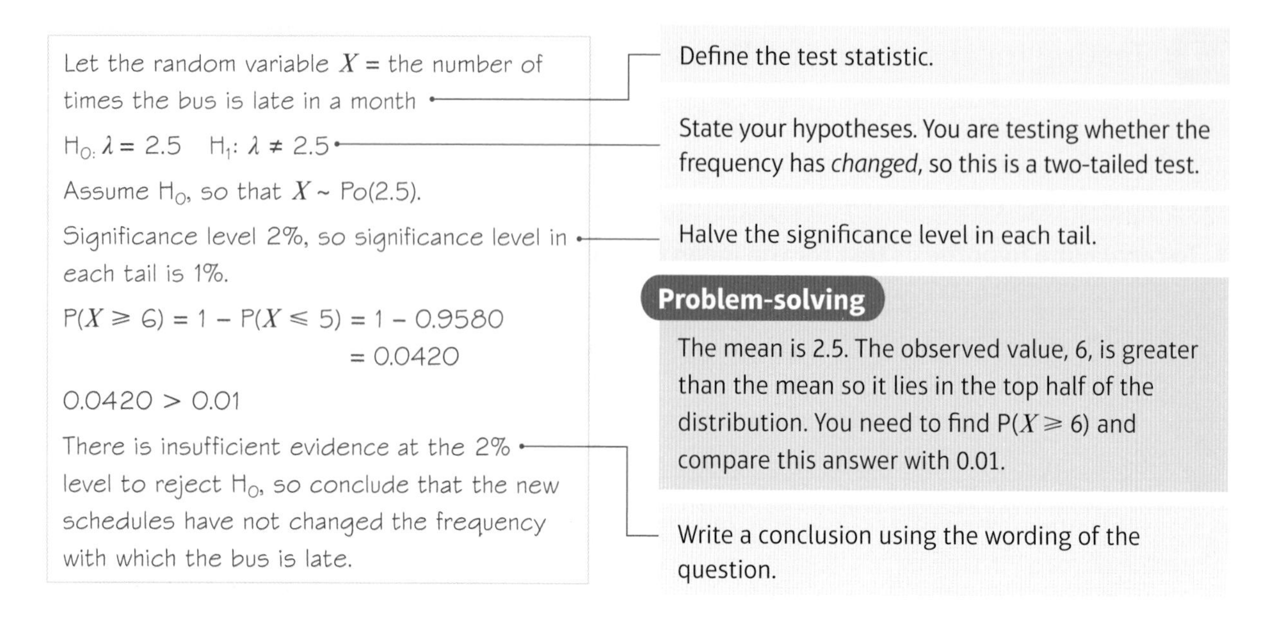

Let the random variable X = the number of times the bus is late in a month

$H_0: \lambda = 2.5 \quad H_1: \lambda \neq 2.5$

Assume H_0, so that $X \sim \text{Po}(2.5)$.

Significance level 2%, so significance level in each tail is 1%.

$P(X \geqslant 6) = 1 - P(X \leqslant 5) = 1 - 0.9580$
$= 0.0420$

$0.0420 > 0.01$

There is insufficient evidence at the 2% level to reject H_0, so conclude that the new schedules have not changed the frequency with which the bus is late.

Define the test statistic.

State your hypotheses. You are testing whether the frequency has *changed*, so this is a two-tailed test.

Halve the significance level in each tail.

Problem-solving

The mean is 2.5. The observed value, 6, is greater than the mean so it lies in the top half of the distribution. You need to find $P(X \geqslant 6)$ and compare this answer with 0.01.

Write a conclusion using the wording of the question.

You might have to use a Poisson distribution as an approximation to a binomial distribution when carrying out a hypothesis test.

Example 3

During an influenza epidemic, 4% of a population of a large city were affected on a given day. The manager of a factory that employs 250 people found that 17 of the employees at his factory were absent, claiming to be suffering from influenza. Using a Poisson approximation to the binomial distribution and a 5% level of significance, test whether or not the proportion of employees suffering from influenza at his factory was larger than that of the whole city.

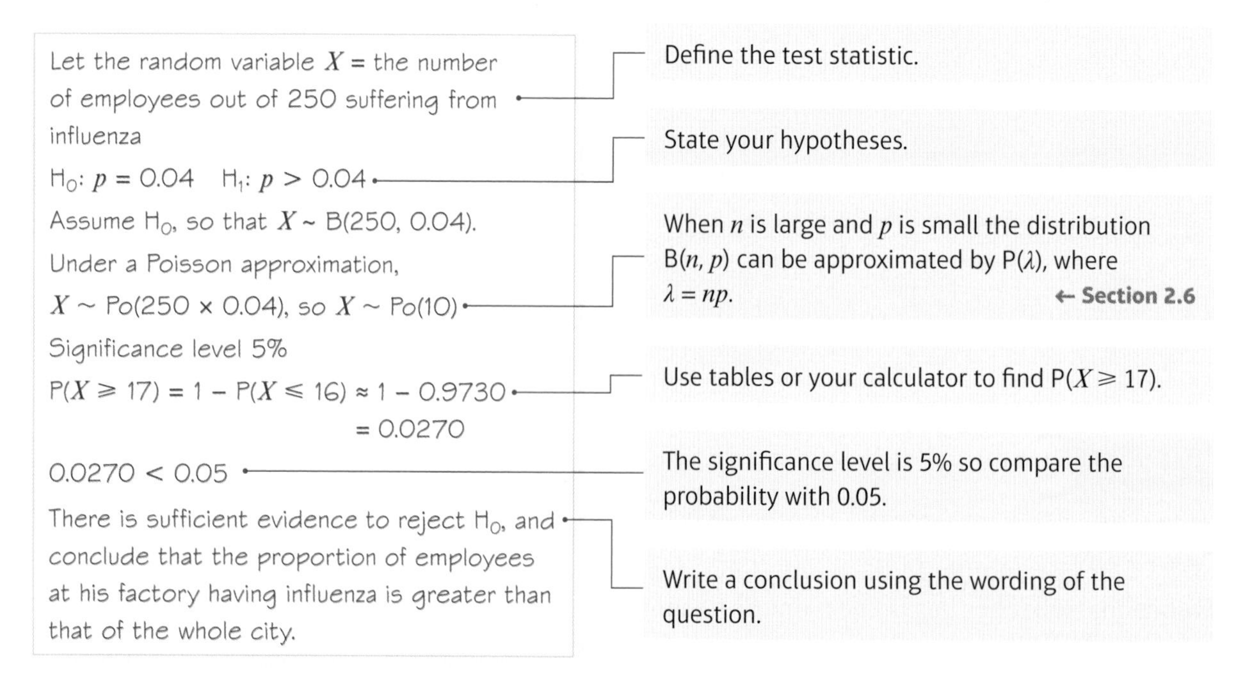

Let the random variable X = the number of employees out of 250 suffering from influenza

$H_0: p = 0.04 \quad H_1: p > 0.04$

Assume H_0, so that $X \sim B(250, 0.04)$.

Under a Poisson approximation,

$X \sim \text{Po}(250 \times 0.04)$, so $X \sim \text{Po}(10)$

Significance level 5%

$P(X \geqslant 17) = 1 - P(X \leqslant 16) \approx 1 - 0.9730$
$= 0.0270$

$0.0270 < 0.05$

There is sufficient evidence to reject H_0, and conclude that the proportion of employees at his factory having influenza is greater than that of the whole city.

Define the test statistic.

State your hypotheses.

When n is large and p is small the distribution $B(n, p)$ can be approximated by $P(\lambda)$, where $\lambda = np$. ← Section 2.6

Use tables or your calculator to find $P(X \geqslant 17)$.

The significance level is 5% so compare the probability with 0.05.

Write a conclusion using the wording of the question.

Exercise 4A

1 A single observation is taken from a Poisson distribution $Po(\lambda)$ and a value of 3 is obtained. Use this observation to test $H_0: \lambda = 8$ against $H_1: \lambda < 8$, using a 5% level of significance.

2 A random variable X has a Poisson distribution $Po(\lambda)$. A single observation of $x = 2$ is taken from the distribution. Test at the 5% level of significance, $H_0: \lambda = 6.5$ against $H_1: \lambda < 6.5$.

3 A single observation is taken from a Poisson distribution $Po(\lambda)$ and a value of 8 is obtained. Use this observation to test $H_0: \lambda = 5.5$ against $H_1: \lambda > 5.5$. using a 5% level of significance.

4 A random variable X has a Poisson distribution $Po(\lambda)$. A single observation of $x = 10$ is taken from the distribution. Test at the 5% level of significance, $H_0: \lambda = 5.5$ against $H_1: \lambda \neq 5.5$.

Hint This is a two-tailed test. $10 > 5.5$ so calculate $P(X \geqslant 10)$ and compare the answer with 0.025.

5 The number of misprints on each page of the *Daily Moaner* is found to have a Poisson distribution with mean 7.5. Soon after a new proofreader is employed, the editor finds one day that there are 13 misprints on a particular page, and claims that the mean number of misprints has increased. Test this claim at the 5% level of significance.

6 On a stretch of road, accidents occur at a rate of 0.8 per month. In the month following new markings on the road, it is found there are 3 accidents. Is there any evidence at the 5% level of significance to suggest an increase in the rate of accidents along this stretch of road?

(P) 7 A restaurant has a coffee machine which seizes up and stops working at a rate of 0.2 times per week. In the 5 weeks following the introduction of a new brand of coffee, the machine seizes up 3 times. Is there any evidence, at the 5% level of significance, to suggest that the rate at which the coffee machine seizes up has increased?

Problem-solving

Your random variable should be the number of times the machine seizes up in 5 weeks, so the mean is $0.2 \times 5 = 1$.

(P) 8 The number of houses sold per week by a firm of estate agents follows a Poisson distribution with a mean of 2.25. The firm appoints a new salesman. In the four-week period following the appointment, the number of sales is 6. Test, at the 5% level of significance, whether or not there is evidence to suggest the rate of sales has changed.

(P) 9 The number of accidents each week at a crossroads controlled by traffic lights may be modelled by a Poisson distribution with mean 1.25. The timings on the lights are changed and in the next 6 weeks there is a total of 4 accidents. Is there any evidence, at a 5% level of significance, of a reduction in the mean weekly number of accidents?

(P) 10 The average number of flaws per 50 m of cloth produced by a machine is found to be 2.3. After the machine is serviced, the number of flaws in the first 150 m is found to be 3. Test, at the 5% level of significance, whether or not the average number of flaws has changed.

(E/P) 11 The number of breakdowns per day in a large fleet of hire cars has a Poisson distribution with mean 0.3.

Find the probability that in a 20-day period the number of breakdowns is:

a exactly 5 **(2 marks)**

b no more than 8. **(2 marks)**

The hire car company introduces new servicing guidelines in order to try to decrease the number of cars that break down. In a randomly chosen 30-day period following the introduction of the new measures, 5 cars break down.

c Test, at the 5% level of significance, whether or not the mean number of breakdowns has decreased. State your hypotheses clearly. **(4 marks)**

E/P **12** A doctor expects to see, on average, 2.25 patients per week with a particular condition. The doctor decides to send information to her patients to try and reduce the number of patients she sees with the condition. In the first four weeks after the information is sent, she sees 4 patients with the condition. Test at the 5% level of significance, whether or not there is reason to believe that sending the information has reduced the number of times the doctor sees patients with the condition. **(5 marks)**

E/P **13** Breakdowns occur on a particular machine at a rate of 1.5 every week. A manager feels that the rate of breakdowns has changed and decides to monitor the machine. Over a 6-week period she finds that there are 13 breakdowns. Test at the 5% level of significance, whether or not the manager's suspicion is correct. **(5 marks)**

E/P **14** A factory produces components, of which 1% are defective. The components are packed in boxes of 1000.

a Using a Poisson approximation to the binomial distribution, estimate the probability that a randomly chosen box contains:

i exactly 9 defective components **(2 marks)**

ii no more than 7 defective components. **(2 marks)**

b Explain why this approximation is suitable. **(2 marks)**

The machinery in the factory is serviced and it is found that in the first box produced following the servicing there are 5 defective components.

c Is there evidence to suggest at the 5% level of significance that the servicing has reduced the number of defective components? State your hypotheses clearly. **(4 marks)**

4.2 Finding critical regions for a Poisson distribution

The critical region is the range of values of the test statistic that would lead to you rejecting H_0. The value(s) on the boundary of the critical region are called critical value(s).

You need to be able to find a critical region for a one- or two-tailed test of the mean of a Poisson distribution.

- **A one-tailed test has an alternative hypothesis $H_1 : \theta < m$ or $H_1 : \theta > m$. There is a single part to the critical region and one critical value.**
- **A two-tailed test has an alternative hypothesis $H_1 : \theta \neq m$. There are two parts to the critical region and two critical values.**

Notation θ is the parameter you are testing for. In this case, it would be the mean of the Poisson distribution, λ.

Example 4

An estate agent has been selling houses at a rate of 9 per month. He believes that the rate of sales will decrease in the next month.

a Using a 5% level of significance, find the critical region for a one-tailed test of the hypothesis that the rate of sales will decrease from 9 per month.

b Write down the actual significance level of the test in part **a**.

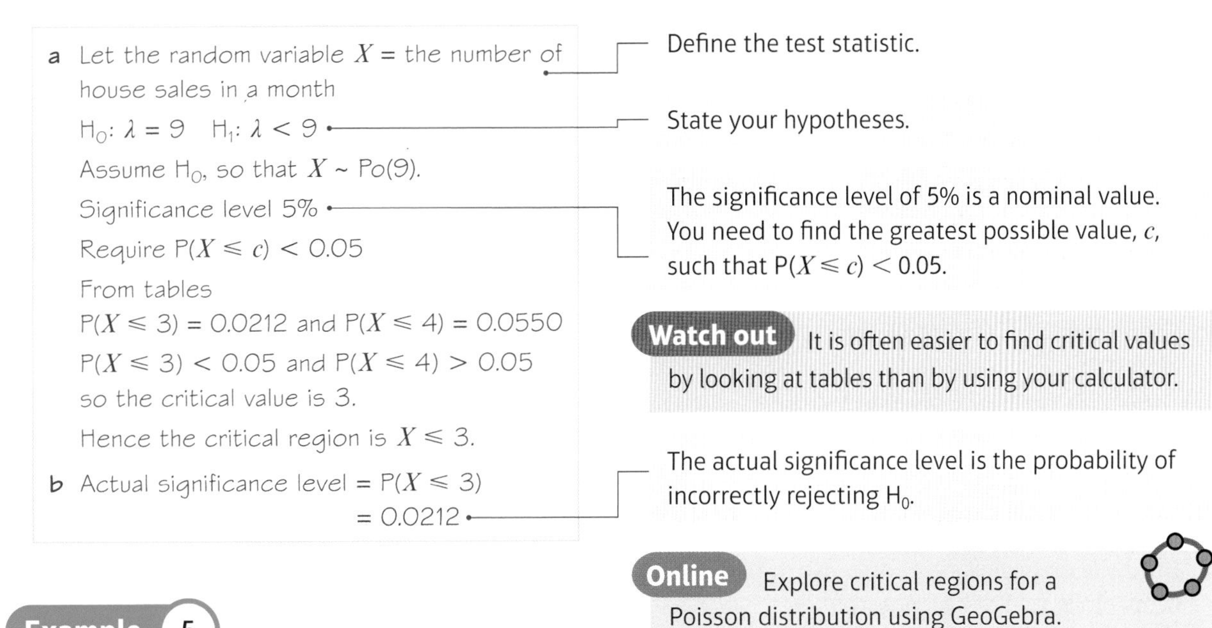

Online Explore critical regions for a Poisson distribution using GeoGebra.

Example 5

Leonora Metti is regarded as a super striker who plays for *Statistics All Stars*. The mean number of goals she has scored over the several years she has been with the club is 0.7 goals per game. She has now been transferred to *Mechanics Ladies*. The management at *Mechanics Ladies* wish to see if the rate at which she scores goals has now increased. They monitor the number of goals that she scores in her first 10 games.

Assuming a Poisson distribution,

a use a 5% level of significance to find the critical region for a one-tailed test of the hypothesis that Leonora has increased her rate of scoring from 0.7 goals per game

b write down the actual significance level of the test in part **a**.

Leonora scores 11 goals in her first 10 games for *Mechanics Ladies*.

c Comment on this observation in light of your critical region.

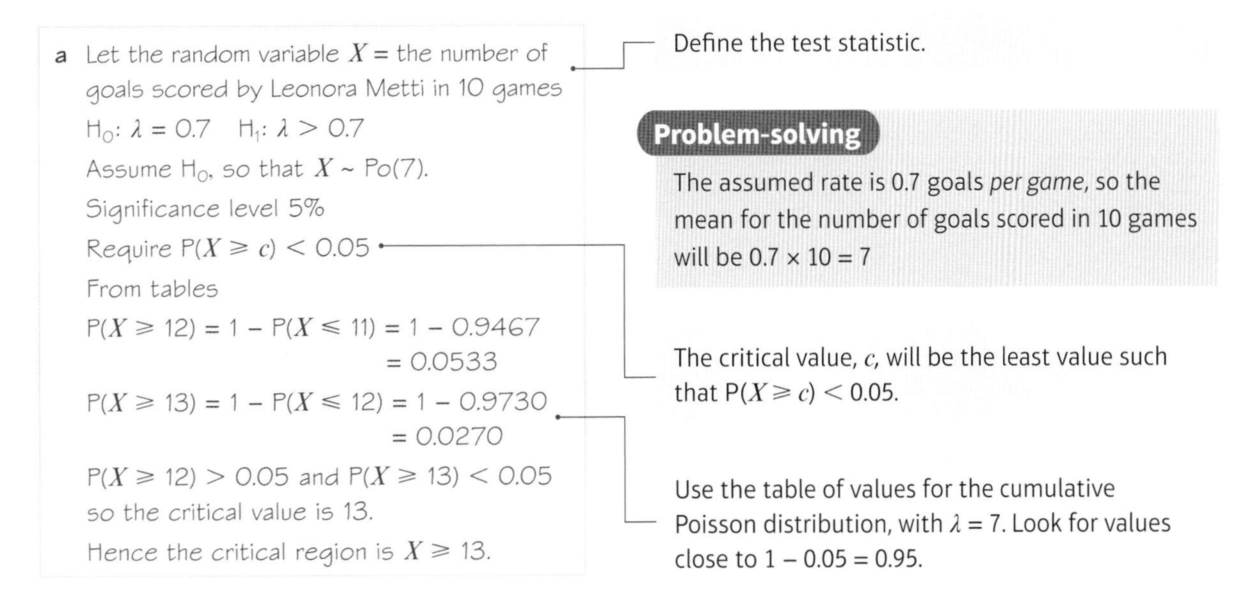

b Actual significance level = P(rejecting H_0)
= $P(X \geqslant 13) = 0.0270$

c $X = 11$ does not lie in the critical region, so there is insufficient evidence to reject H_0. Conclude that Leonora has not increased her goal-scoring rate.

This means that $P(X \geqslant 11) > 0.05$.

Example 6

An office finds that, over a long period of time, incoming calls from customers occur at a rate of 0.325 per minute.

They believe that the rate of calls has changed recently. To test this, the number of incoming calls during a random 20-minute interval is recorded.

a Find the critical region for a two-tailed test of this hypothesis. The probability in each tail should be as close to 2.5% as possible.

b Write down the actual significance level of the test.

The actual number of calls recorded in the 20-minute period was 13.

c Comment on this observation in light of your critical region.

a Let the random variable X = the number of incoming calls in a 20-minute interval

Define the test statistic.

Assumed rate is 0.325 per minute, so in 20 minutes you would expect $20 \times 0.325 = 6.5$ calls, hence

$H_0: \lambda = 6.5 \quad H_1: \lambda \neq 6.5$

You could also state your hypotheses in terms of the rate per minute: $H_0: \lambda = 0.325$ and $H_1: \lambda \neq 0.325$.

Assume H_0, so that $X \sim \text{Po}(6.5)$.

If $X = c_1$ is the upper boundary of the lower critical region, we require $P(X \leqslant c_1)$ to be as close as possible to 2.5%.

From tables

$P(X \leqslant 2) = 0.0430$ and $P(X \leqslant 1) = 0.0113$

Use the table of values for the cumulative Poisson distribution, with $\lambda = 6.5$. Look for values close to 0.025.

0.0113 is closer to 0.025, so $c_1 = 1$, hence lower critical region is $X \leqslant 1$.

If $X = c_2$ is the lower boundary of the upper critical region, we require $P(X \geqslant c_2)$ to be as close as possible to 2.5%.

From tables

$P(X \geqslant 12) = 1 - P(X \leqslant 11) = 1 - 0.9661 = 0.0339$

and $P(X \geqslant 13) = 1 - P(X \leqslant 12) = 1 - 0.9840 = 0.0160$

Since 0.0339 is closer to 0.025, we choose c_2 to be 12, hence upper critical region is $X \geqslant 12$.

Hence critical region is $X \leqslant 1$ or $X \geqslant 12$.

Watch out The probability in each tail has to be as close to 2.5% as possible, but it doesn't necessarily have to be less than 2.5%.

b Actual significance level = $P(X \leqslant 1) + P(X \geqslant 12)$
$= 0.0113 + 0.0339 = 0.0452$

Add together the actual probabilities in each tail.

c $X = 13$ is in the critical region so reject H_0. Conclude that there is evidence to suggest that the rate of incoming calls has changed.

Exercise 4B

1 A single observation is to be taken from a Poisson distribution with parameter λ. This value is used to test H_0 against H_1. In each question part, find the critical region for the test, and write down the actual significance level of the test.

a $H_0: \lambda = 5.5$; $H_1: \lambda < 5.5$ using a 5% level of significance

b $H_0: \lambda = 8$; $H_1: \lambda > 8$ using a 1% level of significance

c $H_0: \lambda = 4$; $H_1: \lambda > 4$ using a 5% level of significance

2 A fisherman is known to catch fish at a mean rate of 5 per hour. The number of fish caught by the fisherman in an hour follows a Poisson distribution. The fisherman buys some new equipment and wants to test whether or not there is an increase in the mean number of fish caught per hour. He records the number of fish he catches in a two-hour period.

Using a 5% level of significance, find the critical region for this test.

(P) **3** The number of sales made by Hans, a telephone sales person, averages 0.8 per day.

Hans is given some extra training and his total sales over a period of 10 days is noted. Find the critical region of a test at the 5% level of significance to determine whether the daily sales achieved by Hans have increased.

(P) **4** In the manufacture of cloth in a factory, defects occur randomly in the production process at a rate of 1.3 per $5\,m^2$. The factory introduces a new procedure to manufacture the cloth. After the introduction of this new procedure, the manager takes a random sample of $25\,m^2$ of cloth from the next batch produced to test if there has been any decrease in the rate of defects.

Using a 5% level of significance, find the critical region for this test.

(P) **5** Accidents occur randomly at a crossroads at a rate of 0.5 per month. A new system is introduced at the crossroads. The number of accidents in the next 12 months is recorded. Find the critical region of a test at the 5% level of significance to determine whether the rate of accidents has decreased.

(P) **6** An online shop sells a computer game at an average rate of 0.35 per day. In an attempt to increase sales of the computer game, the price is reduced for 20 days. Find the critical region of a test at the 5% level of significance to determine whether the rate of sales has increased.

(P) **7** A single observation is to be taken from a Poisson distribution with parameter λ. This value is used to test H_0 against H_1. Using a 5% significance level, find the critical region for this test. The probability of rejection in either tail should be as close as possible to 2.5%. Write down the actual significance level of each test.

a $H_0: \lambda = 4$; $H_1: \lambda \neq 4$ **b** $H_0: \lambda = 8$; $H_1: \lambda \neq 8$ **c** $H_0: \lambda = 9.5$; $H_1: \lambda \neq 9.5$

(E/P) **8** During term time, incoming calls to a school are thought to occur at a rate of 0.25 per minute. To test this, the number of calls during a random 30-minute interval is recorded.

a Find the critical region for a two-tailed test of the hypothesis that the number of incoming calls occurs at a rate other than 0.25 per minute. The probability in each tail should be as close to 2.5% as possible. **(3 marks)**

b Find the actual significance level of the above test. **(1 mark)**

The actual number of calls recorded in this 30-minute period was 11.

c Comment on this observation in light of the critical region. **(2 marks)**

E/P **9** Millie manufactures printed material. She knows that defects occur randomly in the manufacturing process at a rate of 1 every 7 metres. Once a week the machinery is cleaned and reset. Millie then takes a random sample of 35 metres of material from the next batch produced to test if there has been any change in the rate of defects.

a Stating your hypotheses clearly and using a 10% level of significance, find the critical region for this test. You should choose your critical region so that the probability of rejection is less than 0.05 in each tail. **(3 marks)**

b State the actual significance level of this test. **(1 mark)**

E/P **10** A company claims that it receives emails at a mean rate of 3 every 5 minutes.

a Give two conditions under which a Poisson distribution would be a suitable model for the number of emails received. **(2 marks)**

To test its claim, the company records the number of emails received in a 15-minute period.

b Using a 5% level of significance, find the critical region for a two-tailed test of the hypothesis that the mean number of emails received in a 15-minute period is different from 9. The probability of rejection in each tail should be as close as possible to 0.025. **(3 marks)**

c Find the actual level of significance of this test. **(1 mark)**

The actual number of emails received in this period was 13.

d Comment on the company's claim in the light of this value. Justify your answer. **(2 marks)**

E/P **11** A single observation x is to be taken from a Poisson distribution with parameter λ.

This observation is to be used to test, at a 5% level of significance, H_0: $\lambda = c$ against H_1: $\lambda \neq c$, where c is a positive integer. The probability in each tail is less than 0.025.

Given that the critical region for this test is $X \leqslant 2$ or $X \geqslant 15$,

a find the value of c, justifying your answer. **(3 marks)**

b Find the actual significance level of this test. **(2 marks)**

4.3 Hypothesis testing for the parameter p of a geometric distribution

A

The geometric distribution can be used to model the number of trials until a single successful trial is achieved.

Links For a geometric distribution to be valid, the probability of success must be constant on each trial and each trial must be independent. The parameter p used in a geometric distribution is the probability of success on each trial.
If $X \sim \text{Geo}(p)$ then the mean of X is $\frac{1}{p}$.
← Sections 3.1, 3.2

- **To carry out a hypothesis test for the parameter p of geometric distribution, you form two hypotheses.**
  - **The null hypothesis, $H_0 : p = m$ is the value of p that you assume to be true.**
  - **The alternative hypothesis, H_1, tells you about the value of p if your assumption is shown to be wrong.**

Testing for p allows you to answer questions such as:

- Is a dice biased because it takes me lots of rolls to get a 6?
- Does the introduction of a new drug reduce the probability of suffering a particular symptom of my condition on any one day?

Unlike the parameters p of a binomial distribution and λ of a Poisson distribution, when the parameter p of a geometric distribution *increases* the mean of the distribution *decreases*. So when you set up a hypothesis test to see if the parameter p of a geometric distribution has increased you need to look to see if the test statistic falls in the lower end of the distribution and vice versa.

The following results are particularly useful when carrying out a hypothesis test for the mean of a geometric distribution:

- **If $X \sim \text{Geo}(p)$**
  - $\mathbf{P(X = x) = p(1 - p)^{x-1}}$
  - $\mathbf{P(X \leqslant x) = 1 - (1 - p)^{x}}$
  - $\mathbf{P(X \geqslant x) = (1 - p)^{x-1}}$

Example 7

A company claims that 1 in 8 of their packets of crisps contains a prize ticket. Lee decides to test this claim and he buys a packet of crisps each day until he finds a prize ticket. Lee finds a prize ticket for the first time on the 24th day. Test, at the 5% significance level, whether there is any evidence to suggest that the company is overstating the proportion of packets containing a prize ticket.

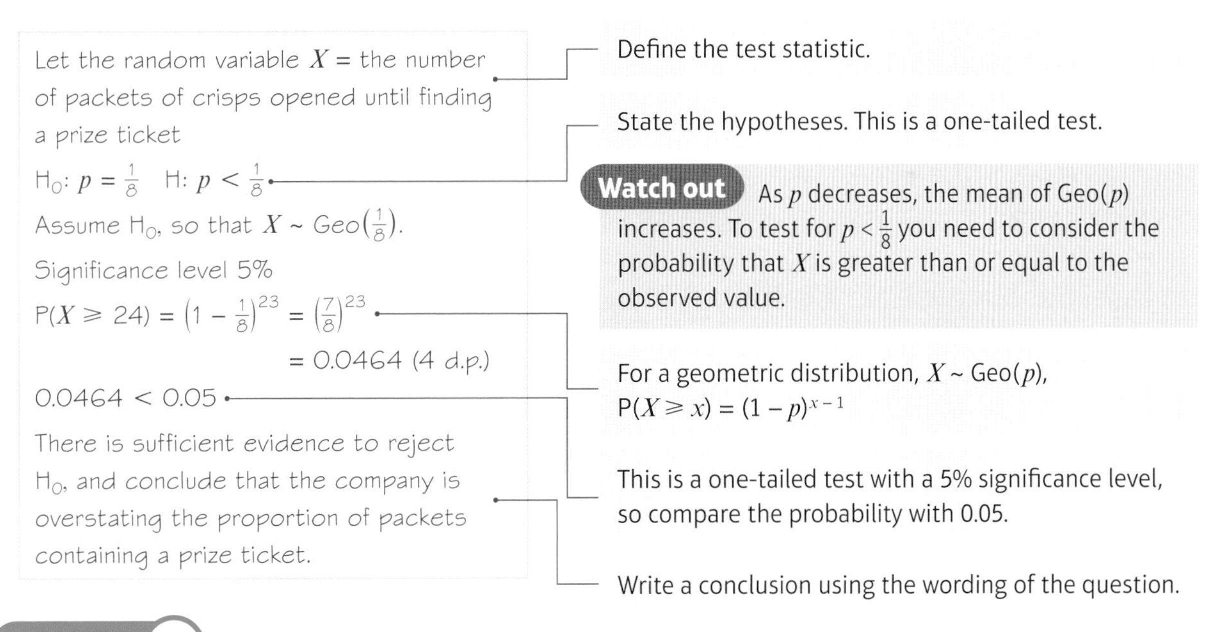

Let the random variable X = the number of packets of crisps opened until finding a prize ticket

H_0: $p = \frac{1}{8}$ $\quad$ H_1: $p < \frac{1}{8}$

Assume H_0, so that $X \sim \text{Geo}\left(\frac{1}{8}\right)$.

Significance level 5%

$P(X \geqslant 24) = \left(1 - \frac{1}{8}\right)^{23} = \left(\frac{7}{8}\right)^{23}$

$= 0.0464$ (4 d.p.)

$0.0464 < 0.05$

There is sufficient evidence to reject H_0, and conclude that the company is overstating the proportion of packets containing a prize ticket.

Define the test statistic.

State the hypotheses. This is a one-tailed test.

Watch out As p decreases, the mean of Geo(p) increases. To test for $p < \frac{1}{8}$ you need to consider the probability that X is greater than or equal to the observed value.

For a geometric distribution, $X \sim \text{Geo}(p)$, $P(X \geqslant x) = (1 - p)^{x-1}$

This is a one-tailed test with a 5% significance level, so compare the probability with 0.05.

Write a conclusion using the wording of the question.

Example 8

An electronics company makes small components for use in computers. It claims that the percentage of defective components coming off the production line is 0.05%. The electronics company sells the components to a retailer. The retailer suspects that the percentage defective stated by the electronics company should be higher. From a very large consignment recently purchased, the retailer tests the components until he finds a defective component. He finds a defective component on the 90th component that he tests. Is there any evidence to suggest that the retailer's suspicions are correct? Test at the 5% level of significance.

A

Let the random variable X = the number of components tested until finding a defective

$H_0: p = 0.0005 \quad H_1: p > 0.0005$

Assume H_0, so that $X \sim \text{Geo}(0.0005)$.
Significance level 5%

$P(X \leqslant 90) = 1 - (1 - 0.0005)^{90}$
$= 1 - 0.9995^{90}$
$= 0.0440$ (4 d.p.)

$0.0440 < 0.05$

There is sufficient evidence to reject H_0 and conclude that the retailer's suspicions are correct i.e the percentage defective is greater than 0.05%.

Define the test statistic.

Since the consignment was very large, the non-replacement of a defective component should not greatly affect the probability. Hence a geometric distribution model is still appropriate.

For a geometric distribution, $X \sim \text{Geo}(p)$, $P(X \leqslant x) = 1 - (1 - p)^x$

This is a one-tailed test with a significance level of 5%, so compare the probability with 0.05.

Write a conclusion using the wording of the question.

Exercise 4C

1 A single observation is taken from a geometric distribution Geo(p) and a value of 9 is obtained. Use this observation to test $H_0: p = 0.25$ against $H_1: p < 0.25$, using a 5% level of significance.

2 A random variable X has a geometric distribution Geo(p). A single observation of $X = 6$ is taken from the distribution. Test at the 5% level of significance, $H_0: p = 0.6$ against $H_1: p < 0.6$.

3 A single observation is taken from a geometric distribution Geo(p) and a value of 3 is obtained. Use this observation to test $H_0: p = 0.01$ against $H_1: p > 0.01$ using a 5% level of significance.

4 A random variable has distribution $X \sim \text{Geo}(p)$. A single observation of $X = 18$ is obtained. Use this observation to test $H_0: p = 0.15$ against $H_1: p < 0.15$ using a 5% level of significance.

5 A random variable has distribution $X \sim \text{Geo}(p)$. A single observation of $X = 2$ is obtained. Test, at the 5% level of significance, $H_0: p = 0.02$ against $H_1: p > 0.02$.

6 A dice used in a board game is suspected of not giving the number 6 often enough. A player throws the dice and finds that she gets her first 6 on her 20th throw. Does this give significant evidence at the 5% level of significance, that the probability of getting a 6 is less than $\frac{1}{6}$?

7 It is claimed that a computer program produces at random a letter from the list A, B, C, D, E. It is found that the first A occurs as the 15th letter after the computer is set running. Is there any evidence to suggest at the 5% level of significance, that the probability of getting an A is less than $\frac{1}{5}$?

A E **8** On average Lucy scores a goal in one of every 4 attempts from a free kick.

a State a suitable distribution to model the number of attempts needed to score her first goal. **(2 marks)**

b Find the probability that she scores her first goal on her 5th attempt. **(2 marks)**

After an injury, Lucy scores her first goal from a free kick on her 10th attempt.

c Test, at the 5% level of significance, whether the probability of her scoring a goal from a free kick is now less than $\frac{1}{4}$. **(3 marks)**

9 It is claimed that 1 in 4 scratch cards is a winner. A statistics student decides to test this claim because she suspects the probability is less than this. She buys one scratch card every day and finds that she gets her first win on the 12th day. Use a 5% level of significance to test whether the student's suspicion is valid.

10 It is claimed by *Wisetalk* that 22% of the population own one of their phones. People are selected one at a time, and asked if they own a *Wisetalk* phone. The number of people questioned, up to and including the first person to own a *Wisetalk* phone, was found to be 14. Is there any evidence at the 5% level of significance that *Wisetalk* are overstating the percentage?

E **11** Marie claims that she scores a penalty on 30% of her attempts. One of her rivals claims that she is overstating her ability. In an attempt to prove her case, Marie takes consecutive shots until she scores her first penalty. She scores her first penalty on her 10th shot. Test her rival's claim, using a 5% level of significance and clearly stating your null and alternative hypotheses. **(4 marks)**

E **12** Imelda is a bird watcher. The probability that she will see a robin on any given day is $\frac{1}{6}$.
The random variable X represents the number of days until Imelda first sees a robin.

a Write down a suitable distribution to model X, and give two conditions that are necessary for this model to be valid. **(3 marks)**

b Calculate the probability that Imelda sees her first robin

i on the third day **(2 marks)**

ii after the fourth day. **(2 marks)**

Imelda further claims that the probability that she will see a magpie on any given day is $\frac{1}{4}$. It is decided to test Imelda's claim. It is found from her records that it took 12 days until she saw her first magpie.

c Is there any evidence at the 5% level of significance that Imelda is overstating the probability of seeing a magpie on any given day? State your hypotheses clearly. **(4 marks)**

4.4 Finding critical regions for a geometric distribution

You need to be able to find a critical region for a hypothesis test of the parameter of a geometric distribution.

Example 9

Ayesha wishes to test if a five-sided spinner numbered from 1 to 5 is fair. She spins the spinner and counts the number of trials until she gets the spinner to show the number 1.

a Using a 5% level of significance, find the critical region for a one-tailed test of the hypothesis that the probability of getting the number 1 on a single spin is less than $\frac{1}{5}$.

b Write down the actual significance level of the test in part **a**.

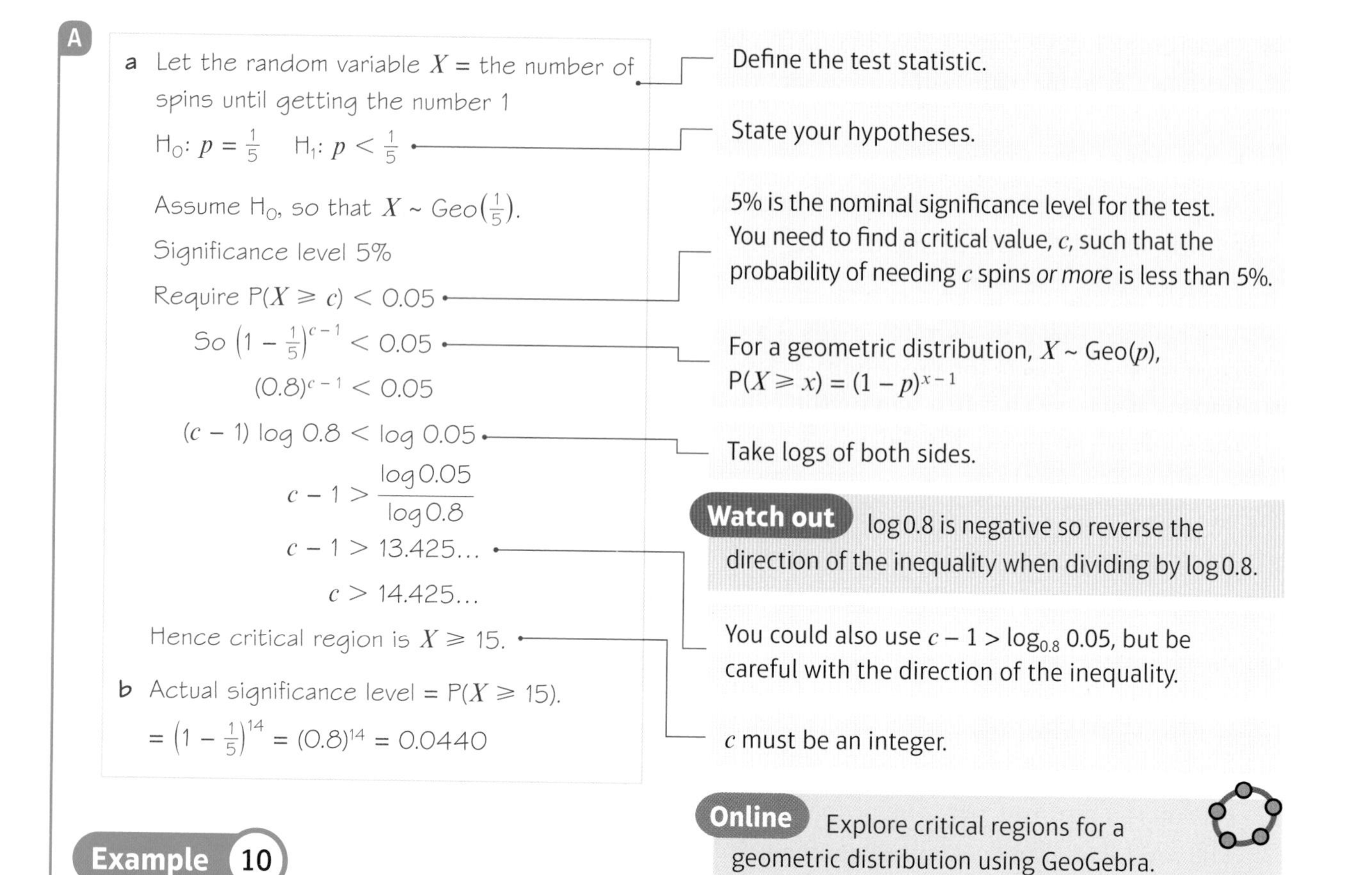

a Let the random variable X = the number of spins until getting the number 1 — Define the test statistic.

$H_0: p = \frac{1}{5} \quad H_1: p < \frac{1}{5}$ — State your hypotheses.

Assume H_0, so that $X \sim \text{Geo}\left(\frac{1}{5}\right)$.

Significance level 5%

Require $P(X \geqslant c) < 0.05$ — 5% is the nominal significance level for the test. You need to find a critical value, c, such that the probability of needing c spins *or more* is less than 5%.

So $\left(1 - \frac{1}{5}\right)^{c-1} < 0.05$ — For a geometric distribution, $X \sim \text{Geo}(p)$, $P(X \geqslant x) = (1 - p)^{x-1}$

$(0.8)^{c-1} < 0.05$

$(c - 1) \log 0.8 < \log 0.05$ — Take logs of both sides.

$c - 1 > \frac{\log 0.05}{\log 0.8}$

$c - 1 > 13.425...$ — You could also use $c - 1 > \log_{0.8} 0.05$, but be careful with the direction of the inequality.

$c > 14.425...$

Hence critical region is $X \geqslant 15$. — c must be an integer.

Watch out $\log 0.8$ is negative so reverse the direction of the inequality when dividing by $\log 0.8$.

b Actual significance level = $P(X \geqslant 15)$.

$= \left(1 - \frac{1}{5}\right)^{14} = (0.8)^{14} = 0.0440$

Online Explore critical regions for a geometric distribution using GeoGebra.

Example 10

In a particular city, a *Lobster Card* is used as a method of payment on trains. The company that administers the card claims that only 1 in 1000 cards will be rejected by the card reader at the train station. A station manager feels that the company is understating the proportion of cards rejected by the card reader, and decides to carry out a test. When he comes on shift at 5:00 am he counts the number of passengers who pass through until he notes a passenger who has a *Lobster Card* rejected.

a Using a 5% level of significance, find the critical region for a one-tailed test of the hypothesis that the proportion of *Lobster Cards* rejected by the card reader is greater than 1 in 1000.

b Write down the actual significance level of the test in part **a**.

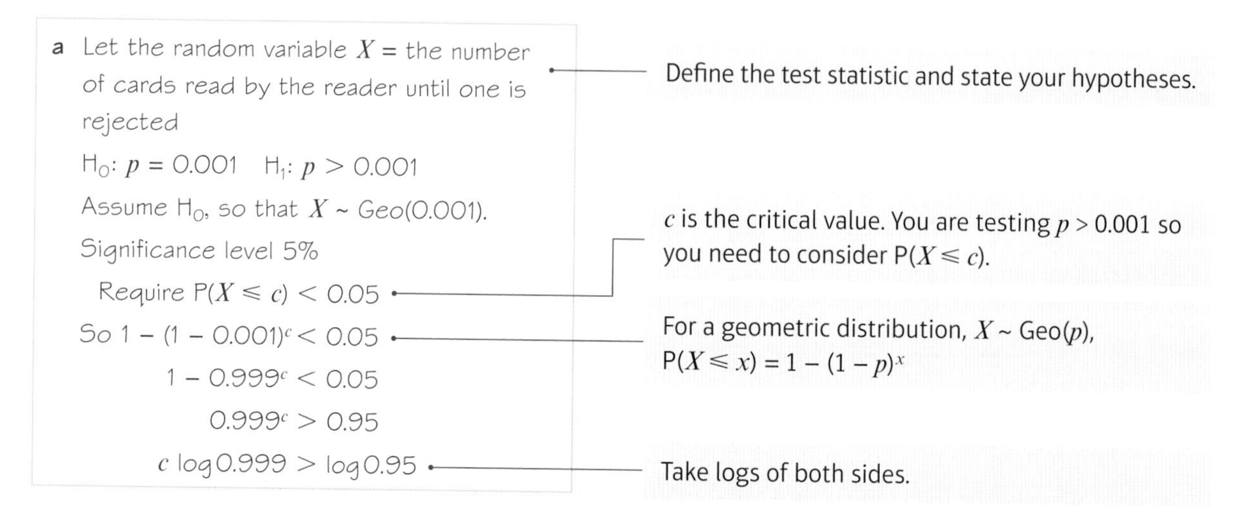

a Let the random variable X = the number of cards read by the reader until one is rejected — Define the test statistic and state your hypotheses.

$H_0: p = 0.001 \quad H_1: p > 0.001$

Assume H_0, so that $X \sim \text{Geo}(0.001)$.

Significance level 5%

Require $P(X \leqslant c) < 0.05$ — c is the critical value. You are testing $p > 0.001$ so you need to consider $P(X \leqslant c)$.

So $1 - (1 - 0.001)^c < 0.05$ — For a geometric distribution, $X \sim \text{Geo}(p)$, $P(X \leqslant x) = 1 - (1 - p)^x$

$1 - 0.999^c < 0.05$

$0.999^c > 0.95$

$c \log 0.999 > \log 0.95$ — Take logs of both sides.

$$c < \frac{\log 0.95}{\log 0.999}$$
$$c < 51.267...$$
Hence critical region is $X \leqslant 51$.

b Actual significance level = $P(X \leqslant 51)$
$= 1 - 0.999^{51} = 0.0497$ (4 d.p.)

log 0.999 is negative so reverse the inequality when you divide.

Choose the largest integer value of c which satisfies this inequality.

Exercise 4D

P **1** A random variable has distribution $X \sim \text{Geo}(p)$. A single observation is used to test H_0: $p = 0.3$ against H_1: $p < 0.3$.

a Using a 5% level of significance, find the critical region for this test.

b Calculate the actual significance level of this test.

P **2** A random variable has distribution $X \sim \text{Geo}(p)$. A single observation is used to test H_0: $p = 0.35$ against H_1: $p < 0.35$.

a Using a 5% level of significance, find the critical region for this test.

b Calculate the actual significance level of this test.

P **3** A random variable has distribution $X \sim \text{Geo}(p)$. A single observation is used to test H_0: $p = 0.05$ against H_1: $p > 0.05$.

a Using a 10% level of significance, find the critical region for this test.

b Calculate the actual significance level of this test.

P **4** Each day Arun enters a ballot for a concert ticket. It is claimed by the concert organisers that the probability of winning a ticket each day is 0.23. Arun decides to test this claim.

a Find the critical region, at the 5% level of significance, for the number of days that Arun has to wait before winning a ticket in order for him to claim that the organisers are overstating the probability of winning a ticket. **(6 marks)**

b Find the probability of incorrectly rejecting the null hypothesis in this test. **(2 marks)**

Arun waits 11 days before winning a ticket.

c Comment on this in light of your critical region. **(2 marks)**

P **5** Dot is a professional darts player. She claims that the probability that she will hit the bullseye with a single dart is $\frac{1}{3}$. Her arch rival Sharon claims that Dot is exaggerating her ability to hit the bullseye. To test this claim, Dot throws darts until she hits a bullseye.

a Find the critical region, at the 5% level of significance, for the number of darts thrown by Dot in order to accept Sharon's claim. **(6 marks)**

b Find the actual significance level for this test. **(2 marks)**

P **6** Rita has a medical condition, a symptom of which is to have a tremor. The probability that Rita has a tremor on any given day is 0.6. A new drug to help treat Rita's condition becomes available. Rita wants to test, at the 5% significance level, whether the new drug has reduced the

A probability that she will have a tremor on any given day. After allowing a period of adjustment, Rita is observed to see how many days it will be before she has a tremor.

Find the critical region for this test. **(6 marks)**

Challenge

A single observation is to be taken from a geometric distribution $X \sim \text{Geo}(p)$. This observation is used to test $\text{H}_0\text{: } p = 0.009$ against $\text{H}_1\text{: } p \neq 0.009$.

a Using a 5% level of significance, find the critical regions for this test. The probability of rejecting either tail should be as close as possible to 2.5%.

b Find the probability of incorrectly rejecting the null hypothesis on this test.

The actual value of X obtained is 5.

c Based on this observation state, with reasons, whether there is sufficient evidence to reject H_0.

Mixed exercise 

E/P **1** Vehicles pass a particular point on a road at a rate of 39 vehicles per hour.

Find the probability that in any randomly selected 10-minute interval

a exactly 6 cars pass this point **(2 marks)**

b at least 8 cars pass this point. **(2 marks)**

After the introduction of a new one-way system, it is suggested that the number of vehicles passing this point has decreased.

During a randomly selected 10-minute interval 2 vehicles pass the point.

c Test, at the 5% level of significance, whether or not there is evidence to support the suggestion that the number of vehicles has decreased. State your hypotheses clearly. **(4 marks)**

E/P **2** An effect of a certain disease is that a small number of the red blood cells are deformed. Francesca has this disease and the deformed blood cells occur randomly at a rate of 3.2 per ml of her blood. Following a course of new treatment, a random sample of 2.5 ml of Francesca's blood is found to contain only 4 deformed red blood cells.

Stating your hypotheses clearly and using a 5% level of significance, test whether or not there has been a decrease in the number of deformed red blood cells in Francesca's blood. **(4 marks)**

A E **3** The probability that Peter completes the crossword successfully each day in his daily newspaper is $\frac{1}{5}$.

A new crossword setter has been appointed. It takes Peter 7 days until he completes his first crossword successfully. Is there evidence to suggest that the crosswords are now more difficult? Test at the 5% level of significance, stating your hypotheses clearly. **(4 marks)**

4 During the winter in the ski resort of Glen Hoe, the probability that snow falls on any one day is 0.45. Roisin starts her winter break in Glen Hoe on 1st December.

a Calculate the probability that the first fall of snow that Roisin sees is on or after 3rd December. **(2 marks)**

A meteorologist feels that due to changes in climate, the probability of seeing a fall of snow on any day in winter in Glen Hoe is now less than 0.45. Roisin eventually sees her first fall of snow on December 7th.

b Test, at the 5% significance level, whether there is sufficient evidence to suggest that the meteorologist is correct. **(4 marks)**

E/P **5** Scoobie is the receptionist at a large company. Records show that over the many years that Scoobie has worked for the company the probability of his connecting to the wrong extension is 0.03. Find, using a Poisson approximation to the binomial distribution, the probability that in a day when Scoobie receives 150 calls he puts through:

a 5 calls to the wrong extension **(3 marks)**

b no more than 3 calls to the wrong extension. **(3 marks)**

Scoobie retires and a new receptionist, Waldo is appointed. The company monitors his first 300 calls and finds that he puts through 4 calls to the wrong extension.

c Test, using a Poisson approximation to the binomial distribution, whether there is any evidence to suggest that Waldo has decreased the rate at which calls are put through to the wrong extension. Test at the 5% level of significance. **(3 marks)**

E/P **6** Arnold is a printer. Breakdowns on his printing press occur at an average rate of 1.75 per month. Assuming a Poisson distribution, find the probability that:

a exactly 3 breakdowns occur in a particular month **(2 marks)**

b more than 5 breakdowns occur in a two-month period **(3 marks)**

c in four consecutive months there are two months in which there are exactly 3 breakdowns. **(3 marks)**

Arnold has his printing press serviced and wants to test whether the rate of breakdowns has been reduced. He records the number of breakdowns in the next four months.

d Find the critical region, at the 5% significance level, for Arnold's test **(3 marks)**

e State the actual significance level of Arnold's test. **(1 mark)**

E/P **7** An electrical goods retailer sells a mean of 3.5 television sets per day on a weekday. The retailer decides to do some advertising in a local paper. In a two-day period following the advertising, the retailer sells 11 television sets, leading him to believe that the advert has increased his sales. Stating your hypotheses clearly, test at the 5% level of significance whether or not there is evidence of an increase in sales following the appearance of the advert. **(4 marks)**

E **8** A company's website is visited on weekdays, at a rate of 8.5 visits per minute. In a random one minute on a Saturday the website is visited 12 times.

a Test, at the 5% level of significance, whether or not there is evidence that the rate of visits is greater on a Saturday than on a weekday. **(3 marks)**

b State the minimum number of visits that would be required in a given minute to obtain a significant result. **(2 marks)**

E/P **9** A manager thinks that 5% of her workforce are absent for at least one day each month. She chooses 200 workers at random and finds that in the last month 15 workers had been absent for at least one day.

Using a Poisson approximation to the binomial distribution, test at the 5% level of significance whether the percentage of workers who are absent for at least one day each month is higher than the manager thinks. **(4 marks)**

10 It is known that 15% of products produced by a machine are defective. Products are tested, one at a time, until the first defective one is encountered.

The random variable X represents the number of products tested until the first defective one is found. Find:

a $\text{P}(X = 5)$ **(2 marks)**

b $\text{P}(X \geqslant 3)$ **(2 marks)**

The machine is serviced and it is hoped that this has reduced the proportion of defective products.

c Find the critical region for a hypothesis test that the proportion of defective products has reduced. Use a 5% level of significance. **(6 marks)**

d Find the actual significance level of this test. **(2 marks)**

11 Over a number of years, the mean number of hurricanes experienced in a certain area during the month of August is 4. A scientist suggests that, due to global warming, the number of hurricanes will have increased, and proposes to do a hypothesis test based on the number of hurricanes this year.

a Suggest suitable hypotheses for this test. **(2 marks)**

b Find to what level the number of hurricanes must increase for the null hypothesis to be rejected at the 5% level of significance. **(3 marks)**

The actual number of hurricanes this year was 8.

c Comment on this observation in light of your answer to part **b**. **(2 marks)**

12 A coin is believed to be biased. Alison and Paul want to test the coin to see if the probability of it landing on heads, p, is significantly less than $\frac{1}{2}$. They both use a 2% significance level.

Alison spins the coin 30 times and records the number of heads.

a Find the critical region for Alison's test. **(2 marks)**

Paul spins the coin until it lands on heads for the first time.

b Find the critical region for Paul's test. **(6 marks)**

Alison and Paul both observe values that lie within their respective critical regions. As a result, they reject the assumption that $p = \frac{1}{2}$.

c Find the probability that Paul and Alison have incorrectly rejected the assumption that $p = \frac{1}{2}$. **(5 marks)**

Challenge

An oil company found that in a certain region there is an 18% chance of striking oil when a well is drilled. The company has now started drilling in a neighbouring region and wishes to test if there is greater chance in this region of striking oil when a well is sunk. It decides to count the number of wells, N, sunk until it strikes oil for the third time.

a Suggest a suitable distribution for N, stating any assumptions that are necessary.

b Using a 5% level of significance find the critical region for this test.

c Write down the actual significance level of the test.

Summary of key points

1 To carry out a **hypothesis test** for a given parameter, θ, you form two hypotheses.

- The **null hypothesis**, $H_0 : \theta = m$ is the value of the parameter that you assume to be true.
- The **alternative hypothesis**, H_1, tells you about the value of the parameter if your assumption is shown to be wrong.

2
- A **one-tailed test** has an alternative hypothesis $H_1 : \theta < m$ or $H_1 : \theta > m$. There is a single part to the critical region and one critical value.
- A **two-tailed test** has an alternative hypothesis $H_1 : \theta \neq m$. There are two parts to the critical region and two critical values.

3 The actual significance level of a test is the probability of incorrectly rejecting H_0.

A

4 If $X \sim \text{Geo}(p)$

- $P(X = x) = p(1 - p)^{x-1}$
- $P(X \leqslant x) = 1 - (1 - p)^x$
- $P(X \geqslant x) = (1 - p)^{x-1}$

5 Central limit theorem

Objectives

After completing this chapter you should be able to:

- Understand and apply the central limit theorem to approximate the sample mean of a random variable, $\overline{X}$ → pages 59–62
- Apply the central limit theorem to other distributions → pages 62–64

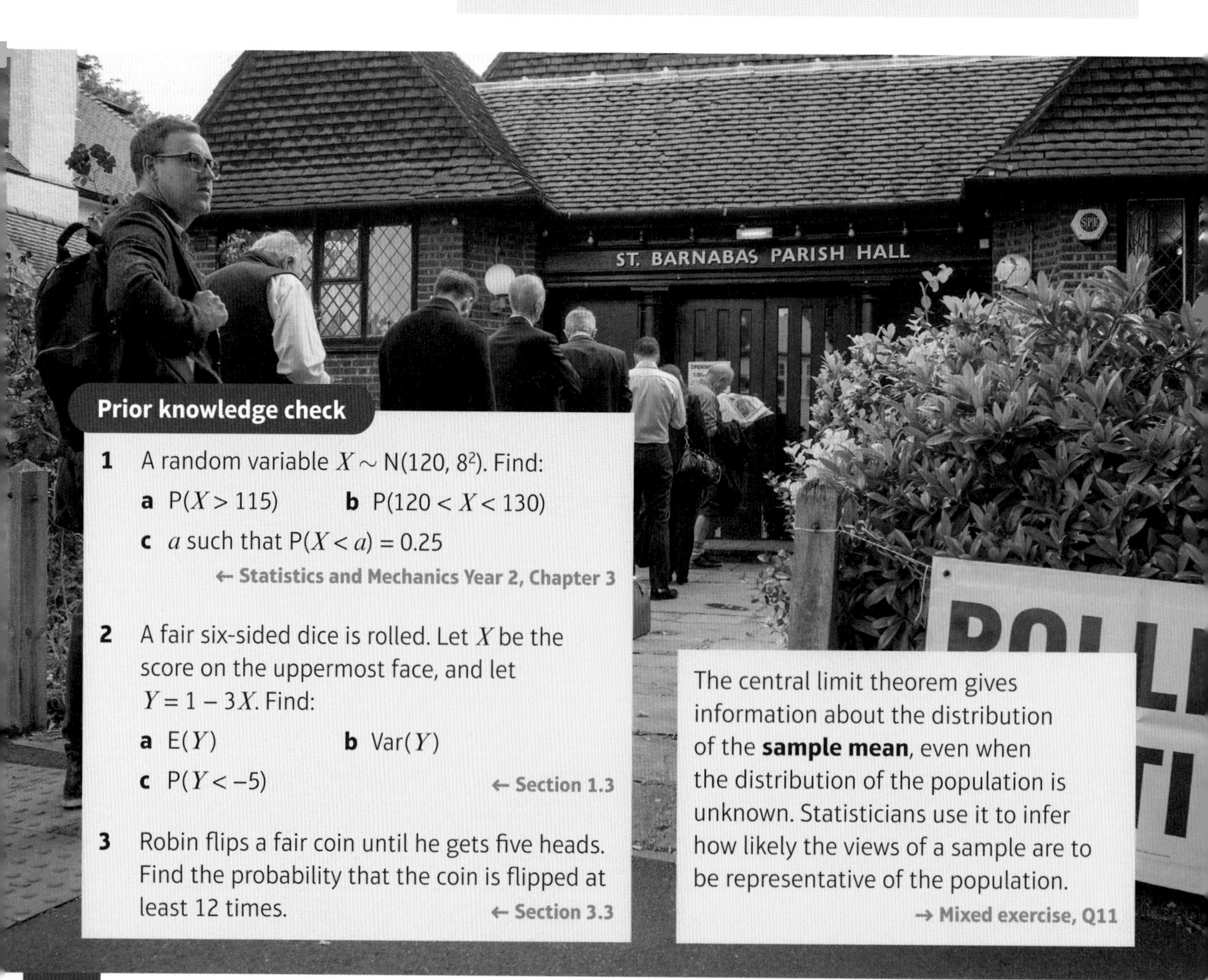

Prior knowledge check

1 A random variable $X \sim N(120, 8^2)$. Find:

a $P(X > 115)$ b $P(120 < X < 130)$

c a such that $P(X < a) = 0.25$

← Statistics and Mechanics Year 2, Chapter 3

2 A fair six-sided dice is rolled. Let X be the score on the uppermost face, and let $Y = 1 - 3X$. Find:

a $E(Y)$ b $Var(Y)$

c $P(Y < -5)$ ← Section 1.3

3 Robin flips a fair coin until he gets five heads. Find the probability that the coin is flipped at least 12 times. ← Section 3.3

The central limit theorem gives information about the distribution of the **sample mean**, even when the distribution of the population is unknown. Statisticians use it to infer how likely the views of a sample are to be representative of the population.

→ Mixed exercise, Q11

5.1 The central limit theorem

A

If you take a random sample of n observations from a normally distributed random variable $X \sim \mathrm{N}(\mu, \sigma^2)$, then the **sample mean**, $\overline{X}$, is also normally distributed with $\overline{X} \sim \mathrm{N}\left(\mu, \frac{\sigma^2}{n}\right)$.

Links This result is useful for hypothesis testing for the mean of a normal distribution.
← Statistics and Mechanics Year 2, Section 3.7

In fact, this result is a special case of a more powerful result called the **central limit theorem**. This states that the mean of a large random sample taken from *any* random variable is always approximately normally distributed. This result is true regardless of the distribution of the original random variable.

- **The central limit theorem says that if $X_1, X_2, \ldots, X_n$ is a random sample of size n from a population with mean μ and variance σ^2, then $\overline{X}$ is approximately $\sim \mathrm{N}\left(\mu, \frac{\sigma^2}{n}\right)$.**

Note that in general the sample mean is only **approximately** distributed with $\mathrm{N}\left(\mu, \frac{\sigma^2}{n}\right)$. As n gets larger, this approximation gets better.

Watch out You can see that this is only an approximation by considering $n = 1$. In this case, each sample is a single observation, so the sample mean will have the same distribution as the original random variable.

The variance of the sample mean also decreases as n gets large. You can say that for a large sample, the sample mean will be very close to the population mean.

Example 1

A sample of size 9 is taken from a population with distribution $\mathrm{N}(10, 2^2)$. Find the probability that the sample mean $\overline{X}$ is more than 11.

The population is normal, so $\overline{X}$ will have a normal distribution despite the small size of the sample.

$$\mathrm{Var}(\overline{X}) = \frac{\sigma^2}{n} = \frac{2^2}{9} = \left(\frac{2}{3}\right)^2$$

So $\overline{X} \sim \mathrm{N}\left(10, \left(\frac{2}{3}\right)^2\right)$

The mean of $\overline{X}$ is μ (= 10) and the variance of $\overline{X}$ is $\frac{\sigma^2}{n}$.

The mean of $\overline{X}$ is 10 and the standard deviation is $\frac{2}{3}$ so:

$$\mathrm{P}(\overline{X} > 11) = 1 - \mathrm{P}(\overline{X} < 11)$$
$$= 1 - 0.9332$$
$$= 0.0668 \text{ (4 d.p.)}$$

Use the normal distribution function on your calculator to find the probability.

Watch out In this case the distribution of the sample mean is not an approximation. This is only true when the population is normally distributed.

Example 2

A

A six-sided dice is relabelled so that there are three faces marked 1, two faces marked 3 and one face marked 6. The dice is rolled 40 times and the mean of the 40 scores is recorded.

a Find an approximate distribution for the mean of the scores.

b Use your approximation to estimate the probability that the mean is greater than 3.

a Let the random variable X = the score on a single roll; then the distribution of X is:

x	1	3	6
$P(X = x)$	$\frac{1}{2}$	$\frac{1}{3}$	$\frac{1}{6}$

So: $\mu = E(X) = \sum xP(X = x)$

$= 1 \times \frac{1}{2} + 3 \times \frac{1}{3} + 6 \times \frac{1}{6}$

$= 2.5$

and $\sigma^2 = \text{Var}(X)$

$= \sum x^2 P(X = x) - \mu^2$

$= 1^2 \times \frac{1}{2} + 3^2 \times \frac{1}{3} + 6^2 \times \frac{1}{6} - \left(\frac{5}{2}\right)^2$

$= \frac{19}{2} - \frac{25}{4} = 3.25$ or $\frac{13}{4}$

Now by the central limit theorem:

$\bar{X} \approx \sim N\left(2.5, \frac{13}{160}\right)$

b $P(\bar{X} > 3) = 1 - P(\bar{X} < 3)$

$\approx 1 - 0.9599$

$= 0.0401$ (4 d.p.)

Problem-solving

Find the mean and variance of the discrete distribution. ← Sections 1.1, 1.2

The population is clearly not normally distributed but the sample size ($n = 40$) is quite large so the central limit theorem can be used.

You can use the normal distribution function on your calculator to find $P(\bar{X} > 3)$.

Watch out You do not need to apply a continuity correction when using the central limit theorem. This is because the underlying distribution is the *mean* of the sample. Although this is a discrete random variable, it does not have to take integer values. It takes fractional values, and the gaps between values get smaller and smaller as n gets larger.

Exercise 5A

E

1 A sample of size 6 is taken from a population that is normally distributed with mean 10 and standard deviation 2.

a Find the probability that the sample mean is greater than 12. **(3 marks)**

b State, with a reason, whether your answer is an approximation. **(1 mark)**

2 A machine fills cartons in such a way that the amount of drink in each carton is distributed normally with a mean of 40 cm^3 and a standard deviation of 1.5 cm^3.

A sample of four cartons is examined.

a Find the probability that the mean amount of drink is more than 40.5 cm^3.

A sample of 49 cartons is examined.

b Find the probability that the mean amount of drink is more than 40.5 cm^3 on this occasion.

E/P

3 The lengths of bolts produced by a machine have an unknown distribution with mean 3.03 cm and standard deviation 0.20 cm.

A sample of 100 bolts is taken.

a Estimate the probability that the mean length of this sample is less than 3 cm. **(3 marks)**

A second sample is taken. The probability that the mean of this sample is less than 3 cm needs to be less than 1%.

b Find the minimum sample size required. **(5 marks)**

E

4 A random variable X has the discrete uniform distribution

$$P(X = x) = \tfrac{1}{5} \qquad x = 1, 2, 3, 4, 5$$

40 observations are taken from X, and their mean $\overline{X}$ is recorded.

Find an estimate for $P(\overline{X} > 3.2)$. **(6 marks)**

P

5 A fair dice is rolled 35 times.

a Find the approximate probability that the mean of the 35 scores is more than 4.

b Find the approximate probability that the total of the 35 scores is less than 100.

6 The 25 children in a class each roll a fair dice 30 times and record the number of sixes they obtain. Find an estimate of the probability that the mean number of sixes recorded for the class is less than 4.5.

E

7 The random variable X has the probability distribution shown in the table.

a Find the value of k. **(2 marks)**

x	0	2	3	5
$P(X = x)$	0.1	$3k$	k	0.3

A random sample of 100 observations of X is taken.

b Use the central limit theorem to estimate the probability that the mean of these observations is greater than 3. **(6 marks)**

c Comment on the accuracy of your estimate. **(1 mark)**

P

8 A fair dice is rolled n times. Given that there is less than a 1% chance that the mean of all the scores differs from 3.5 by more than 0.1, find the minimum sample size.

E/P

9 The annual salaries of employees at a large company have an unknown distribution with mean £28 500 and standard deviation £6800.

A random sample of 5 members of the senior management team is taken.

A researcher suggests that $N(28\,500, \frac{6800^2}{5})$ could be used to model the distribution of the sample mean.

a Give a reason why this is unlikely to be a good model. **(1 mark)**

A second random sample of 15 employees from the whole company is taken.

b Estimate the probability that the mean annual salary of these employees is:

i less than £25 000 **ii** between £25 000 and £30 000. **(4 marks)**

c Comment on the accuracy of your estimate. **(1 mark)**

10 An electrical company repairs very large numbers of television sets and wishes to estimate the mean time taken to repair a particular fault. It is known from previous research that the standard deviation of the time taken to repair this particular fault is 2.5 minutes.
The manager wishes to ensure that the probability that the estimate differs from the true mean by less than 30 seconds is 0.95.

Find how large a sample is required. **(6 marks)**

5.2 Applying the central limit theorem to other distributions

You can use the central limit theorem to solve problems involving the Poisson, geometric, binominal and negative binomial distributions.

Example 3

A supermarket manager is trying to model the number of customers that visit her store each day. She observes that, on average, 20 new customers enter the store every minute.

a Calculate the probability that fewer than 15 customers arrive in a given minute.

b Find the probability that in one hour no more than 1150 customers arrive.

c Use the central limit theorem to estimate the probability that in one hour no more than 1150 customers arrive. Compare your answer to part **b**.

a Let X denote the number of customers that arrive in a minute. Then $X \sim \text{Po}(20)$.
$P(X < 15) = 0.1049$ (4 d.p.)

It's reasonable to assume that customers arrive independently of each other at a constant rate, so the number of customers arriving each minute will have a Poisson distribution. ← Chapter 2

b Let T denote the number of customers that arrive in an hour.
Then $T \sim \text{Po}(60 \times 20)$
so $T \sim \text{Po}(1200)$.
$P(T \leqslant 1150) = 0.07578$ (4 d.p.)

c Consider a sample of 60 observations taken from X.

You could also consider the number of customers who arrive in one hour as a sample of 60 observations from X, the number who arrive in one minute.

By the central limit theorem $\overline{X}$ is approximately $\sim N\left(20, \frac{20}{60}\right)$, or $N\left(20, \frac{1}{3}\right)$.
If $T \leqslant 1150$ then $\overline{X} \leqslant \frac{1150}{60} = 19.1666...$
So $P(T \leqslant 1150) = P(\overline{X} \leqslant 19.1666...)$
≈ 0.0745 (4 d.p.)

The two answers are close, so the approximation from the central limit theorem is quite good.

Problem-solving

If $\sum X_i$ is the sum of the observations from a sample of size n, then the sample mean is given by $\overline{X} = \frac{\sum X_i}{n}$.

Example 4

Billy is the captain of a football team. Each week he gets a team together by calling his friends one by one and asking if they would like to play. The probability of each friend agreeing to play is $\frac{2}{3}$. Once he has 10 other players he stops calling.

a Calculate the number of friends Billy expects to have to call to find 10 other players.

b Find the probability that Billy has to call exactly 12 friends.

In a season, Billy's team plays 25 matches.

c Estimate the probability that the mean number of calls per match Billy had to make was less than 15.5.

a Let X be the number of friends Billy calls.
Then $X \sim$ Negative B$(10, \frac{2}{3})$, so
$E(X) = \frac{30}{2} = 15$

The number of friends Billy calls is the number of trials required for 10 successes with probability $\frac{2}{3}$ of success, which has a **negative binomial** distribution. If $X \sim$ Negative B(r, p), then $E(X) = \mu = \frac{r}{p}$ and $\text{Var}(X) = \sigma^2 = \frac{r(1-p)}{p^2}$.

← Section 3.3

b $P(X = 12) = \binom{11}{9} \times \left(\frac{2}{3}\right)^{10} \times \left(\frac{1}{3}\right)^2$
$= 0.1060$ (4 d.p.)

Use $P(X = x) = \binom{x-1}{r-1} p^r (1-p)^{x-r}$.

c $E(X) = 15$, and $\text{Var}(X) = \frac{10\left(\frac{1}{3}\right)}{\left(\frac{2}{3}\right)^2} = 7.5$

For a sample of size 25, the sample mean $\overline{X}$ is approximately $\sim N\left(15, \frac{7.5}{25}\right)$, or $N(15, 0.3)$, by the central limit theorem.

$P(\overline{X} < 15.5) \approx 0.8193$ (4 d.p.)

Use the normal distribution function on your calculator.

Exercise 5B

1 A random sample of 10 observations is taken from a Poisson distribution with mean 3.

a Find the exact probability that the sample mean does not exceed 2.5.

b Estimate the probability that the sample mean does not exceed 2.5 using the central limit theorem, and compare your answer to part **a**.

2 A random sample of 12 observations is taken from a random variable $X \sim \text{Geo}(0.25)$.

a Find the mean and variance of X.

b Estimate the probability that the sample mean is greater than 5 using the central limit theorem.

Hint X has the **geometric distribution** with parameter 0.25, so $E(X) = \mu = \frac{1}{p}$ and $\text{Var}(X) = \frac{1-p}{p^2}$.

← Section 3.2

3 A sample of size 20 is taken from a binomial distribution with $n = 10$ and $p = 0.2$. Estimate the probability that the sample mean does not exceed 2.4. **(4 marks)**

A E/P **4** There are 20 children in a class. Each flips a fair coin until they get heads 5 times.

a Write down the expected number of times each student will have to flip the coin. **(2 marks)**

b Find an estimate of the probability that the mean number of flips is at most 9. **(3 marks)**

E/P **5** A town is hit by three thunderstorms per month, on average.

a Find the probability that there are four thunderstorms next month. **(2 marks)**

b Use the central limit theorem to estimate the probability that over the course of a year, the average number of thunderstorms each month is at most 2.5. **(4 marks)**

E/P **6** A patient is awaiting a liver transplant. The probability that a randomly selected donor is a match is 0.2.

a Find the expected number of donors that will have to be tested before finding a match. **(2 marks)**

A random sample of 20 patients awaiting liver transplants was selected, and the number of donors tested for each patient before a match was found was recorded.

b Estimate the probability that the average number of donors to be tested per patient is more than 5.5. **(3 marks)**

E/P **7** David is selling raffle tickets from door to door to raise money for charity. To reach his daily fundraising goal, he needs to sell 10 tickets. He observes that, on average, an occupant in one in every three houses he visits will buy a ticket.

a Find the probability that on a given day he reaches his daily goal after visiting exactly 35 houses. **(2 marks)**

b In one month, David worked on 20 days, and met his daily goal on each day. Estimate the probability that the average number of houses he visited per day was 35 or fewer. **(4 marks)**

E/P **8** Telephone calls arrive at an exchange at an average rate of two per minute. Over a period of 30 days a telephonist records the number of calls each day that arrive in the five-minute period before her break.

a Find an approximation for the probability that the total number of calls recorded is more than 350. **(2 marks)**

b Estimate the probability that the mean number of calls received in this period each day is less than 9.0. **(4 marks)**

Mixed exercise 5

E **1** A random sample of 100 observations is taken from a probability distribution with mean 5 and variance 1. Estimate the probability that the mean of the sample is greater than 5.2. **(3 marks)**

E/P **2** A fair six-sided dice numbered 1, 2, 4, 5, 7, 8 is rolled 20 times. Estimate the probability that the average score is less than 4. **(4 marks)**

A E/P **3** A sample of size n is taken from a normal distribution with $\mu = 1$ and $\sigma = 1$. Find the minimum sample size such that the probability of the sample mean being negative is less than 5%. **(3 marks)**

E/P **4** In a group of 20 students, each rolls a fair six-sided dice 10 times and records the number of sixes. Estimate the probability that the average number of sixes rolled by each student is greater than 2. **(4 marks)**

E **5** Buses arrive at a bus stop on average once every 5 minutes.

a Find the probability that exactly 3 buses arrive in the next 10 minutes. **(2 marks)**

b Use the central limit theorem to estimate the probability that at least 25 buses arrive in the next 2 hours. **(3 marks)**

E/P **6** A married couple plan to have children, and are desperate to have a daughter. They decide they will keep having children until they have a daughter and then stop. You can assume that giving birth to a girl or boy is equally likely, and independent of the gender of any other children the couple have had.

a Find the probability that they will have more than 2 children. **(2 marks)**

Suppose a group of 10 couples all decide on the same plan.

b Estimate the probability that between them, the 10 couples have more than 24 children. **(4 marks)**

E/P **7** The masses of eggs are normally distributed with mean 60 g and standard deviation 5 g.

A crate contains 48 randomly chosen eggs.

a Calculate the probability that the mean mass of an egg in a randomly chosen crate is greater than 59 g. **(3 marks)**

b State, with a reason, whether your answer to part **a** is an estimate. **(1 mark)**

The probability that an egg has a double yolk is 0.1. A sample of 30 crates is taken.

c Estimate the probability that the sample will contain fewer than 150 double-yolk eggs in total. **(5 marks)**

E/P **8** An automatic coffee machine uses milk powder. The mass, S grams, of milk powder used in one cup of coffee is modelled by $S \sim \text{N}(4.9, 0.8^2)$.

'Semi-skimmed' milk powder is sold in 500 g packs. Find the probability that one pack will be sufficient for 100 cups of coffee. **(4 marks)**

E/P **9** A random sample of size n is to be taken from a population with mean 40 and variance 9. Find the minimum sample size such that the probability of the sample mean being greater than 42 is less than 5%. **(5 marks)**

E **10** A sample of size 20 is taken from a population with an unknown distribution, with mean 35 and variance 9. Find the probability that the sample mean will be greater than 37. **(3 marks)**

11 A nationwide poll asked 500 people whether they prefer white chocolate or milk chocolate. The polling company wants to determine whether the proportion of people who prefer milk chocolate differs significantly from 60%.

The polling company assumes that in the population 60% of people prefer milk chocolate, and defines the random variable X to take the value 1 if a randomly selected member of the population prefers milk chocolate, and 0 otherwise.

a Describe the distribution of X and state its mean and variance. **(3 marks)**

Modelling the poll as a random sample of size 500 from the distribution in part **a**:

b estimate the probability that the sample mean differs from 0.6 by 0.03 or more. **(3 marks)**

c How many people should be polled in order for there to be a greater than 95% chance that the sample mean differs from 0.6 by at most 0.03? **(5 marks)**

Challenge

Let $X_1, \ldots, X_n$ be a random sample from a population with distribution $N(\mu, \sigma^2)$. Show that $\overline{X} \sim N\left(\mu, \frac{\sigma^2}{n}\right)$.

Hint You can use the fact that if $X_1 \sim N(\mu_1, \sigma_1^2)$ and $X_2 \sim N(\mu_2, \sigma_2^2)$ are independent, then $X_1 + X_2 \sim N(\mu_1 + \mu_2, \sigma_1^2 + \sigma_2^2)$.

Summary of key points

1 The **central limit theorem** states that given a random sample of size n from any distribution with mean μ and variance σ^2, the sample mean $\overline{X}$ is approximately distributed as $N\left(\mu, \frac{\sigma^2}{n}\right)$.

Review exercise

(E) **1** The random variable X has probability function

$$P(X = x) = \frac{(2x - 1)}{36} \quad x = 1, 2, 3, 4, 5, 6$$

a Construct a table giving the probability distribution of X. **(2)**

Find:

b $P(2 < X \leqslant 5)$, **(1)**

c the exact value of $E(X)$. **(2)**

d Show that $Var(X) = 1.97$ to three significant figures. **(3)**

e Find $Var(2 - 3X)$. **(2)**

← Sections 1.1, 1.2, 1.3

(E) **2** The random variable X has probability function

$$P(X = x) = \begin{cases} kx & x = 1, 2, 3, \\ k(x + 1) & x = 4, 5 \end{cases}$$

where k is a constant.

a Find the value of k. **(2)**

b Find the exact value of $E(X)$. **(2)**

c Show that, to three significant figures, $Var(X) = 1.47$. **(3)**

d Find, to one decimal place, $Var(4 - 3X)$. **(2)**

← Sections 1.1, 1.2, 1.3

(E/P) **3** The random variable X has probability distribution given by

x	1	2	3	4	5
$P(X = x)$	0.1	p	0.20	q	0.30

a Given that $E(X) = 3.5$, write down two equations involving p and q. **(3)**

Find:

b the value of p and the value of q **(2)**

c $Var(X)$ **(3)**

d $Var(3 - 2X)$ **(2)**

← Sections 1.1, 1.2, 1.3

(E/P) **4** The random variable X has probability distribution given by

x	1	3	5	7	9
$P(X = x)$	0.2	p	0.2	q	0.15

a Given that $E(X) = 4.5$, write down two equations involving p and q. **(3)**

Find:

b the value of p and the value of q **(2)**

c $P(4 < X \leqslant 7)$. **(1)**

Given that $E(X^2) = 27.4$, find:

d $Var(X)$ **(2)**

e $E(19 - 4X)$ **(1)**

f $Var(19 - 4X)$. **(2)**

← Sections 1.1, 1.2, 1.3

(E/P) **5** The discrete random variable X has probability distribution given by

x	−2	−1	0	1
$P(X = x)$	0.2	0.3	a	b

The random variable Y is defined as $Y = 2 - 3X$. Given that $E(Y) = 2.9$,

a find the values of a and b **(5)**

b calculate $E(X^2)$ and $Var(X)$ **(3)**

c write down the value of $Var(Y)$ **(1)**

d find $P(Y + 1 < X)$. **(2)**

← Section 1.4

E/P **6** The discrete random variable X has probability distribution given by

x	-3	-2	0	1	3
$P(X = x)$	a	b	b	a	c

The random variable Y is defined as $Y = \frac{1 - 2X}{4}$. Given that $E(Y) = -0.05$ and $P(Y > 0) = 0.5$, find:

a the probability distribution of X **(7)**

b $P(-3X < 5Y)$. **(2)**

← Section 1.4

E/P **7** Accidents on a particular stretch of motorway occur at an average rate of 1.5 per week.

a Write down a suitable model to represent the number of accidents per week on this stretch of motorway. **(1)**

Find the probability that

b there will be 2 accidents in the same week **(2)**

c there is at least one accident per week for 3 consecutive weeks **(3)**

d there are more than 4 accidents in a two-week period. **(2)**

← Sections 2.1, 2.2

E/P **8** **a** State two conditions under which a Poisson distribution is a suitable model to use in statistical work. **(2)**

The number of cars passing an observation point in a 10-minute interval is modelled by a Poisson distribution with mean 1.

b Find the probability that in a randomly chosen 60-minute period there will be

i exactly 4 cars passing the observation point **(2)**

ii at least 5 cars passing the observation point. **(2)**

The number of other vehicles (i.e. other than cars), passing the observation point in a 60-minute interval is modelled by a Poisson distribution with mean 12.

c Find the probability that exactly 1 vehicle, of *any type*, passes the observation point in a 10-minute period. **(4)**

← Sections 2.1, 2.2, 2.3

E/P **9** Two garden machinery firms hire out equipment independently of each other.

Quikmow hire out lawn-mowers at a rate of 1.5 mowers per hour.

Easitrim hire out lawn-mowers at a rate of 2.2 mowers per hour.

a In a one-hour period, find the probability that each company hires exactly 1 lawn-mower. **(2)**

b In a one-hour period, find the probability that between them, the two companies hire out 4 lawn-mowers. **(3)**

c In a three-hour period, find the probability that the total number of lawn-mowers hired out by the two companies is less than 12. **(3)**

← Sections 2.2, 2.3

E **10** A manufacturer places toys in cereal boxes. A random sample of 200 cereal boxes is taken, and the number of toys, x, in each box is observed. The data is summarised as follows:

$$\sum x = 290 \quad \sum x^2 = 702.$$

a Calculate the mean and the variance of these data. **(2)**

b Explain why the results in part **a** suggest that a Poisson distribution may be a suitable model for the number of toys in each box of cereal. **(1)**

c Use a suitable Poisson distribution to estimate the probability that a randomly chosen box of cereal will contain at least 2 toys. **(3)**

← Section 2.4

E/P **11** **a** Write down the conditions under which the Poisson distribution may be used as an approximation to the binomial distribution. **(2)**

A call centre routes incoming telephone calls to agents who have specialist knowledge to deal with the call. The probability of the caller being connected to the wrong agent is 0.01.

b Find the probability that 2 consecutive calls will be connected to the wrong agent. **(1)**

c Find the probability that more than 1 call in 5 consecutive calls are connected to the wrong agent. **(2)**

The call centre receives 1000 calls each day.

d Find the mean and variance of the number of wrongly connected calls. **(2)**

e Use a Poisson approximation to find, to three decimal places, the probability that more than 6 calls each day are connected to the wrong agent. **(3)**

← Sections 2.5, 2.6

E **12** The random variable X has a binomial distribution $X \backsim B(150, 0.02)$.

a Find $P(X = 3)$ **(2)**

A Poisson random variable, $Y \backsim Po(\lambda)$ is used to approximate X.

b Write down the value of λ and justify the use of a Poisson approximation in this instance. **(2)**

← Section 2.6

E **13** In a manufacturing process, 1.5% of the articles produced are defective. A random sample of 200 articles is selected, and the number of defective articles, X, is recorded.

a Write down the distribution of X. **(2)**

b Find $P(X = 4)$ **(2)**

c Explain why a Poisson distribution could be used as an approximation for X, and write down the parameter for this approximation. **(2)**

d Use your answer to part **c** to find an estimate for $P(X = 4)$, and calculate the percentage error in your estimate. **(3)**

← Section 2.6

A E **14** A computer software engineer is checking the coding of a number of apps. The percentage of apps with defective code is thought to be 5%. X represents the number of apps checked up to and including the first one with defective code.

a State the distribution than can be used to model X. **(1)**

b Find the mean and variance of X. **(2)**

c Find the probability that the engineer has to check the coding of at least 15 apps before finding a defective one. **(2)**

← Sections 3.1, 3.2

15 Anne-Marie is practising basketball and she continues to throw the ball until she gets it in the basket. The random variable Y represents the number of throws she needs.

a State a suitable distribution to model Y. **(1)**

Given that the mean of Y is 10,

b find the probability that Anne-Marie gets her first basket on her seventh attempt **(2)**

c find the variance of Y. **(2)**

d State any assumptions you have made in using this model. **(2)**

← Sections 3.1, 3.2

16 A fair twelve-sided spinner has one side coloured red and eleven sides coloured blue. In a fairground game, players spin the spinner until it lands on red, at which point they win a prize.

A

a Find the probability of winning a prize if the player is limited to 10 spins. **(3)**

Keisha wants the probability of winning a prize to be greater than 0.75.

b Find the minimum number of spins to which she should limit players. **(4)**

← Sections 3.3

E/P **17** Matt is playing a game at a school fete where his probability of winning a prize is 0.18. He plays the game several times.

a Find the probability that he wins his third prize on his thirteenth game. **(2)**

b State two assumptions that have to be made for the model used in part **a** to be valid. **(2)**

c Find the mean and standard deviation of the number of times Matt needs to play to win his fourth prize. **(3)**

Naomi plays a different game until she has won r prizes. Given that Y represents the number of games Naomi plays and that $E(Y) = 20$ and $Var(Y) = 113\frac{1}{3}$,

d find the probability of Naomi winning a prize. **(3)**

← Sections 3.3, 3.4

E/P **18** The random variable X is the number of times a biased dice is rolled until 3 threes have occurred. The variance of X is $23\frac{1}{3}$.

a Find the probability of rolling a three on the biased dice. **(3)**

b Find $P(X = 9)$. **(2)**

c Find $P(X = 11 | \text{a three occurs on the first roll})$. **(3)**

← Sections 3.3, 3.4

E **19** **a** Explain what you understand by

i an hypothesis test,

ii a critical region. **(2)**

During term time, incoming calls to a school are thought to occur at a rate of 0.45 per minute. To test this, the number of calls during a random 20-minute interval is recorded.

b Find the critical region for a two-tailed test of the hypothesis that the number of incoming calls occurs at a rate of 0.45 per 1-minute interval. The probability in each tail should be as close to 2.5% as possible. **(5)**

c Write down the actual significance level of the above test. **(1)**

In the school holidays, 1 call occurs in a 10-minute interval.

d Test, at the 5% level of significance, whether or not there is evidence that the rate of incoming calls is less during the school holidays than in term time. **(5)**

← Sections 4.1, 4.2

E **20** A telesales agent claims she makes a sale on 2.5% of her calls. During her last 300 calls, she made 11 sales.

a Using a Poisson approximation to the binomial distribution, test the claim at the 5% level of significance. **(4)**

b To the nearest multiple of 5%, what level of significance would need to be set in order to reject the claim of the telesales agent? **(1)**

← Sections 4.1, 4.2

E/P **21** Counter staff at a take-away shop claim to receive orders at a mean rate of 4 every 10 minutes.

To test the claim, the shop owner records the number of orders received in a 60-minute period.

a Using a 10% level of significance, find the critical region for a two-tailed test of the hypothesis that the mean number of orders received in a 60-minute period is 24. The probability of rejection in each tail should be as close as possible to 0.05. **(3)**

b Find the actual level of significance of this test. **(1)**

The actual number of orders received in this period was 18.

c Comment on the counter staff's claim in the light of this value. Justify your answer. **(2)**

← Section 4.2

A E/P **22** At a school tombola, it is claimed that 1 in 5 tickets is a winning ticket.

a State a suitable distribution to model the number of attempts needed before finding a winning ticket. **(2)**

Mr Taylor, a maths teacher, decides to test this claim because he suspects the probability is less than this. He buys tickets one after the other and finds that he gets his first win on the 15th ticket.

b Use a 5% level of significance to test whether Mr Taylor's suspicion is valid. **(3)**

← Section 4.3

E/P **23** Xander is playing basketball. He claims that he makes a basket on 50% of his attempts. His team-mate claims that Xander is overstating his ability.

Xander takes consecutive shots until he makes his first basket. He scores his first basket on his 5th shot.

Test the team-mate's claim, using a 10% level of significance and clearly stating your null and alternative hypotheses. **(4)**

← Section 4.3

E/P **24** Anne and Brian work in a microchip manufacturing plant. Brian claims that the chance of a chip being faulty is 1 in 2000. Anne suspects that this is an underestimate and that it is more likely that a randomly chosen chip is faulty.

Anne decided to test Brian's claim at the 5% significance level by sampling from a large batch of chips until she finds a faulty one.

A **a** Find the critical region for Anne's test. **(5)**

b Anne finds the first faulty chip after selecting her 115th. Comment on this in light of your answer to part **a**. **(2)**

← Section 4.4

E/P **25** A report on the health and nutrition of a population stated that the mean height of three-year-old children is 90 cm and the standard deviation is 5 cm. A sample of 100 three-year-old children was chosen from the population.

a Write down the distribution of the sample mean height. **(2)**

b Hence find the probability that the sample mean height is at least 91 cm. **(3)**

← Section 5.1

E/P **26** A sample of size 5 is taken from a population that is normally distributed with mean 10 and standard deviation 3.

Find the probability that the sample mean lies between 7 and 10. **(4)**

← Section 5.1

E/P **27** The random variable X has the probability distribution shown in the table:

x	1	2	3	4
$\mathrm{P}(X = x)$	0.4	$2k$	0.3	k

a Find the value of k. **(2)**

A random sample of 200 observations of X are taken.

b Use the central limit theorem to estimate the probability that the mean of these observations is greater than 2.09. **(6)**

c Comment on the accuracy of your estimate. **(1)**

← Section 5.1

28 A busy call centre receives, on average, 15 calls every minute.

a Calculate the probability that fewer than 10 calls come in a given minute. **(1)**

b Find the probability that in one 30-minute period no more than 420 calls come in. **(2)**

c Use the central limit theorem to estimate the probability that in one 30-minute period no more than 420 calls come in. Compare your answer to part **b**. **(5)**

← Section 5.2

29 A bag contains a large number of coloured balls, red and green, in the ratio 3 : 1. A group of 20 students each repeatedly select a ball from the bag and then replace it, continuing until a green ball is selected.

Use the central limit theorem to estimate the probability that the mean number of attempts needed to select a green ball is more than 4.5. **(5)**

← Section 5.2

30 A group of students are completing a multiple choice quiz where there are five answers to each question.

One student is chosen at random. Given that they guess each answer,

a find the probability that they get their 4th question right on their 12th attempt. **(2)**

b Find the expected number of questions they must answer to get 4 right. **(2)**

There are 15 students in the group. Each student continues answering questions until they have achieved four correct answers.

c Estimate the probability that the mean number of questions answered per student is less than 19. **(5)**

← Section 5.2

Challenge

1 Three fair four-sided dice are rolled. The discrete random variable X represents the difference between the highest score and the lowest score on the three dice.

a Write down the probability distribution of X.

b Show that $\text{E}(X) = \frac{15}{8}$

← Sections 1.1, 1.2, 1.3

2 a If $X \sim \text{B}(4, p)$, show that the probability distribution of X can be written as

x	0	1	2	3	4
$\text{P}(X = x)$	q^4	$4q^3p$	$6p^2q^2$	$4p^3q$	p^4

where $q = 1 - p$.

b Hence show that $\text{E}(X) = 4p$ and $\text{Var}(X) = 4p(1 - p)$.

← Section 1.4

3 A metal detectorist has a 12% chance of finding something valuable when he searches a particular quadrant of a field. He starts searching in another field and wishes to test if there is greater chance in this field of finding something valuable. He decides to count the number of quadrants searched until he finds his fourth item of value.

a Using a 2.5% level of significance find the critical region for this test. Use a negative binomial distribution.

b Write down the actual significance level of the test.

← Sections 3.3, 4.4

Chi-squared tests

6

Objectives

After completing this chapter you should be able to:

- Form hypotheses about how well a distribution fits as a model for an observed frequency distribution and measure goodness of fit of a model to observed data → pages 92–96
- Understand degrees of freedom and use the chi-squared (χ^2) family of distributions → pages 96–99
- Be able to test a hypothesis → pages 99–103
- Apply goodness-of-fit tests to discrete data → pages 103–113
- Use contingency tables → pages 113–119
- Apply goodness-of-fit tests to geometric distributions → pages 119–122

Prior knowledge check

1. Suppose that $X \sim \text{Po}(5)$, find $\text{P}(X < 2)$.
← Section 2.1

2. Suppose that $X \sim \text{Geo}(\frac{1}{3})$, find $\text{P}(X > 5)$.
← Section 3.1

3. David claims 60% of the students in his school like baked beans. He takes a random sample of 100 students and finds that 70 of them say that they like baked beans. Test David's claim, at the 5% significance level.
← Statistics and Mechanics Year 1, Chapter 7

The chi-squared test is used in genetics to help determine whether an experiment was fair and unbiased, and to provide a level of confidence for whether the results were obtained by chance. → Exercise 6A, Q4

6.1 Goodness of fit

Goodness of fit is concerned with measuring how well an observed frequency distribution fits to a known distribution.

Suppose you take a dice and throw it 120 times. You might get results like these:

Number, n	1	2	3	4	5	6
Observed frequency	23	15	25	18	21	18

If the dice is *unbiased* you would, in theory, expect each of the numbers 1 to 6 to appear the same number of times.

For 120 throws the expected frequencies would each be:

$$P(X = x) \times 120 = \frac{1}{6} \times 120 = 20$$

You would expect results like these:

Number, n	1	2	3	4	5	6
Expected frequency	20	20	20	20	20	20

The expected results fit a uniform discrete probability distribution:

x:	1	2	3	4	5	6
$P(X = x)$:	$\frac{1}{6}$	$\frac{1}{6}$	$\frac{1}{6}$	$\frac{1}{6}$	$\frac{1}{6}$	$\frac{1}{6}$

Since you are taking a sample, you should not be surprised that the observed frequency for each number doesn't match the expected frequency exactly.

However, suppose now the dice was biased, you would also not expect the observed frequency of each number to be exactly 20.

Although both the results from the biased and unbiased dice would differ from the predicted results, the results from the unbiased dice should be better modelled by the discrete uniform distribution than those from a biased dice.

We form the **hypothesis** that the *observed frequency distribution does not differ from a theoretical one*, and that any differences are due to natural variations. Because this assumes no difference, it is called the **null hypothesis**.

The **alternative hypothesis** is that the observed frequency distribution *does differ from the theoretical one* and that any differences are due to not only natural variations but the bias of the dice as well.

- **H_0: There is no difference between the observed and the theoretical distribution.**
- **H_1: There is a difference between the observed and the theoretical distribution.**

In order to tell how closely the model fits the observed results you need to have a measure of the **goodness of fit** between the observed frequencies and the expected frequencies.

This measure used for goodness of fit may be understood by looking further at the results of the dice-throwing experiment.

The results and the expected frequencies are:

Number on dice, n	1	2	3	4	5	6
Observed frequency, O_i	23	15	25	18	21	18
Expected frequency, E_i	20	20	20	20	20	20

You can show this as a bar chart:

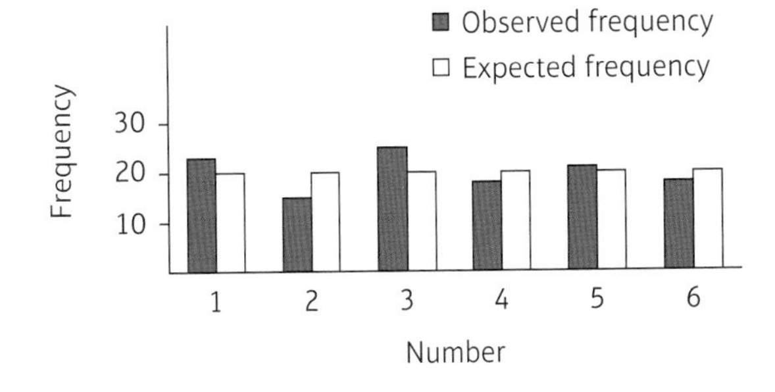

The thing you instinctively look at is the difference between the observed and the expected values.

As a measure of the size of these differences you take the sum of the squares of the differences, divided by the expected frequency:

$$\sum \frac{(O_i - E_i)^2}{E_i} \text{ where}$$

O_i = an observed frequency

E_i = an expected (theoretical) frequency, asserted by the null hypothesis

This gives a positive number that gets larger as the differences between the observed and the expected frequencies get larger, and smaller as the differences get smaller.

- **The measure of goodness of fit is:**

$$X^2 = \sum_{i=1}^{n} \frac{(O_i - E_i)^2}{E_i}$$

The symbol X^2 is used rather than just X because it shows that the value is never going to be negative.

You can see that the less good the fit, the larger the difference between each observed and expected value, and the greater the value of X^2.

- **Here is another way of calculating X^2:**

$$X^2 = \sum \frac{(O_i - E_i)^2}{E_i} = \sum \frac{O_i^2 - 2O_iE_i + E_i^2}{E_i}$$

Multiply out the bracket.

$$= \sum \frac{O_i^2}{E_i} - \sum \frac{2O_iE_i}{E_i} + \sum \frac{E_i^2}{E_i}$$

$$= \sum \frac{O_i^2}{E_i} - \sum 2O_i + \sum E_i$$

$$= \sum \frac{O_i^2}{E_i} - \sum O_i$$

$\sum E$ and $\sum O$ are both equal to the total number of trials, or observations. So $\sum E = \sum O = N$.

$$= \sum \frac{O_i^2}{E_i} - N$$

This formula is not given in the formulae booklet, but is easier to use.

Example 1

Billy and Mel each have two 4-sided spinners numbered 1–4. They each carry out experiments, where they spin their spinners at the same time, and add the scores together. After each student has carried out 160 experiments, the frequency distributions are as follows:

Number, n	2	3	4	5	6	7	8
Observed by Billy (O_i)	12	15	22	41	33	21	16
Observed by Mel (O_i)	6	12	21	37	35	29	20
Expected (E_i)	10	20	30	40	30	20	10

Both Billy and Mel believe that their spinners are fair.

a State the null and alternative hypotheses for the experiment.

One of the students has a biased spinner.

b Calculate the goodness of fit for both students, and determine which of them is most likely to have the biased spinner.

a H_0: the observed distribution is the same as the theoretical distribution. (The spinner is unbiased.)
H_1: the observed distribution is different to the theoretical distribution. (The spinner is biased.)

The null hypothesis is what we believe unless the evidence tells us otherwise. In this case, we think the spinners are fair.

b

	n	2	3	4	5	6	7	8	Total
Billy	$\frac{(O_i - E_i)^2}{E_i}$	0.4	1.25	2.133	0.025	0.3	0.05	3.6	7.755
	$\frac{O_i^2}{E_i}$	14.4	11.25	16.133	42.025	36.3	22.05	25.6	167.755
Mel	$\frac{(O_i - E_i)^2}{E_i}$	1.6	3.2	2.7	0.225	0.833	4.05	10	22.608
	$\frac{O_i^2}{E_i}$	3.6	7.2	14.7	34.225	40.833	42.05	40	182.608

Results for Billy: $X^2 = \sum_{i=2}^{8} \frac{(O_i - E_i)^2}{E_i} = 7.755$

or using the alternative method:

$$X^2 = \sum_{i=2}^{8} \frac{O_i^2}{E_i} - N = 167.755 - 160 = 7.755$$

Calculate $\sum \frac{(O_i - E_i)^2}{E_i}$ or $\sum \frac{O_i^2}{E_i} - N$, with $N = 160$.

It is often easier to calculate using the alternative method.

Results for Mel: $X^2 = \sum_{i=2}^{8} \frac{(O_i - E_i)^2}{E_i} = 22.608.$

or using the alternative method:

$$X^2 = \sum_{i=2}^{8} \frac{O_i^2}{E_i} - N = 182.608 - 160 = 22.608$$

Mel's goodness of fit is higher, so she is more likely to have the biased spinner.

Watch out The higher the value of X^2, the **less similar** the observed distribution is to the theoretical distribution.

Exercise 6A

1 An octagonal dice is thrown 500 times and the results are noted. It is assumed that the dice is unbiased. A test is to be done to see whether the observed results differ from the expected ones. Write down a null hypothesis and an alternative hypothesis that can be used.

2 A six-sided dice is rolled 180 times to try to establish whether or not it is fair. The results of the rolls are as follows:

Number, n	1	2	3	4	5	6
Observed rolls (O_i)	27	33	31	28	34	27

a State the null and alternative hypotheses for the experiment.

b Calculate X^2 for the observed data.

3 A random sample of 750 UK secondary school students is taken, and the year group they are each in is recorded:

Year	7	8	9	10	11
Observed (O_i)	190	145	145	140	130

A researcher wants to test to see whether UK secondary school students are uniformly distributed across each year group.

a State suitable null and alternative hypotheses.

b Calculate the expected number of students in each year group assuming your null hypothesis is true.

c Calculate X^2 for the observed data.

4 A particular genetic mutation is believed to have a 75% chance of being passed from parent to child. In an experiment, 160 adults with the mutation each had one of their children tested to see if the child had inherited the mutation. The results were as follows:

Mutation present	Yes	No
Observed (O_i)	117	43

a Calculate the expected frequencies.

b State the null and alternative hypotheses.

c Calculate the goodness of fit of the data to the expected result.

(P) **5** John has two coins that he can't tell apart. One is fair. The other is biased and will land on heads with probability 0.6. He flips one of the coins 50 times and records the results in the frequency table given below.

Result	H	T
Observed (O_i)	28	22

a Calculate the expected frequencies for each coin.

b Calculate the goodness of fit between the observed results and the expected results for each coin.

c Which coin is John more likely to have been using? Give a reason for your answer.

(P) **6** The BMI profile of English adults is given below.

Country	Underweight	Normal	Overweight	Obese	Total
England	2%	35%	36%	27%	100%

Obesity Statistics, House of Commons Briefing Paper, Number 3336, 2017

You may assume that these percentages reflect the true distribution. A sample is taken of adults in Wales, and the results are recorded in the table below.

Country	Underweight	Normal	Overweight	Obese	Total
Wales (Men)	4	70	80	46	200
Wales (Women)	6	81	65	48	200

By calculating the goodness-of-fit statistic for both Welsh men and women, determine which group more closely matches the English distribution.

6.2 Degrees of freedom and the chi-squared (χ^2) family of distributions

An important consideration when deciding goodness of fit is the number of **degrees of freedom**.

In general, degrees of freedom are calculated from the size of the sample. They are a measure of the amount of information from the sample data that has not been used up. Every time a statistic is calculated from a sample, one degree of freedom is used up.

In this chapter, in order to create a model for the observed frequency distribution, you must use the information about the data in order to select a suitable model. To begin with you have n observed frequencies, and your model has to have the same total frequency as the observed distribution. The requirement that the totals have to agree is called a **constraint**, or **restriction**, and uses up one of your degrees of freedom.

The number of constraints will also depend on the number of parameters needed to describe the distribution and whether or not these parameters are known. If you do not know a parameter you have to estimate it from the observed data and this uses up a further degree of freedom.

Problem-solving

If you flip a coin 100 times and observe the results, there are **two observed frequencies**: the number of heads, and the number of tails. There is **one constraint**: the fact that the total frequency must be 100. Therefore the number of degrees of freedom is $2 - 1 = 1$. If you know one frequency, x, then you can calculate the other, $100 - x$. Similarly, if a dice is rolled 120 times, there are **six observed frequencies** and **one constraint** (the total number of rolls). So there are $6 - 1 = 5$ degrees of freedom. Setting values for any 5 of the frequencies uniquely determines the 6th. → Section 6.4

It is usual to refer to each rectangle of a table that contains an observation as a cell. You sometimes have to combine frequencies from different cells of the table. (The reason for this is given on the next page.) If cells are combined in this way then there are fewer expected values, so when you calculate the number of degrees of freedom you have to count the number of cells after any such combination and subtract the number of constraints from this.

Watch out If the estimate of a parameter is calculated then it is a restriction. If it is guessed by using an estimate that seems sensible from observations then it is not a restriction.

- **Number of degrees of freedom = number of cells (after any combining) − number of constraints**

The χ^2 (pronounced kye-squared) family of distributions can be used as approximations for the statistic X^2. We write this:

- $X^2 = \sum \frac{(O_i - E_i)^2}{E_i} = \sum \frac{O_i^2}{E_i} - N \sim \chi^2$

X^2 is approximated well by χ^2 as long as none of the expected values (E_i) fall below 5.

- **If any of the expected values are less than 5, then you have to combine frequencies in the data table until they are greater than 5.**

Usually frequencies adjacent to each other in the table are joined together because if one value is low the next one is also likely to be low.

The χ^2 family of distributions are theoretical ones. The probability distribution function of each member of the family depends on the number of degrees of freedom.

Notation

The number of degrees of freedom in a χ^2 distribution is written using the greek letter ν, which is pronounced 'nu'.

To distinguish which member of the family of distributions you are talking about you write χ^2_ν. Thus χ^2_4 is the χ^2 distribution with $\nu 4$.

- **When selecting which of the χ^2 family to use as an approximation for X^2 you select the distribution which has ν equal to the number of degrees of freedom of your expected values.**

Example 2

In a sample of 100 households, the expected number of dogs is as follows:

Dogs	0	1	2	3	4	5	>5	Total
Expected	55	20	10	7	4	3	1	100

Select an appropriate chi-squared distribution to model the goodness of fit, X^2, for these data.

χ^2 is only a good approximation for X^2 if none of the expected values fall below 5, so we combine the expected values for 4, 5 and >5 dogs into a single value.

Combining the frequencies:

Dogs	0	1	2	3	>3	Total
Expected	55	20	10	7	8	100

Degrees of freedom = 5 − 1 = 4

There are 5 values. You know the total of the frequencies must be 100 so there is one constraint.

Therefore $X^2 \sim \chi^2_4$ is the correct approximation.

You began this chapter by forming a null hypothesis that there was no difference between the observed and the theoretical distributions.

Next, you found a measure of goodness of fit.

The question which arises is, 'could the value of $X^2 = \sum \frac{(O_i - E_i)^2}{E_i}$ calculated for your sample come from a population for which X^2 is equal to zero?'

As with all hypothesis tests, you will only reject the null hypothesis if, by accepting the alternative hypothesis, you have only a small chance of being wrong. Typically this figure is probability at 5%.

To find the value of X^2 that is only exceeded with probability of 5% (the critical value), we use the appropriate χ^2 distribution.

Notation For a given value of ν, the critical value of χ^2 which is exceeded with probability 5% is written χ^2_ν (5%) or χ^2_ν (0.05).

Example 3

With $\nu = 5$ find the value of χ^2 that is exceeded with 0.05 probability.

Look across the table to get column 0.050.

Look down the table to get row $\nu = 5$.

χ^2_5 (5%) = 11.070

This is shown on the probability diagram below.

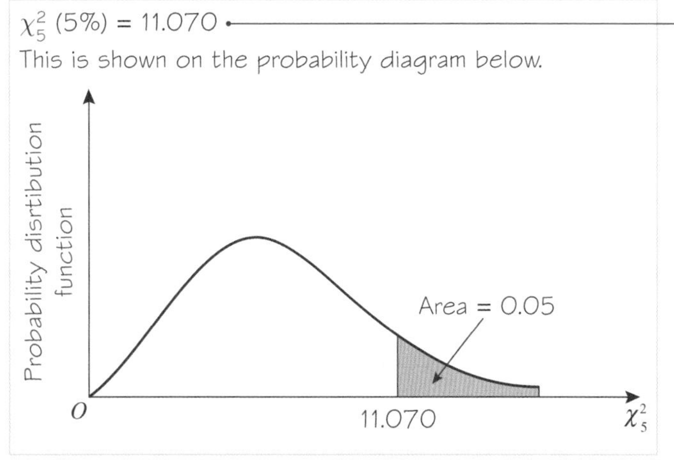

Use the table of values on page 192 for the percentage points of the χ^2 distribution, with $\nu = 5$.

ν	0.995	0.100	0.050	0.025
1	0.000	2.705	3.841	5.024
2	0.010	4.605	5.991	7.378
3	0.072	6.251	7.815	9.348
4	0.207	7.779	9.488	11.143
5	0.412	9.236	11.070	12.832
6	0.676	0.645	12.592	14.449
7	0.989	2.017	14.067	16.013

Read off where row and column cross: the value is 11.070.

Also from the table, χ^2_5 (10%) = 9.236 and χ^2_5 (2.5%) = 12.832.
For each other value of ν the critical values may be looked up in the same way.

You can also find values for χ^2 distributions on some graphical calculators.

Example 4

a Find the following critical values:
i χ^2_3 (95%) **ii** χ^2_4 (10%)

b Find the smallest values of y such that:
i $P(\chi^2_2 > y) = 0.95$ **ii** $P(\chi^2_4 > y) = 0.99$

Online Explore the χ^2-squared distribution and use it to determine critical values for goodness of fit using GeoGebra.

a i $\nu = 3$
Level of significance = 0.95
From the table on page 192, the critical value of χ^2 is 0.352.

ii $\nu = 4$
Level of significance = 0.1
From the table on page 192, the critical value of χ^2 is 7.779.

b i $P(\chi^2_2 > 0.103) = 95\%$

ii $P(\chi^2_4 > 0.297) = 99\%$

The significance level is the probability that the distribution exceeds the critical value.
χ^2_2 (95%) = 0.103
so $P(\chi^2_2 > 0.103) = 95\%$

Exercise 6B

1 A group of 50 students record the days (Monday–Sunday) that their birthdays will fall on this year. How many degrees of freedom are there in the frequency distribution?

2 For 5 degrees of freedom find the critical value of χ^2 which is exceeded with a probability of 5%.

3 Find the following critical values:

a $\chi^2_5(5\%)$ b $\chi^2_8(1\%)$ c $\chi^2_{10}(10\%)$

4 With $\nu = 10$ find the value of χ^2 that is exceeded with 0.05 probability.

5 With $\nu = 8$ find the value of χ^2 that is exceeded with 0.10 probability.

6 The random variable Y has a χ^2 distribution with 8 degrees of freedom.
Find y such that $P(Y > y) = 0.99$.

7 The random variable X has a χ^2 distribution with 5 degrees of freedom.
Find x such that $P(X > x) = 0.95$.

(P) 8 The random variable Y has a χ^2 distribution with 12 degrees of freedom.
Find:

Notation The χ^2 distribution is continuous, so $P(Y < y) = 1 - P(Y > y)$.

a y such that $P(Y < y) = 0.05$ b y such that $P(Y < y) = 0.95$.

(E) 9 The frequency table below shows 50 samples from what is believed to be a Geo(0.5) distribution.

x	1	2	3	4	5
$P(X = x)$	24	12	6	6	2

a State, with a reason, which cells should be combined before using a χ^2 distribution to model the goodness of fit for these data. **(1 mark)**

b State a suitable χ^2 distribution and find the critical value for a 1% significance level. **(2 marks)**

6.3 Testing a hypothesis

By using a suitable χ^2 distribution to model the goodness of fit you can carry out a **hypothesis test** for the null hypothesis that the observed data fits the theoretical distribution.

Notation α is used to denote the significance level to which the data is being tested.

You need to choose a significance level, α, for the hypothesis test. This is often 5%, and will be given in the question.

You can then compute the critical value, $\chi^2_\nu(\alpha)$, which will depend on the significance level, α, and the number of degrees of freedom, ν. The probability of observing data with a goodness of fit exceeding the critical value is α.

- **If X^2 exceeds the critical value, it is unlikely that the null hypothesis is correct, so you reject it in favour of the alternative hypothesis.**

Watch out A hypothesis test for goodness of fit is always **one-tailed**. This means the **critical region** is always the set of values **greater than** the critical value.

Example 5

In an experiment where a dice is rolled 120 times, the frequency distribution is to be compared to a discrete uniform distribution as shown:

Number on dice, n	1	2	3	4	5	6
Observed frequency, O_i	23	15	25	18	21	18
Expected frequency, E_i	20	20	20	20	20	20

Test, at the 5% significance level, whether or not the observed frequencies could be modelled by a discrete uniform distribution.

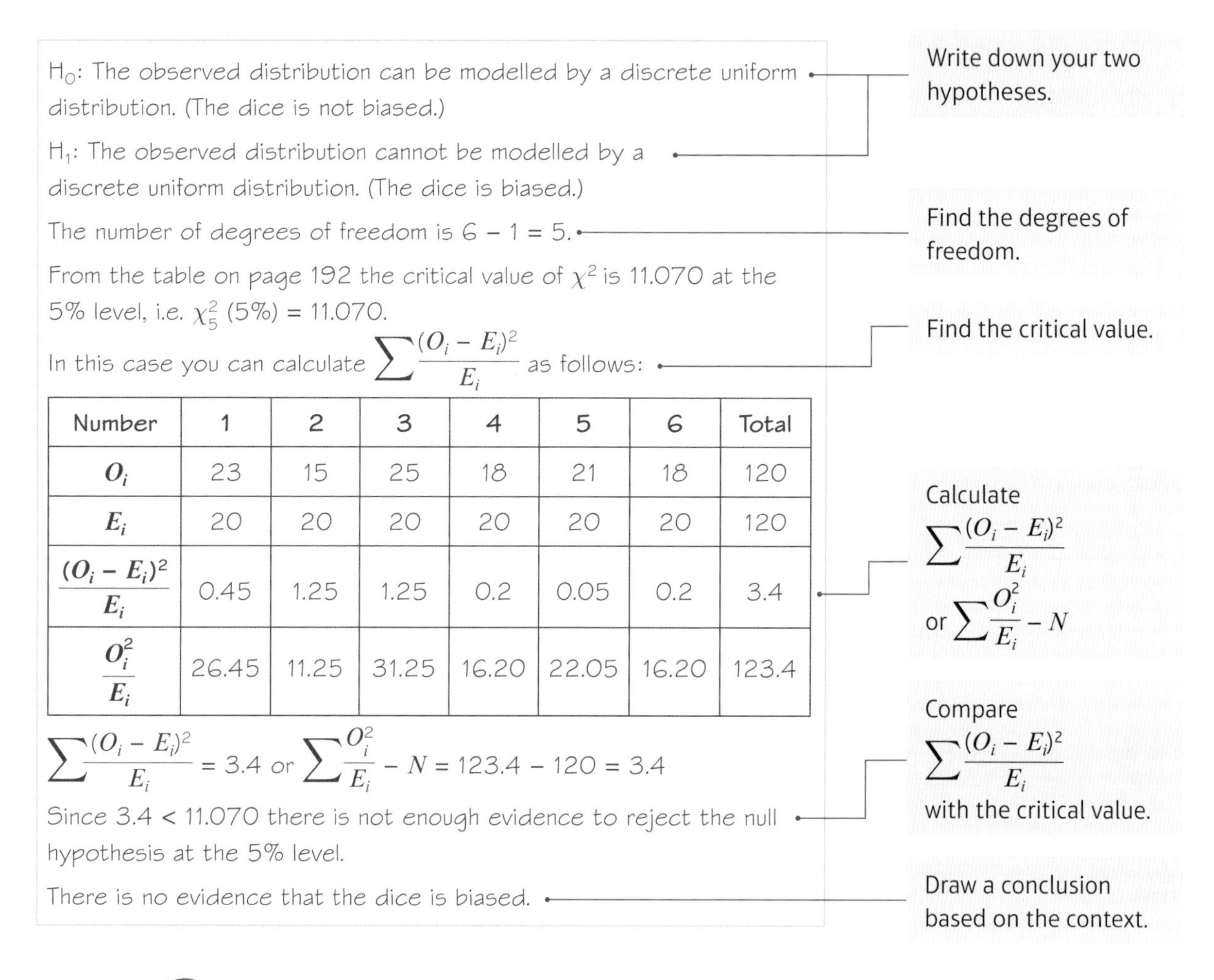

H_0: The observed distribution can be modelled by a discrete uniform distribution. (The dice is not biased.)

H_1: The observed distribution cannot be modelled by a discrete uniform distribution. (The dice is biased.)

Write down your two hypotheses.

The number of degrees of freedom is 6 − 1 = 5.

Find the degrees of freedom.

From the table on page 192 the critical value of χ^2 is 11.070 at the 5% level, i.e. χ^2_5 (5%) = 11.070.

Find the critical value.

In this case you can calculate $\sum \frac{(O_i - E_i)^2}{E_i}$ as follows:

Number	1	2	3	4	5	6	Total
O_i	23	15	25	18	21	18	120
E_i	20	20	20	20	20	20	120
$\frac{(O_i - E_i)^2}{E_i}$	0.45	1.25	1.25	0.2	0.05	0.2	3.4
$\frac{O_i^2}{E_i}$	26.45	11.25	31.25	16.20	22.05	16.20	123.4

Calculate $\sum \frac{(O_i - E_i)^2}{E_i}$ or $\sum \frac{O_i^2}{E_i} - N$

$\sum \frac{(O_i - E_i)^2}{E_i} = 3.4$ or $\sum \frac{O_i^2}{E_i} - N = 123.4 - 120 = 3.4$

Since 3.4 < 11.070 there is not enough evidence to reject the null hypothesis at the 5% level.

Compare $\sum \frac{(O_i - E_i)^2}{E_i}$ with the critical value.

There is no evidence that the dice is biased.

Draw a conclusion based on the context.

Example 6

Alan has two identical 4-sided spinners with the numbers 1–4 written on each of them. He carries out experiments, where he spins both of his spinners at the same time, and adds the scores together. After 160 experiments, the frequency distribution is as follows:

Number, n	2	3	4	5	6	7	8
Observed by Alan (O_i)	14	11	26	33	42	18	16
Expected (E_i)	10	20	30	40	30	20	10

The table also shows the expected distribution of the scores if both spinners are unbiased. Test, at the 2.5% significance level, whether the observed frequencies could be modelled by the expected distribution shown.

H_0: The observed distribution can be modelled by the expected distribution shown. (The spinners are not biased.)

H_1: The observed distribution cannot be modelled by the expected distribution shown. (The spinners are biased.)

The null hypothesis is what we believe unless the evidence suggests otherwise. In this case, we assume the spinners are equally likely to land on any number.

The number of degrees of freedom is 7 − 1 = 6.

There are 7 observed frequencies, and 1 constraint (that the total is 160).

From the table on page 192, the critical value of χ^2 is 14.449 at the 2.5% level i.e. $\chi^2_6(2.5\%) = 14.449$.

We calculate X^2 as follows:

n	2	3	4	5	6	7	8	Total
O_i	14	11	26	33	42	18	16	160
E_i	10	20	30	40	30	20	10	160
$\frac{(O_i - E_i)^2}{E_i}$	1.600	4.050	0.533	1.225	4.800	0.200	3.600	16.008
$\frac{O_i^2}{E_i}$	19.600	6.050	22.533	27.225	58.800	16.200	25.600	176.008

Watch out You only need to use one of the formulae to calculate the goodness of fit in your exam – they both give the same answer.

So $X^2 = 16.008$

Since 16.008 is greater than 14.449, we reject the null hypothesis at the 2.5% level.

X^2 is greater than the critical value. This means that if the spinners were unbiased, there would be less than a 2.5% probability that the observed frequencies would differ from the expected frequencies by this amount.

There is evidence, at the 2.5% level, that the spinners are biased.

Example 7

A school conducted a survey into the impact that a new exercise club was having on students. Prior to the new club starting, 60% of students said they had no regular exercise, 30% reported exercising once a week and 10% reported exercising more than once a week. After the new club started, they surveyed the 150 students to find out how often they exercised.

	No regular exercise	Once a week	More than once a week	Total
Frequency	73	57	20	150

Based on these data, is there evidence of a change in attitude to exercise following the introduction of the new club? Test the data at a 5% significance level.

H_0: The observed distribution has not changed from the original distribution. (The new club has had no effect on the number of times students exercise each week.)

H_1: The observed distribution has changed from the original distribution. (The new club has had an effect on the number of times students exercise each week.)

The number of degrees of freedom is 3 − 1 = 2.

From the table on page 192, the critical value of χ^2 is 5.991 at the 5% level, i.e. $\chi^2_2(5\%) = 5.991$.

We calculate X^2 as follows:

	No exercise	Once a week	More than once a week	Total
Observed	73	57	20	150
Expected	0.6 × 150 = 90	0.3 × 150 = 45	0.1 × 150 = 15	150
$\frac{(O_i - E_i)^2}{E_i}$	3.211	3.200	1.667	8.078
$\frac{O_i^2}{E_i}$	59.211	72.200	26.667	158.078

So $X^2 = 8.078$.

Since 8.078 is greater than 5.991, we reject the null hypothesis at the 5% significance level.

Therefore, at the 5% significance level, there is evidence that the new club has had an effect on the number of times students exercise each week.

Unless there is evidence to the contrary, we should not assume that the exercise club has changed people's attitudes either way, so the null hypothesis is that nothing has changed.

There are 3 observed frequencies, and 1 constraint, so 2 degrees of freedom.

The expected values are based on the proportions before the club was started.

Watch out A test for goodness of fit only tells you how closely the observed data matches the theoretical (or assumed) distribution. You cannot conclude that the new club has increased the amount of exercise students do – only that the amount has changed.

Exercise 6C

(E) **1** In an experiment where a dice is rolled 72 times, the frequency distribution is to be compared to a discrete uniform distribution as shown:

Number on dice, n	**1**	**2**	**3**	**4**	**5**	**6**
Observed frequency, O_i	16	11	13	15	8	9
Expected frequency, E_i	12	12	12	12	12	12

Test, at the 5% significance level, whether or not the observed frequencies could be modelled by a discrete uniform distribution. **(6 marks)**

(E) **2** In a tombola, tickets ending in a 0 or a 5 are guaranteed a prize. All other tickets will lose. At a fair, 120 tickets were drawn, and the numbers of winning tickets were as follows.

	Winning	Losing	Total
Observed ticket draws	15	105	120

Test, at the 5% significance level, whether or not the tombola was fair. **(6 marks)**

Hint If the tickets were fairly distributed, it would be expected that 2 in every 10 would be winning tickets.

(E) **3** A local travel agent has made a prediction as to how many trips abroad his customers make. He surveys a sample of 100 customers and compares the results to his expectations.

Trips abroad	None	One	Two or more
Expected	10%	60%	30%
Sample	4	73	23

Test, at the 2.5% significance level, whether the travel agent's prediction fits the observed data. **(6 marks)**

(E/P) **4** In a sample of 100 households, the actual and expected numbers of dogs is as follows:

Dogs	0	1	2	3	4	5	>5	Total
Observed	45	19	11	8	7	6	4	100
Expected	55	20	10	7	4	3	1	100

a Explain why there are 4 degrees of freedom in this case. **(2 marks)**

b Test, at the 5% significance level, whether the observed data fits the expected distribution given. **(5 marks)**

Problem-solving

When using a χ^2 distribution to approximate the distribution of X^2, if any of the expected frequencies is less than 5 you need to combine cells.

(E/P) **5** In the year 2000, the birth weights of babies were distributed as follows:

Weight (g)	Under 1500	1500–1999	2000–2499	2500–2999	3000–3499	3500 and over	Total
Percentage	1.3%	1.5%	5%	16.5%	35.7%	40%	100%

In the year 2015, the birth weights of babies were as follows:

Weight (g)	Under 1500	1500–1999	2000–2499	2500–2999	3000–3499	3500 and over	Total
Frequency	7286	9304	32 121	112 535	244 472	281 942	687 660

Using a 5% significance level, decide whether the distribution of birth weights from 2000 can be used as a model for the weights in 2015. **(8 marks)**

6.4 Testing the goodness of fit with discrete data

The steps you need to take to test goodness of fit with discrete data can be summarised as follows.

1 Determine which distribution is likely to be a good model by examining the conditions applying to the observed data.

2 Set the significance level, for example, 5%.

These will often be given in the question.

3 Estimate parameters (if necessary) from your observed data.

4 Form your hypotheses.
5 Calculate expected frequencies.
6 Combine any expected frequencies so that none are less than 5.
7 Find ν using ν = number of cells after combining – number of constraints or restrictions.
8 Find the critical value of χ^2 from the table.
9 Calculate $\sum \frac{(O_i - E_i)^2}{E_i}$ or $\sum \frac{O_i^2}{E_i} - N$.
10 See if your value is significant.
11 Draw the appropriate conclusion and interpret in the context of the original problem.

Testing a discrete uniform distribution as a model

You have already seen an example of this. The conditions under which a discrete uniform distribution arises are:

- the discrete random variable X is defined over a set of k distinct values
- each value is equally likely

The probability of each value is given by

$$\mathrm{P}(X = x_r) = \frac{1}{k} \qquad r = 1, 2, \ldots, k$$

The frequencies for a sample size of N are given by

$$\text{Frequency} = \mathrm{P}(X = x_r) \times N = \frac{1}{k} \times N \qquad r = 1, 2, \ldots, k$$

In a discrete uniform distribution, the probability of each outcome is only dependent on the size of the sample space. This means that there are no additional parameters to estimate, so the only restriction is that the expected frequencies add up to N. The number of degrees of freedom is **one less** than the number of cells, after any cells have been combined.

Example 8

100 digits between 0 and 9 are selected from a table with the frequencies as shown below.

Digit	0	1	2	3	4	5	6	7	8	9
Frequency	11	8	8	7	8	9	12	9	13	15

Could the digits be from a random number table? Test at the 0.05 level.

Each digit should have an equal chance of selection, so the appropriate model is the discrete uniform distribution.

Determine the distribution. The significance level is given as 5% and no parameters need estimating.

H_0: A discrete uniform distribution is a suitable model. (The digits are random.)
H_1: A discrete uniform distribution is not a suitable model. (The digits are not random.)

State your hypotheses.

$$\mathrm{P}(X = x_r) = \frac{1}{10} \qquad r = 0, 1, \ldots, 9$$

The number of degrees of freedom is:

$$\nu = 10 - 1 = 9$$

Find ν.

From the table on page 192, χ^2_9 (5%) = 16.919 — Find the critical value.

Digit	0	1	2	3	4	5	6	7	8	9
Observed, O_i	11	8	8	7	8	9	12	9	13	15
Expected, E_i	10	10	10	10	10	10	10	10	10	10
$\frac{(O_i - E_i)^2}{E_i}$	0.1	0.4	0.4	0.9	0.4	0.1	0.4	0.1	0.9	2.5

$$\sum \frac{(O_i - E_i)^2}{E_i} = 6.2$$

Calculate $\sum \frac{(O_i - E_i)^2}{E_i}$ or $\sum \frac{O_i^2}{E_i} - N$.

So: $\sum \frac{(O_i - E_i)^2}{E_i} < 16.919$

See if your value is significant.

Do not reject H_0: there is no evidence to suggest the digits are not random.

Draw a conclusion in the context of the original problem.

Testing a binomial distribution as a model

The conditions under which a binomial distribution arises are:

- there must be a fixed number (n) of trials in each observation
- the trials must be independent
- the trials have only two outcomes: success and failure
- the probability of success (p) is constant

For a binomial random variable:

$$P(X = r) = \binom{n}{r} p^r (1 - p)^{n-r} \qquad r = 0, 1, 2, \ldots, n$$

The frequency f_r with which each r occurs when the number of observations is N is given by

$$f_r = P(X = r) \times N$$

The binomial distribution has two parameters, n and p. You have the usual restriction that the expected frequencies have to have the same total as observed frequencies, while p may be known or it may be estimated from the observed values by using frequencies of success.

$$p = \frac{\text{total number of successes}}{\text{number of trials} \times N} = \frac{\Sigma(r \times f_r)}{n \times N}$$

If p is not estimated by calculation: ν = number of cells – 1

If p is estimated by calculation: ν = number of cells – 2

Watch out For a binomial distribution, each observation is of the number of successes in n trials. So for N observations there are $n \times N$ trials in total.

Example 9

The data in the table are thought to be modelled by a binomial distribution B(10, 0.2). Use the table for the binomial cumulative distribution function to find expected values, and conduct a test to see if this is a good model. Use a 5% significance level.

x	0	1	2	3	4	5	6	7	8
Frequency	12	28	28	17	7	4	2	2	0

H_0: A B(10, 0.2) distribution is a suitable model for the results.
H_1: The results cannot be modelled by a B(10, 0.2) distribution.

State your hypotheses.

x	0	1	2	3	4	5	6	7	8
Probability of x	0.1074	0.2684	0.3020	0.2013	0.0881	0.0264	0.0055	0.0008	0.0001
Expected frequencies	10.74	26.84	30.20	20.13	8.81	2.64	0.55	0.08	0.01

Use your calculator to find the probabilities for a binomial random variable $X \sim$ B(10, 0.2). For example P(X = 0) = 0.1074. There are no frequencies for $X > 8$ and the probabilities are very small for these values, so you don't need to calculate them.

There are 7 + 4 + 2 + 2 = 15 observed values when $x \geqslant 4$.

Expected frequency = 8.81 + 2.64 + 0.55 + 0.08 + 0.01 = 12.09

Combine any expected frequencies < 5. (You will have to combine the last five cells in the table since 2.64 + 0.55 + 0.08 + 0.01 < 5.)

x	0	1	2	3	$\geqslant 4$
O_i	12	28	28	17	15
E_i	10.74	26.84	30.20	20.13	12.09
$\frac{(O_i - E_i)^2}{E_i}$	0.1478	0.0501	0.1603	0.4867	0.7004

Number of degrees of freedom = number of cells − 1 = 5 − 1 = 4.
(p was not estimated by calculation this time.)

Find the number of degrees of freedom.

From the table on page 192 the critical value χ_4^2 (5%) is 9.488.

Find the critical value.

$$\sum \frac{(O_i - E_i)^2}{E_i} = 1.5453$$

Calculate $\sum \frac{(O_i - E_i)^2}{E_i}$ or $\sum \frac{O_i^2}{E_i} - N$.

1.545 < 9.488

See if the value is significant.

Do not reject H_0: B (10, 0.2) is a possible model for the data.

Draw a conclusion.

Example 10

A study of the number of girls in families with five children was done on 100 such families. The results are summarised in the following table.

Number of girls (r)	0	1	2	3	4	5
Frequency (f)	13	18	38	20	10	1

It is suggested that the distribution may be modelled by a binomial distribution with $p = 0.5$.

a Give reasons why this might be so.

b Test, at the 5% significance level, whether or not a binomial distribution is a good model.

a There is a fixed number of children in the family so $n = 5$. The trials are independent. (Assume no multiple births.) There are two outcomes to each trial: success (a girl), failure (a boy). The assumption that a girl is as likely as a boy is reasonable.

Compare the conditions with the known conditions for a binomial distribution.

b H_0: B(5, 0.5) is a suitable model.
H_1: B(5, 0.5) is not a suitable model.

State your hypotheses. Note that the value of p is given in the hypotheses.

r	0	1	2	3	4	5
O_i	13	18	38	20	10	1
E_i	3.12	15.63	31.25	31.25	15.63	3.12

Calculate the expected frequencies using tables or your calculator.

Since 3.12 < 5 you must combine cells.

r	0 or 1	2	3	4 or 5
O_i	31	38	20	11
E_i	18.75	31.25	31.25	18.75

Combine cells so that no $E_i < 5$.

$$\sum \frac{(O_i - E_i)^2}{E_i} = 16.715$$

Calculate $\sum \frac{(O_i - E_i)^2}{E_i}$.

You have $4 - 1 = 3$ degrees of freedom.

Find the degrees of freedom.

From the tables: $\chi^2_3(0.5) = 7.815$

$16.715 > 7.815$

Find the critical value.
See if the result is significant.

Reject H_0: the number of girls in families of 5 children cannot be modelled by B(5, 0.5).

Draw a conclusion based on the context.

Example 11

Look at the data in the previous example. By estimating a suitable value of the parameter, p, carry out a test, at the 5% significance level, to determine whether a binomial distribution is a suitable model for the data.

Problem-solving

In the previous example you determined that B(5, 0.5) was not a suitable model for the data, at the 5% level. However, B(5, p) may be a suitable model for some different value of p.

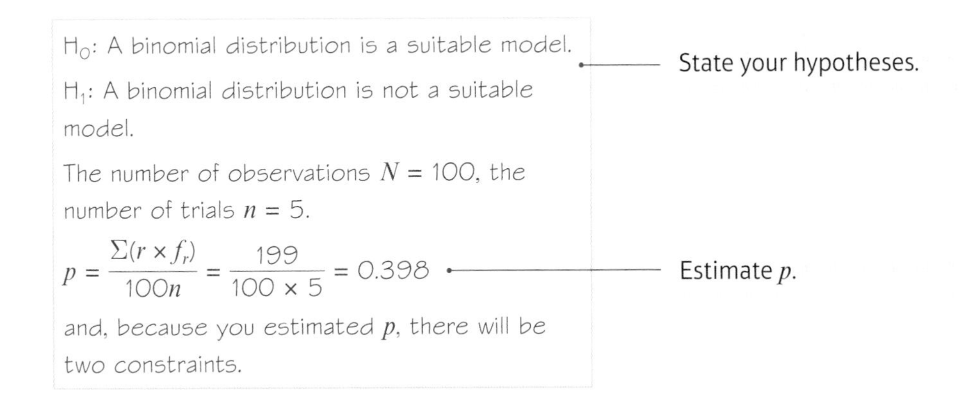
H_0: A binomial distribution is a suitable model.
H_1: A binomial distribution is not a suitable model.

The number of observations $N = 100$, the number of trials $n = 5$.

$$p = \frac{\Sigma(r \times f_r)}{100n} = \frac{199}{100 \times 5} = 0.398$$

and, because you estimated p, there will be two constraints.

State your hypotheses.

Estimate p.

r	P(r)	E_i
0	$(0.602)^5 = 0.0791$	7.91
1	$5(0.602)^4(0.398) = 0.2614$	26.14
2	$10(0.602)^3(0.398)^2 = 0.3456$	34.56
3	$10(0.602)(0.398)^4 = 0.2285$	22.85
4	$5(0.602)^1(0.398)^4 = 0.0755$	7.55
5	$(0.398)^5 = 0.0099$	0.99

Calculate the probabilities using your calculator or the formula for a binomial probability.

r	O_i	E_i	$\frac{O_i^2}{E_i}$
0	13	7.91	21.37
1	18	26.14	12.39
2	38	34.56	41.78
3	20	22.85	17.51
>3	11	8.54	14.17
Total			107.22

Combine cells if $E_i < 5$.

There are 5 − 2 = 3 degrees of freedom.

Calculate the degrees of freedom.

The critical value is $\chi_3^2 = 7.815$.

Find the critical value.

$$\sum \frac{(O_i)^2}{E_i} - N = 107.22 - 100 = 7.22$$

Calculate $\sum \frac{O_i^2}{E_i} - N$ or $\sum \frac{(O_i - E_i)^2}{E_i}$.

$7.22 < 7.815$

See if your value is significant.

Do not reject H_0. A binomial distribution is a suitable model.

Draw a conclusion.

Testing a Poisson distribution as a model

The conditions under which a Poisson distribution is likely to arise are:

- the events occur *independently* of each other
- the events occur *singly* and at random in continuous space or time
- the events occur at a *constant rate*, in the sense that the mean number in an interval is proportional to the length of the interval
- the mean and the variance are *equal*

For a Poisson distribution with mean λ:

$$P(X = r) = \frac{e^{-\lambda}\lambda^r}{r!} \qquad r = 0, 1, 2, \ldots$$

Although, theoretically, r can be any one of an infinite number of integer values, in practice, all those values greater than or equal to some number n are put together and the probability $P(X > n)$ is found from:

$$P(X \geqslant n) = 1 - P(X \leqslant n - 1)$$

You choose n equal to the highest value of r for which the observed frequency is > 0.
In Example 12, n is chosen to be 7 since all telephone calls for $r \geqslant 8$ have zero frequencies.

The frequency f_r with which each r occurs is given by $P(X = r) \times N$.

The Poisson distribution has a single parameter λ, which may be known or which may be estimated from the observed data using:

$$\lambda = \frac{\Sigma(r \times f_r)}{N}$$

There is the usual restriction on the total of the expected frequencies being equal to the total of the observed frequencies.

If λ is not estimated by calculation: ν = number of cells – 1

If λ is estimated by calculation: ν = number of cells – 2

Example 12

The numbers of telephone calls arriving at an exchange in six-minute periods were recorded over a period of 8 hours, with the following results.

Number of calls, r	0	1	2	3	4	5	6	7	8
Frequency, f_r	8	19	26	13	7	5	1	1	0

Can these results be modelled by a Poisson distribution? Test at the 5% significance level.

H_0: A Poisson distribution Po(λ) is a suitable model.

H_1: The calls cannot be modelled by a Poisson distribution.

Total number of observations $= N = \frac{8 \times 60}{6}$

$= 80$

$$\lambda = \frac{\Sigma(r \times f_r)}{N} = \frac{176}{80} = 2.2$$

Since you do not know the value of λ you must estimate it from the observed frequencies.

r	$P(X = r)$	Expected frequency of r
0	0.1108	8.864 (0.1108 × 80)
1	0.2438	19.504
2	0.2681	21.448
3	0.1966	15.728
4	0.1082	8.656
5	0.0476	3.808
6	0.0174	1.392
7 or more	0.0075	0.6

Use the Poisson distribution function on your calculator, or the formula for a Poisson probability, to calculate the probabilities and expected frequencies for each value of r.

$\lambda = 2.2$ is not in the table so you must calculate the expected frequencies.

The value for $P(r = 7 \text{ or more})$ is obtained by subtracting the sum of the other probabilities from 1.

r	O_i	E_i	$\frac{(O_i - E_i)^2}{E_i}$
0	8	8.864	0.0842
1	19	19.504	0.0130
2	26	21.448	0.9661
3	13	15.728	0.4732
4	7	8.656	0.3168
5 or more	7	5.8	0.2483

In this case you have to combine cells to give expected frequencies of more than 5.

$$\sum\frac{(O_i - E_i)^2}{E_i} = 2.1016$$

You have 6 − 2 = 4 degrees of freedom.

Since λ was estimated from the observed frequencies there are 2 constraints.

From the table on page 192, χ_4^2 (5%) = 9.488

2.1016 < 9.488

See if your value is significant.

There is not enough evidence to reject H_0. The calls may be modelled by a Poisson distribution.

Draw a conclusion.

Exercise 6D

1 The following table shows observed values for a distribution which it is thought may be modelled by a Poisson distribution.

x	0	1	2	3	4	5	>5
Frequency of x	12	23	24	24	12	5	0

A possible model is thought to be Po(2). From tables, the expected values are found to be as shown in the following table.

x	0	1	2	3	4	5	>5
Expected frequency of x	13.53	27.07	27.07	18.04	9.02	3.61	1.66

a Conduct a goodness-of-fit test at the 5% significance level.

b It is suggested that the model could be improved by estimating the value of λ from the observed results. What effect would this have on the number of constraints placed upon the degrees of freedom?

2 A mail-order firm receives packets every day through the mail.

They think that their deliveries are uniformly distributed throughout the week. Test this assertion, given that their deliveries over a four-week period were as follows. Use a 0.05 significance level.

Day	Mon	Tues	Wed	Thurs	Fri	Sat
Frequency	15	23	19	20	14	11

(P) **3** Over a period of 50 weeks the numbers of road accidents reported to a police station were as shown.

Number of accidents	0	1	2	3	4
Number of weeks	15	13	9	13	0

a Find the mean number of accidents per week.

b Using this mean and a 0.10 significance level, test the assertion that these data are from a population with a Poisson distribution.

(P) **4** A marksman fires 6 shots at a target and records the number r of bullseye hits. After a series of 100 such trials he analyses his scores, the frequencies being as follows.

r	0	1	2	3	4	5	6
Frequency	0	26	36	20	10	6	2

a Estimate the probability of hitting a bullseye.

b Use a test at the 0.05 significance level to see if these results are consistent with the assumption of a binomial distribution.

(E) **5** The table below shows the numbers of employees, in thousands, at five factories and the numbers of accidents in 3 years.

Factory	A	B	C	D	E
Employees (thousands)	4	3	5	1	2
Accidents	22	14	25	8	12

Using a 0.05 significance level, test the hypothesis that the number of accidents per 1000 employees is constant at each factory. **(6 marks)**

(E/P) **6** In a test to determine the red blood cell count in a patient's blood sample, the number of cells in each of 80 squares is counted with the following results.

Number of cells per square, x	0	1	2	3	4	5	6	7	8
Frequency, f	2	8	15	18	14	13	7	3	0

It is assumed that these will fit a Poisson distribution. Test this assertion at the 0.05 significance level. **(10 marks)**

(E/P) **7** A factory has a machine. The number of times it broke down each week was recorded over 100 weeks with the following results.

Number of times broken down	0	1	2	3	4	5
Frequency	50	24	12	9	5	0

It is thought that the distribution is Poisson.

a Give reasons why this assumption might be made. **(2 marks)**

b Conduct a test at the 0.05 level of significance to see if the assumption is reasonable. **(8 marks)**

(E) **8** In a lottery there are 505 prizes, and it is assumed that they will be uniformly distributed throughout the numbered tickets. An investigation gave the following:

Ticket number	1–1000	1001–2000	2001–3000	3001–4000	4001–5000	5001–6000	6001–7000	7001–8000	8001–9000	9001–10000
Frequency	56	49	35	47	63	58	44	52	51	50

Using a suitable test with a 0.05 significance level, and stating your null and alternative hypotheses, see if the assumption is reasonable. **(6 marks)**

(E) **9** Data were collected on the numbers of female puppies born in 200 litters of 8 puppies. It was decided to test whether or not a binomial model with parameters $n = 8$ and $p = 0.5$ is a suitable model for the data. The following table shows the observed frequencies and the expected frequencies, to 2 decimal places, obtained in order to carry out this test.

Number of females	**Observed number of litters**	**Expected number of litters**
0	1	0.78
1	9	6.25
2	27	21.88
3	46	R
4	49	S
5	35	T
6	26	21.88
7	5	6.25
8	2	0.78

a Find the values of R, S and T. **(3 marks)**

b Carry out the test to determine whether or not this binomial model is a suitable one. State your hypotheses clearly and use a 5% level of significance. **(5 marks)**

An alternative test might have involved estimating p rather than assuming $p = 0.5$.

c Explain how this would have affected the test. **(2 marks)**

(E) **10** A random sample of 300 football matches was taken and the numbers of goals scored in each match was recorded. The results are given in the table below.

Number of goals	0	1	2	3	4	5	6	7
Frequency	33	55	80	56	56	11	5	4

a Show that an estimate of the mean number of goals scored in a football match is 2.4 and find an estimate of the variance. **(3 marks)**

It is thought that a Poisson distribution might provide a good model for the number of goals per match.

b Give one reason why the observed data might support this model. **(1 mark)**

Using a Poisson distribution, with mean 2.4, expected frequencies were calculated as follows:

Number of goals	0	1	2	3	4	5	6	7
Expected frequency	s	65.3	t	62.7	37.6	18.1	7.2	2.5

c Find the values of s and t. **(2 marks)**

d State clearly the hypotheses required to test whether or not a Poisson distribution provides a suitable model for these data. **(1 mark)**

In order to carry out this test, the class for 7 goals is redefined as 7 or more goals.

e Find the expected frequency for this class. **(1 mark)**

The test statistic for the test in part **d** is 15.7 and the number of degrees of freedom used is 5.

f Explain fully why there are 5 degrees of freedom. **(1 mark)**

g Stating clearly the critical value used, carry out the test in part **d**, using a 5% level of significance. **(3 marks)**

(E) **11** A student of botany believed that a certain species of wild orchid plants grow in random positions in grassy meadowland. He recorded the number of plants in one square metre of grassy meadow, and repeated the procedure to obtain the 148 results in the table.

Number of plants	0	1	2	3	4	5	6	7 or greater
Frequency	9	24	43	34	21	15	2	0

a Show that, to two decimal places, the mean number of plants in one square metre is 2.59. **(2 marks)**

b Give a reason why the Poisson distribution might be an appropriate model for these data. **(1 mark)**

Using the Poisson model with mean 2.59, expected frequencies corresponding to the given frequencies were calculated, to two decimal places, and are shown in the table below.

Number of plants	0	1	2	3	4	5	6	7 or greater
Expected frequencies	11.10	28.76	s	32.15	20.82	10.78	4.65	t

c Find the values of s and t to two decimal places. **(2 marks)**

d Stating clearly your hypotheses, test at the 5% level of significance whether or not this Poisson model is supported by these data. **(5 marks)**

6.5 Using contingency tables

So far in this chapter you have considered the frequency with which a single event occurs. For example, you might count the number of times each of the numbers 1 to 6 appears when a dice is thrown 100 times. Sometimes however, we may be interested in the frequencies with which two criteria are fulfilled at the same time. If you study the frequency with which A-Level Maths passes at grades A, B and C occur you may also be interested in which of two schools the students attended. Here you have two criteria: the pass level and the school. You can show these results by means of a **contingency table**, which shows the frequency with which each of the results occurred at each school separately.

		Pass (criterion 1)			
		A	**B**	**C**	**Totals**
School (criterion 2)	X	18	12	20	50
	Y	26	12	32	70
Totals		44	24	52	120

18 students at school X got a grade A pass.

32 students at school Y got a grade C pass.

A total of 44 students out of a total of 120 got a grade A pass.

This is called a **2 × 3** contingency table since there are two rows and three columns.

Setting the hypotheses

What we are interested in is whether there is any association between the two schools' sets of results. We pose the hypothesis 'are the two criteria independent?'.

H_0: School and grade of pass are independent.

H_1: School and grade of pass are not independent.

Selecting a model

If the hypothesis H_0 is true then you would expect school X to get $\frac{50}{120}$ of each grade and school Y to get $\frac{70}{120}$ of each grade.

Now, overall: $\text{P(A grade)} = \frac{44}{120}$

$\text{P(school } X) = \frac{50}{120}$

So

$$\text{P(A grade and school } X) = \text{P(A grade)} \times \text{P(school } X) = \frac{44}{120} \times \frac{50}{120}$$

The expected frequency of passes at A from school X is therefore

$$\frac{44}{120} \times \frac{50}{120} \times 120 = \frac{44 \times 50}{120} = 18.33$$

Notice that:

- **Expected frequency = $\dfrac{\textbf{row total} \times \textbf{column total}}{\textbf{grand total}}$**

The expected frequency is calculated on the assumption that the criteria are independent. You can find the other expected frequencies in the same way. These are shown in the table.

		Pass (criterion 1)			
		A	**B**	**C**	**Totals**
School (criterion 2)	X	$\frac{50 \times 44}{120} = 18.33$	$\frac{50 \times 24}{120} = 10$	$\frac{50 \times 52}{120} = 21.67$	50
	Y	$\frac{70 \times 44}{120} = 25.67$	$\frac{70 \times 24}{120} = 14$	$\frac{70 \times 52}{120} = 30.33$	70
Totals		44	24	52	120

Degrees of freedom

When calculating expected values you need not calculate the last value in each row because the sum of the values in each row has to equal the row total.

This creates one constraint on the number of degrees of freedom.

For example, the expected frequency of students who obtain a grade C from school X would be $50 - (18.33 + 10) = 21.67$

In the same way the last value in each of the columns is fixed by the column total once the other values in the column are known.

This creates another constraint on the number of degrees of freedom.

For example, the expected frequency of students who obtain a grade A from school Y would be $44 - 18.33 = 25.67$.

In general, if there are h rows, then once $(h - 1)$ expected frequencies have been calculated the last value in the row is *fixed* by the row total. If there are k columns, once $(k - 1)$ columns have been calculated the last column value is fixed by the column total.

The number of independent variables is given therefore by $(h - 1)(k - 1)$. That is to say:

- **The number of degrees of freedom**
  $\nu = (h - 1)(k - 1)$

Watch out If the expected frequency in any column is < 5, you will need to **combine columns**. Make sure you use the new number of columns after combining as your value of k when working out the number of degrees of freedom.

Example 13

Conduct a goodness-of-fit test, at the 5% significance level, on the data given on pages 113–114 for the two schools X and Y.

H_0: School and grade of pass are independent.
H_1: School and grade of pass are not independent.
Form your hypotheses.

$\nu = (h - 1)(k - 1) = (2 - 1)(3 - 1) = 2$
Calculate the number of degrees of freedom.

From tables the critical value at the 0.05 significance level is 5.991.
Find the critical value.

O_i	E_i	$\frac{(O_i - E_i)^2}{E_i}$
18	18.33	0.0059
12	10.00	0.4000
20	21.67	0.1287
26	25.67	0.0042
12	14.00	0.2857
32	30.33	0.0920

The expected values have already been calculated on a page 114.

$$\sum \frac{(O_i - E_i)^2}{E_i} = 0.9165$$

Calculate $\sum \frac{(O_i - E_i)^2}{E_i}$ or $\sum \frac{O_i^2}{E_i} - N$.

So: $\sum \frac{(O_i - E_i)^2}{E_i} < 5.991$
See if the value is significant.

Do not reject H_0: there is insufficient evidence to suggest an association between the school and the grades of pass.
School and grade of pass are independent.
Draw a conclusion.

Example 14

During the trial of a new drug, 60 volunteers out of 200 were treated with the drug.
Those who experienced relief of their symptoms and those who did not were recorded as in the table.

	Relief	No relief	Totals
Treated	10	50	60
Not treated	40	100	140
Totals	50	150	200

Use a suitable test to see if there is any association between treatment with the drug and relief of symptoms. Use a 5% significance level.

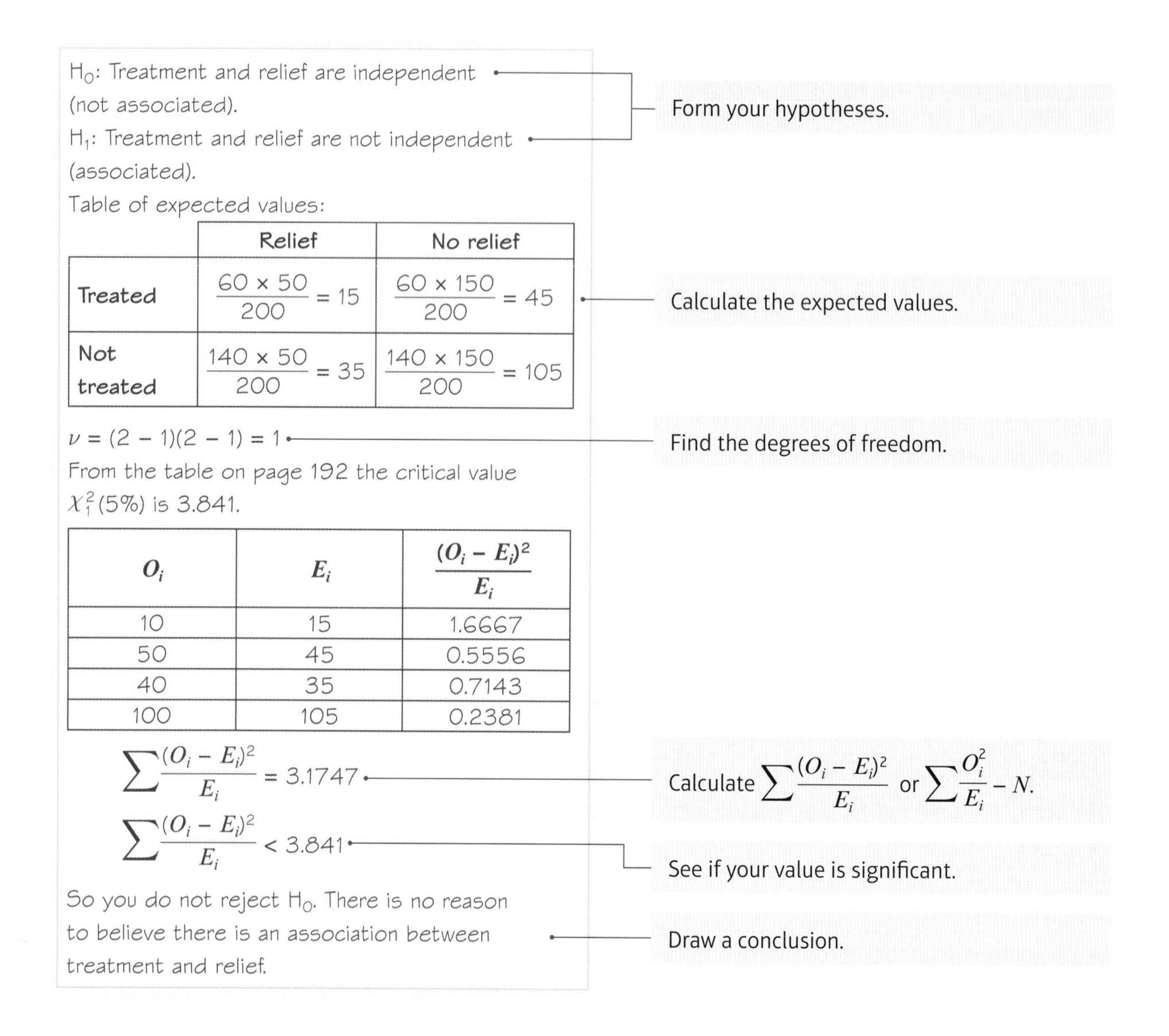

H_0: Treatment and relief are independent (not associated).
H_1: Treatment and relief are not independent (associated).

Form your hypotheses.

Table of expected values:

	Relief	No relief
Treated	$\frac{60 \times 50}{200} = 15$	$\frac{60 \times 150}{200} = 45$
Not treated	$\frac{140 \times 50}{200} = 35$	$\frac{140 \times 150}{200} = 105$

Calculate the expected values.

$\nu = (2 - 1)(2 - 1) = 1$

Find the degrees of freedom.

From the table on page 192 the critical value $\chi_1^2(5\%)$ is 3.841.

O_i	E_i	$\frac{(O_i - E_i)^2}{E_i}$
10	15	1.6667
50	45	0.5556
40	35	0.7143
100	105	0.2381

$$\sum \frac{(O_i - E_i)^2}{E_i} = 3.1747$$

Calculate $\sum \frac{(O_i - E_i)^2}{E_i}$ or $\sum \frac{O_i^2}{E_i} - N$.

$$\sum \frac{(O_i - E_i)^2}{E_i} < 3.841$$

See if your value is significant.

So you do not reject H_0. There is no reason to believe there is an association between treatment and relief.

Draw a conclusion.

Exercise 6E

1 When analysing the results of a 3 × 2 contingency table it was found that

$$\sum_{i=1}^{6} \frac{(O_i - E_i)^2}{E_i} = 2.38$$

Write down the number of degrees of freedom and the critical value appropriate to these data in order to carry out a χ^2 test of significance at the 5% level.

2 Three different types of locality were studied to see if the ownership, or non-ownership, of a television was or was not related to the locality. $\sum \frac{(O_i - E_i)^2}{E_i}$ was evaluated and found to be 13.1.

Using a 5% level of significance, carry out a suitable test and state your conclusion.

3 In a college, three different groups of students sit the same examination. The results of the examination are classified as Credit, Pass or Fail. In order to test whether or not there is an association between the group, and exam results, the statistic $\sum \frac{(O_i - E_i)^2}{E_i}$ is evaluated and found to be equal to 10.28.

a Explain why there are 4 degrees of freedom in this situation.

b Using a 5% level of significance, carry out the test and state your conclusions.

(E) **4** The grades of 200 students in both Mathematics and English were studied with the following results.

		English grades		
		A	**B**	**C**
Maths grades	**A**	17	28	18
	B	38	45	16
	C	12	12	14

Using a 0.05 significance level, test these results to see if there is an association between English and Mathematics results. State your conclusions. **(6 marks)**

(E) **5** The number of trains on time and the number of trains that were late were observed at three different London stations. The results were:

		Observed frequency	
		On time	**Late**
Station	A	26	14
	B	30	10
	C	44	26

Using the χ^2 statistic and a significance test at the 5% level, decide if there is any association between station and lateness. **(6 marks)**

6 In addition to being classed into grades A, B, C, D and E, 200 students are classified as male or female and their results are summarised in a contingency table.

Assuming all expected values are 5 or more, the statistic $\sum \frac{(O_i - E_i)^2}{E_i}$ was 14.27.

Stating your hypotheses and using a 1% significance level, investigate whether or not gender and grade are associated.

7 In a random sample of 60 articles made in factory A, 13 were defective. In factory B, 12 out of 40 similar articles were defective.

a Draw up a contingency table.

b Test at the 0.05 significance level the hypothesis that quality was independent of the factory involved.

8 During an influenza epidemic, 15 boys and 8 girls became ill out of a year group of 22 boys and 28 girls. Assuming that this group may be treated as a random sample of the age group, test at the 5% significance level the hypothesis that there is no connection between gender and susceptibility to influenza.

(E) **9** In a study of marine organisms, a biologist collected specimens from three beaches and counted the number of males and females in each sample, with the following results:

		Beach		
		A	*B*	*C*
Gender	**Male**	46	80	40
	Female	54	120	160

Using a significance level of 5%, test these results to see if there is any association between the choice of beach and the gender of the organisms. **(6 marks)**

(E) **10** A research worker studying the ages of adults and the number of credit cards they possess obtained the results shown below:

		Number of cards	
		$\leqslant 3$	> 3
Age	$\leqslant 30$	74	20
	> 30	50	35

Use the χ^2 statistic and a significance test at the 5% level to decide whether or not there is an association between age and number of credit cards possessed. **(6 marks)**

(E) **11** Members of four local gyms were surveyed to find out if they had injured themselves while working out in the last month. The results are summarised in the table below:

	Gym				
	A	*B*	*C*	*D*	**Total**
Injured	15	4	8	7	34
Uninjured	222	254	167	188	831
Total	237	258	175	195	865

A test is carried out at the 5% significance level to determine whether or not there is an association between injuries and choice of gym.

a State the null hypothesis for this test. **(1 mark)**

b Show that the expected frequency of members injured at gym *C* is 6.88 **(1 mark)**

c Calculate the test statistic for this test, and state with reasons whether or not the null hypothesis is rejected. **(5 marks)**

12 Millie wants to investigate whether students who studied different sciences at university get paid the same when they get a job. She surveys science graduates who graduated from university in the last 5 years, and records their salary information. The results are recorded in the table below.

		Salary					
		£0–£20k	**£20k–£40k**	**£40k–£60k**	**£60k–£80k**	**>£80k**	**Total**
Science studied	**Biology**	4	69	23	5	3	104
	Chemistry	3	72	27	4	2	108
	Physics	2	68	32	5	4	111
	Total	9	209	82	14	9	323

She tests at the 5% significance level whether there is an association between the science studied and pay.

a State the null and alternative hypotheses for this test. **(1 mark)**

b Calculate the test statistic for this test, and state with reasons whether or not the null hypothesis is rejected. **(5 marks)**

Problem-solving

If any of the expected frequencies are less than 5 you have to pool columns before calculating X^2.

6.6 Apply goodness-of-fit tests to geometric distributions

Recall that the conditions under which a geometric distribution is likely to arise are:

- Trials have two outcomes: success and failure
- Trials are independent
- Trials are performed until the first success
- The probability of success on each trial is constant (p)
- The measured quantity is the number of trials until the first success

For a geometric distribution with probability of success p we have:

$$P(X = r) = p(1 - p)^{r-1}, \text{ for } r = 1, 2, 3, \ldots$$

As with the Poisson distribution, there is an infinite number of possible values for r, so we once again group together all those values greater than or equal to some cut-off, n, which can be chosen to be the largest value of r for which the observed frequency is non-zero.

If we take a sample of size N from a geometric random variable X, then the expected frequency of observations of the value r is $f_r = P(X = r) \times N$.

The geometric distribution has a single parameter p which can be estimated from the observed frequencies f_r using the formula:

$$p = \frac{\text{Total number of successes}}{\text{Total number of trials}} = \frac{N}{\sum r \times f_r}$$

Because each success represents one observation from the distribution, the total number of successes is equal to the total frequency, N.

Each observation of r from the distribution contributes r trials to the total.

As before, if you estimate the parameter p from the observed frequencies, then that is a constraint, and so reduces the number of degrees of freedom by 1.

Example 15

A

Sarah has a large DVD collection. Every week she picks DVDs off the shelf at random until she finds one that she would like to watch. Sarah thinks that there is about a 50% chance she will be in the mood to watch any particular DVD. Over the course of a year she records the number of DVDs she picks off the shelf before finding one she would like to watch. The results are recorded in the frequency table below.

Number of DVDs	1	2	3	4	Total
Observed frequency O_i	33	12	5	2	52

a Calculate the expected frequencies if the number of DVDs considered is modelled as a Geo(0.5) random variable.

Sarah wants to check if her guess that there is a 50% chance she'll watch any particular DVD is supported by the data.

b Formulate the null and alternative hypotheses.

c Is Sarah right in her assumption? Test at the 5% significance level.

a The expected frequencies are:

Number of DVDs	1	2	3	$\geqslant 4$	Total
Observed frequency O_i	33	12	5	2	52
Expected frequency E_i	26	13	6.5	6.5	52

b H_0: $X \sim$ Geo(0.5) is a suitable model.
H_1: $X \sim$ Geo(0.5) is not a suitable model.

c We calculate the goodness of fit.

Number of DVDs	1	2	3	$\geqslant 4$	Total
Observed Frequency O_i	33	12	5	2	52
Expected Frequency E_i	26	13	6.5	6.5	52
$\frac{(O_i - E_i)^2}{E_i}$	1.8846	0.0769	0.3462	3.1154	5.4231

So $X^2 = 5.4231$. We can model X^2 with a χ^2_3 random variable.

The critical value at the 5% significance level is $\chi^2_3(5\%) = 7.815$. Since $5.4231 < 7.815$, we do not have enough evidence to reject the null hypothesis, so we can model the number of DVDs by a geometric random variable with $p = 0.5$.

Problem-solving

The expected frequency is calculated by
$E_i = P(X = i) \times N = \left(\frac{1}{2}\right)^i \times 52.$

Since there are no observations greater than 4, we combine all larger values into a single observation. The expected number of observations $\geqslant 4$ is

$P(X \geqslant 4) \times N = \frac{1}{8} \times 52$

There are 4 observations and one constraint (that they sum to 52), hence 3 degrees of freedom.

Exercise 6F

A E **1** The following table shows observed values for what is thought to be a geometric distribution with $p = 0.6$.

k	1	2	3	4	5	6	**Total**
Observed frequency O_k	207	66	13	9	3	2	300

Calculate the expected frequencies and, using a 1% significance level, conduct a goodness-of-fit test. **(6 marks)**

E **2** The following table shows observed values for what is thought to be a geometric distribution with $p = 0.4$.

k	1	2	3	4	5	6	**Total**
Observed frequency O_k	42	26	10	8	10	4	100

Calculate the expected frequencies and, using a 5% significance level, conduct a goodness-of-fit test. **(6 marks)**

E/P **3** The following table shows observed values from a distribution which is thought to be modelled by a geometric distribution.

k	1	2	3	4	5	6	**Total**
Observed frequency O_k	61	24	11	1	2	1	100

a Use the observed data to estimate p (to 3 d.p.). **(2 marks)**

b Conduct a goodness-of-fit test at the 5% significance level to determine whether a geometric distribution is a good fit for the data. **(5 marks)**

Problem-solving

Use $\frac{N}{\sum k \times O_k}$ to estimate the parameter.

E/P **4** Each day after work Katie flags down a taxi to take her home. She records how many taxis she tries to flag down before one stops, over the course of 100 days.

Attempts	1	2	3	4	5	**Total**
Frequency	76	17	4	2	1	100

Katie thinks she can model the number of attempts each day using a geometric random variable $X \sim \text{Geo}(p)$.

a Using the observed frequencies, find an estimate for p (to 3 d.p.). **(2 marks)**

b Conduct a goodness-of-fit test at the 2.5% significance level, and say whether a geometric random variable is a good model for the data. **(5 marks)**

E/P **5** Michael has a pet monkey and wants to test the theory that, given a typewriter and enough time, the monkey will eventually type out the complete works of Shakespeare. Unfortunately the experiment is a total disaster, and the monkey has succeeded only in producing a seemingly random string of alphabetic characters!

Michael decides instead he will see how random the string of characters is. He reads the string looking for vowels. Each time he finds a vowel, he counts the number of letters until the next vowel. The results are recorded in the table below.

Characters to next vowel	1	2	3	4	5	6	7	8 or more	Total
Frequency	12	14	11	10	5	9	7	7	75

Michael believes that every letter of the alphabet has an equal chance of being typed by the monkey.

a Assuming Michael's belief is accurate, suggest a suitable distribution to model the number of letters typed before the next vowel. **(2 marks)**

b Conduct a goodness-of-fit test, at the 5% significance level, to see if Michael's belief is supported by the data. **(5 marks)**

c Describe one limitation of using this experiment to test Michael's belief. **(1 mark)**

Challenge

Ellen is trying to make some money, so on the weekends she runs a lemonade stand. Each day she makes 10 cups of lemonade and counts how many people walk past before she sells all of them. The results are recorded in the table below.

Number of people	10	11	12	13	14	15
Frequency	10	25	29	15	15	10

You may assume that every passer-by has an equal probability of buying a cup of lemonade, and that the probabilities do not change from day to day. Suggest an appropriate distribution to model the number of people that go past before Ellen sells out, and test the fit of your model at the 5% significance level.

Mixed exercise 6

1 The random variable Y has a χ^2 distribution with 10 degrees of freedom. Find y such that $P(Y < y) = 0.99$.

2 The random variable X has a chi-squared distribution with 8 degrees of freedom. Find x such that $P(X > x) = 0.05$.

3 As part of an investigation into visits to a Health Centre, a 5×3 contingency table was constructed. A χ^2 test of significance at the 5% level is to be carried out on the table.

Write down the number of degrees of freedom and the critical region appropriate to this test.

4 Data are collected in the form of a 4×4 contingency table.

To carry out a χ^2 test of significance one of the rows is amalgamated with another row and the resulting value of $\sum \frac{(O - E)^2}{E}$ was calculated.

Write down the number of degrees of freedom and the critical value of χ^2 appropriate to this test, assuming a 5% significance level.

(E) **5** A new drug to treat the common cold was used with a randomly selected group of 100 volunteers. Each was given the drug and their health was monitored to see if they caught a cold. A randomly selected control group of 100 volunteers was treated with a dummy pill. The results are shown in the table below.

	Cold	**No cold**
Drug	34	66
Dummy pill	45	55

Using a 5% significant level, test whether or not the chance of catching a cold is affected by taking the new drug. State your hypotheses carefully. **(6 marks)**

(E/P) **6** Breakdowns on a certain stretch of motorway were recorded each day for 80 consecutive days. The results are summarised in the table below.

Number of breakdowns	0	1	2	>2
Frequency	38	32	10	0

It is suggested that the number of breakdowns per day can be modelled by a Poisson distribution.

Using a 5% significant level, test whether or not the Poisson distribution is a suitable model for these data. State your hypotheses clearly. **(9 marks)**

(E) **7** A survey in a college was commissioned to investigate whether or not there was any association between gender and passing a driving test. A group of 50 males and 50 females were asked whether they passed or failed their driving test at the first attempt. All the students asked had taken the test. The results were as follows.

	Pass	**Fail**
Male	23	27
Female	32	18

Stating your hypotheses clearly test, at the 10% level, whether or not there is any evidence of an association between gender and passing a driving test at the first attempt. **(6 marks)**

8 Successful contestants in a TV game show were allowed to select from one of five boxes, four of which contained prizes, and one of which contained nothing. The boxes were numbered 1 to 5, and, when the show had run for 100 weeks, the choices made by the contestants were analysed with the following results:

Box number	1	2	3	4	5
Frequency	20	16	25	18	21

a Explain why these data could possibly be modelled by a discrete uniform distribution.

b Using a significance level of 5%, test to see if the discrete uniform distribution is a good model in this particular case.

(E/P) **9** A pesticide was tested by applying it in the form of a spray to 50 samples of five flies. The numbers of dead flies after 1 hour were then counted with the following results:

Number of dead flies	0	1	2	3	4	5
Frequency	1	1	5	11	24	8

a Calculate the probability that a fly dies when sprayed. **(2 marks)**

b Using a significance level of 5%, test to see if these data could be modelled by a binomial distribution. **(5 marks)**

(E/P) **10** The number of accidents per week at a certain road junction was monitored for four years. The results obtained are summarised in the table.

Number of accidents	0	1	2	>2
Frequency	112	56	40	0

Using a 5% level of significance, carry out a χ^2 test of the hypothesis that the number of accidents per week has a Poisson distribution. **(9 marks)**

(E/P) **11** Samples of stones were taken at two sites on a beach which were 1 mile apart. The rock types of the stones were found and classified as igneous, sedimentary or other types, with the following results.

		Site	
		A	B
Rock type	Igneous	30	10
	Sedimentary	55	35
	Other	15	15

A scientist believes that the distribution of rock types at site A can be used as a model for the distribution at site B. Test this belief, using a 5% significance level. **(6 marks)**

(E/P) **12** A small shop sells a particular item at a fairly steady yearly rate. When looking at the weekly sales it was found that the number sold varied. The results for the 50 weeks the shop was open were as shown in the table.

Weekly sales	0	1	2	3	4	5	6	7	8	>8
Frequency	0	4	7	8	10	6	7	4	4	0

a Find the mean number of sales per week. **(2 marks)**

b Using a significance level of 5%, test to see if these can be modelled by a Poisson distribution. **(8 marks)**

(E) **13** A study was done of how many students in a college were left-handed and how many were right-handed. As well as left- or right-handedness the gender of each person was also recorded with the following results.

	Left-handed	Right-handed
Male	100	600
Female	80	800

Use a significance test at the 0.05 level to see if there is an association between gender and left- and right-handedness. **(6 marks)**

(E) **14** A school science department collected data on which science subject students found the most interesting. A random sample of 300 students gave the following results.

		Subject		
		Physics	**Biology**	**Chemistry**
Gender	**Male**	74	28	68
	Female	45	40	45

A test is carried out at the 1% level of significance to determine whether or not there is an association between gender and preferred subject.

a State the null hypothesis for this test. **(1 mark)**

b Show that the expected frequency for females choosing biology is 29.47 (to 2 d.p.). **(1 mark)**

c Calculate the remaining expected frequencies, and the test statistic for this test. **(3 marks)**

d State whether or not the null hypothesis should be rejected. Justify your answer. **(2 marks)**

e Would the test be rejected if instead the test were carried out at the 5% level of significance? **(1 mark)**

(E) **15** The discrete random variable X follows a Poisson distribution with mean 2.15.

a Write down the values of:

i $P(X = 1)$

ii $P(X > 2)$ **(2 marks)**

The manager at a call centre recorded the number of calls coming in each minute between noon and 1 pm.

Number of calls	0	1	2	3	4	5	6	**Total**
Frequency	10	12	14	12	8	3	1	60

b Show that the average number of calls received in a minute is 2.15. **(1 mark)**

The manager believes that the Poisson distribution may be a good model for the number of calls arriving each minute. She uses a Poisson distribution with mean 2.15 to calculate expected frequencies as follows.

Number of calls	0	1	2	3	4	5	6 or more
Expected frequency	6.99	15.03	a	11.58	6.22	2.67	b

c Find the values of a and b to two decimal places. **(2 marks)**

The manager will test, at the 5% level of significance, if the data can be modelled by a Poisson distribution as she suspects.

d State the null and alternative hypotheses for this test. **(1 mark)**

e Explain why the last two cells in the expected frequency table should be combined when calculating the test statistic for this test. **(1 mark)**

f Calculate the test statistic and state the conclusion for this test. State clearly the critical value used in the test. **(4 marks)**

16 David doesn't have a car. If it's raining in the morning when he has to go to work, he calls his friends one by one to see if anyone can give him a lift. Since starting his job it has rained on 255 mornings. David has recorded the number of calls he had to make on each of these mornings before finding a willing driver.

Number of calls	1	2	3	4	5	6	7	**Total**
Frequency	130	54	24	28	13	5	1	255

Assume that each of David's friends is equally likely to offer him a lift, and that David will never run out of friends to call. (David is extremely popular.)

a Suggest a suitable distribution to model the number of calls made by David on a rainy morning. **(1 mark)**

b Use the observed data to estimate any parameters necessary for your chosen distribution. **(2 marks)**

c Carry out a goodness-of-fit test at the 5% significance level for your chosen distribution. **(6 marks)**

E/P **17** Wilfred calls his parents every weekend to tell them about his week. Unfortunately Wilfred is very forgetful, and can't remember if the last digit of his parents' phone number is 2, 5 or 7. When he wants to call his parents, he simply guesses the last number, and waits to see if his parents answer. If they don't he tries again.

The number of times Wilfred dials before he gets through to his parents throughout the year is recorded in the table below.

Number of attempts	1	2	3	4	7	**Total**
Frequency observed	27	10	10	4	1	52

Wilfred claims that the number of attempts made each time he calls his parents can be modelled as a geometric distribution with probability of success $\frac{1}{3}$.

a Give one criticism of Wilfred's model. **(1 mark)**

b Test Wilfred's claim at the 5% significance level. **(6 marks)**

Challenge

A random sample of 500 phone calls to a call centre revealed the following distribution of call length (in minutes).

Length of call	$0 \leqslant l < 5$	$5 \leqslant l < 10$	$10 \leqslant l < 15$	$15 \leqslant l < 20$	$20 \leqslant l < 25$
Frequency	7	63	221	177	32

a Estimate the mean and variance of the call lengths.

b Using the mean and variance calculated in part **a**, test at the 5% level of significance whether call length can be modelled by a normal distribution.

Summary of key points

1 The **null** and **alternative hypotheses** generally take the following form:

H_0: There is no difference between the observed and the theoretical distribution.

H_1: There is a difference between the observed and the theoretical distribution.

2 **Goodness of fit** is concerned with measuring how well an observed frequency distribution fits to a known distribution.

3 The measure of **goodness of fit** is $X^2 = \sum \frac{(O_i - E_i)^2}{E_i}$ or $X^2 = \sum \frac{O_i^2}{E_i} - N$

4 The χ^2 family of distributions can be used to approximate X^2 as long as none of the expected values is below 5.

5 When calculating **degrees of freedom:**

ν = number of cells after combining – number of constraints

6 When using chi-squared tests, if any of the expected values are less than 5, then you have to combine frequencies in the data table until they are greater than 5.

7 When selecting which of the χ^2 family to use as an approximation for X^2, you have to select the distribution which has ν equal to the number of degrees of freedom of your expected values.

8 If X^2 exceeds the critical value, it is unlikely that the null hypothesis is correct so you reject it in favour of the alternative hypothesis.

9 If n is the number of cells after combining:

Distribution	**Degrees of freedom**	
	Parameters known	**Parameters not known**
Discrete uniform	$n - 1$	
Binomial	$n - 1$	$n - 2$
Poisson	$n - 1$	$n - 2$
Geometric	$n - 1$	$n - 2$

10 For contingency tables:

$$\textbf{expected frequency} = \frac{\text{row total} \times \text{column total}}{\text{grand total}}$$

for an $h \times k$ table, the number of degrees of freedom $\nu = (h - 1)(k - 1)$

7 Probability generating functions

Objectives

After completing this chapter you should be able to:

- Understand the use of probability generating functions → pages 129–131
- Use probability generating functions for standard distributions → pages 132–135
- Use probability generating functions to find the mean and variance of a distribution → pages 135–138
- Know the probability generating function of the sum of independent random variables → pages 139–142

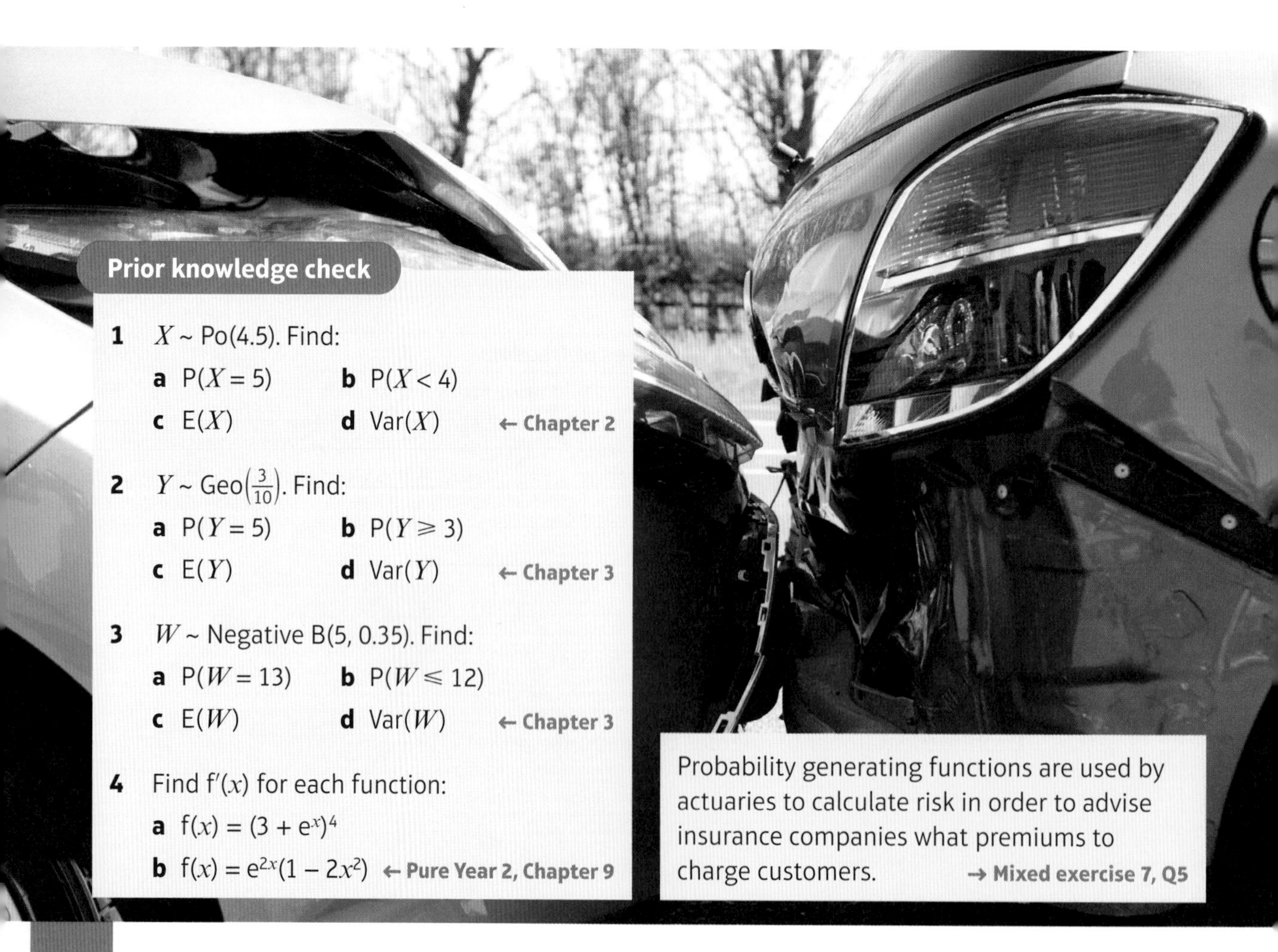

Prior knowledge check

1 $X \sim \text{Po}(4.5)$. Find:

a $\text{P}(X = 5)$ **b** $\text{P}(X < 4)$

c $\text{E}(X)$ **d** $\text{Var}(X)$ ← Chapter 2

2 $Y \sim \text{Geo}\left(\frac{3}{10}\right)$. Find:

a $\text{P}(Y = 5)$ **b** $\text{P}(Y \geqslant 3)$

c $\text{E}(Y)$ **d** $\text{Var}(Y)$ ← Chapter 3

3 $W \sim \text{Negative B}(5, 0.35)$. Find:

a $\text{P}(W = 13)$ **b** $\text{P}(W \leqslant 12)$

c $\text{E}(W)$ **d** $\text{Var}(W)$ ← Chapter 3

4 Find $\text{f}'(x)$ for each function:

a $\text{f}(x) = (3 + e^x)^4$

b $\text{f}(x) = e^{2x}(1 - 2x^2)$ ← Pure Year 2, Chapter 9

Probability generating functions are used by actuaries to calculate risk in order to advise insurance companies what premiums to charge customers. → Mixed exercise 7, Q5

7.1 Probability generating functions

A **probability generating function (p.g.f.)** is a mathematical function that stores details of a probability distribution. It can only be used with a discrete probability distribution that takes non-negative integer values, such as the binomial or Poisson distributions.

- **If a discrete random variable X has probability mass function $P(X = x)$, then the probability generating function of X is given by $G_X(t) = \sum P(X = x)t^x$ where t is a dummy variable.**

Hint You sum $t^x P(X = x)$ over all possible values in the sample space of X.

You can see how this function works by considering an example:
The discrete random variable X has the probability distribution shown in the table below:

x	0	1	2	3
$P(X = x)$	0.2	0.3	0.3	0.2

The probability generating function of X is

$G_X(t) = 0.2t^0 + 0.3t^1 + 0.3t^2 + 0.2t^3$

where the coefficients of t^x are the probabilities $P(X = x)$.

When $t = 1$, the terms of the generating function add up to 1 since $\sum P(X = x) = 1$. This is an important property of any probability generating function.

- **For any probability generating function $G_X(1) = 1$.**

The probability generating function can also be defined in terms of the expectation of a function of the random variable. This definition is given in the formulae booklet in your exam.

- **The probability generating function of a discrete random variable X is given by**

$$G_X(t) = E(t^X)$$

Links $E(g(X)) = \sum g(x)P(X = x)$, so $E(t^X) = \sum t^x P(X = x)$ ← Section 1.3

Example 1

X is the discrete random variable that denotes the absolute difference of the scores when two fair dice are thrown. Construct the probability distribution of X and write down the probability generating function.

Use a sample space diagram to find the possible outcomes:

	1	2	3	4	5	6
1	0	1	2	3	4	5
2	1	0	1	2	3	4
3	2	1	0	1	2	3
4	3	2	1	0	1	2
5	4	3	2	1	0	1
6	5	4	3	2	1	0

Now write down the probability distribution in a table:

x	0	1	2	3	4	5
$P(X = x)$	$\frac{6}{36}$	$\frac{10}{36}$	$\frac{8}{36}$	$\frac{6}{36}$	$\frac{4}{36}$	$\frac{2}{36}$

Check that $\sum P(X = x) = 1$ and do not simplify each probability.

$G_X(t) = \sum P(X = x)t^x$ so

$G_X(t) = \frac{6}{36}t^0 + \frac{10}{36}t^1 + \frac{8}{36}t^2 + \frac{6}{36}t^3 + \frac{4}{36}t^4 + \frac{2}{36}t^5$

$= \frac{1}{18}(3 + 5t + 4t^2 + 3t^3 + 2t^4 + t^5)$

This is the probability generating function for X.

Example 2

The probability generating function of the discrete random variable X is given by

$G_X(t) = k(1 + t)^2$

a Find the value of k. **b** Write down the probability distribution of X.

a $G_X(1) = 1$

$4k = 1$

$k = \frac{1}{4}$

Problem-solving

Use the property that $G_X(1) = 1$. Substitute $t = 1$ into the expression for $G_X(t)$ and set it equal to 1, then solve to find k.

b $G_X(t) = \frac{1}{4}(1 + 2t + t^2)$

$= \frac{1}{4} + \frac{1}{2}t + \frac{1}{4}t^2$

Expand the right-hand side with $k = \frac{1}{4}$

x	0	1	2
$P(X = x)$	$\frac{1}{4}$	$\frac{1}{2}$	$\frac{1}{4}$

The x values are the powers of t and the probabilities are the coefficients of each term.

Exercise 7A

1 The discrete random variable X has probability generating function

$G_X(t) = 0.3 + 0.2t + 0.5t^2$

a Write down the sample space of X.

b Find:

i $P(X = 0)$ **ii** $P(X \geqslant 0)$

2 The discrete random variable X has probability generating function

$G_X(t) = \frac{1}{8}(1 + t)^3$

a Write down the sample space of X.

b Find:

i $P(X = 1)$ **ii** $P(X \leqslant 2)$

3 The discrete random variable Y has probability generating function

$$G_Y(t) = 0.7 + 0.1(t^2 + t^3 + t^5)$$

Find:

a $P(Y = 1)$ **b** $P(Y < 3)$ **c** $P(3 \leqslant Y \leqslant 6)$

4 X is the discrete random variable representing the score when a fair dice is rolled. Write down the probability generating function of X.

5 X is the discrete random variable representing the score on a tetrahedral dice. The dice is biased such that $P(X = 1) = 0.4$. Given that the other three outcomes are equally likely, write down the probability generating function of X.

6 Find the probability generating function for each of these distributions:

a $P(X = x) = \frac{x}{10}$ $x = 1, 2, 3, 4$ **b** $P(X = x) = \frac{x^2}{14}$ $x = 1, 2, 3$

E/P 7 The probability generating function of a discrete random variable Y is given by

$$G_Y(t) = k(2 + t + t^2)^2$$

a Find the value of k. **(2 marks)**

b Find $P(Y = 1)$. **(2 marks)**

E/P 8 The probability generating function of a discrete random variable X is given by

$$G_X(t) = k(1 + 2t + 2t^2)^2$$

a Find the value of k. **(2 marks)**

b Write down the probability distribution of X. **(3 marks)**

9 X is the discrete random variable that denotes the sum of the scores when two fair four-sided dice are rolled.

a Construct the probability distribution for X.

b Write down the probability generating function for X.

10 A student writes the following probability generating function for a discrete random variable X

$G_X(t) = 0.1(2t + 5t^2 + 4t^3)$

Explain why this is not a probability generating function.

P 11 The discrete random variable Y has probability generating function:

$$G_Y(t) = k(1 + t)^{10}$$

a Find the value of k.

b Write down the largest value that Y can take and the probability that it takes this value.

c Find $P(Y = 5)$.

d State the name of the distribution of Y.

7.2 Probability generating functions of standard distributions

The standard probability distributions that you have studied have particularly simple probability generating functions. These are given in the formulae booklet but you should also be able to derive them in simple cases.

Example 3

An archer hits the bullseye with probability 0.6. She fires three shots at a target. The random variable X represents the number of times she hits the bullseye.

Find the probability generating function of X.

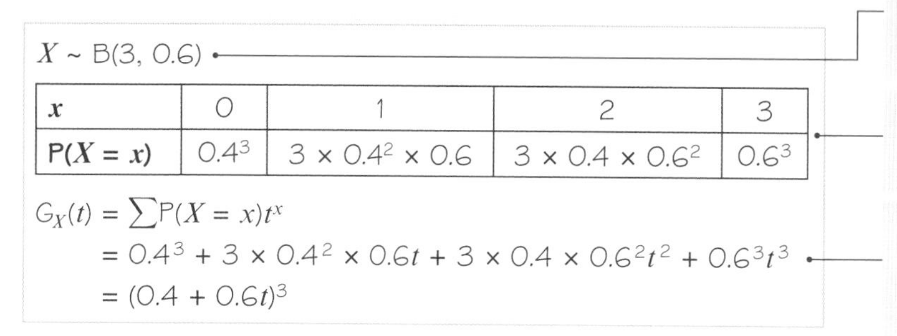

$X \sim B(3, 0.6)$

x	0	1	2	3
$P(X = x)$	0.4^3	$3 \times 0.4^2 \times 0.6$	$3 \times 0.4 \times 0.6^2$	0.6^3

$G_X(t) = \sum P(X = x)t^x$

$= 0.4^3 + 3 \times 0.4^2 \times 0.6t + 3 \times 0.4 \times 0.6^2t^2 + 0.6^3t^3$

$= (0.4 + 0.6t)^3$

This is a binomial model with $n = 3$ and $p = 0.6$. You are assuming that each shot is an independent event.

Write out the probability distribution of X.

The expansion of $G_X(t)$ is a binomial expansion in the form $(a + b)^n$ with $a = 0.4$ and $b = 0.6t$.

- **If a discrete random variable $X \sim B(n, p)$ the probability generating function for X is given by**
  $G_X(t) = (1 - p + pt)^n$

The Poisson distribution is theoretically an infinite distribution but the probability generating function can be derived using the idea of an infinite series.

Example 4

X is a discrete random variable such that $X \sim Po(1.1)$.

Show, from first principles, that the probability generating function of X is given by

$G_X(t) = e^{1.1(t-1)}$

Watch out 'From first principles' means that you cannot quote the standard result for the p.g.f. of a Poisson random variable given in the formulae booklet.

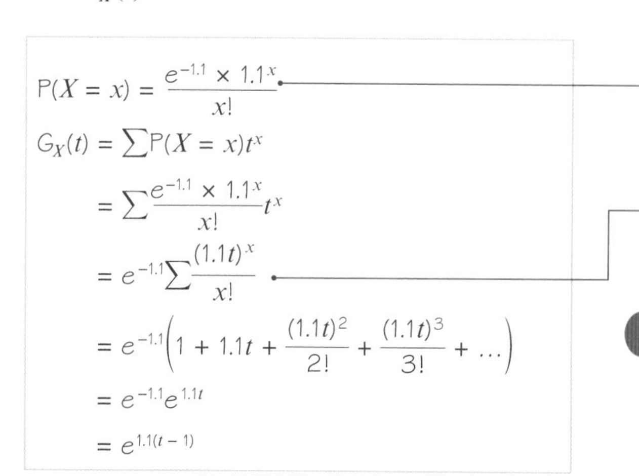

$P(X = x) = \dfrac{e^{-1.1} \times 1.1^x}{x!}$

$G_X(t) = \sum P(X = x)t^x$

$= \sum \dfrac{e^{-1.1} \times 1.1^x}{x!} t^x$

$= e^{-1.1} \sum \dfrac{(1.1t)^x}{x!}$

$= e^{-1.1}\left(1 + 1.1t + \dfrac{(1.1t)^2}{2!} + \dfrac{(1.1t)^3}{3!} + \ldots\right)$

$= e^{-1.1}e^{1.1t}$

$= e^{1.1(t-1)}$

Write down the probability distribution of X.

$e^{-1.1}$ is constant so you can write it outside the summation.

Problem-solving

The bracketed expression is the Maclaurin expansion of e^x where $x = 1.1t$.

← Core Pure Book 2, Chapter 2

- **If a discrete random variable $X \sim \text{Po}(\lambda)$ the probability generating function for X is given by**

  $$G_X(t) = e^{\lambda(t-1)}$$

Example 5

A fair tetrahedral dice is rolled until the dice shows a four. The discrete random variable X represents the number of rolls up to and including the first instance of a four.

Show, from first principles, that the probability generating function for X is given by

$$G_X(t) = \frac{\frac{1}{4}t}{1 - \frac{3}{4}t}$$

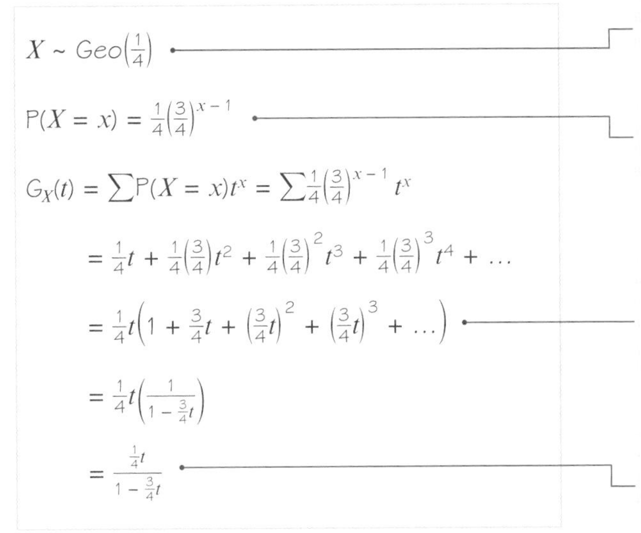

$$X \sim \text{Geo}\left(\tfrac{1}{4}\right)$$

The distribution is geometric with parameter $\frac{1}{4}$.

$$P(X = x) = \tfrac{1}{4}\left(\tfrac{3}{4}\right)^{x-1}$$

Write down the probability distribution of X.

← Chapter 3

$$G_X(t) = \sum P(X = x)t^x = \sum \tfrac{1}{4}\left(\tfrac{3}{4}\right)^{x-1} t^x$$

$$= \tfrac{1}{4}t + \tfrac{1}{4}\left(\tfrac{3}{4}\right)t^2 + \tfrac{1}{4}\left(\tfrac{3}{4}\right)^2 t^3 + \tfrac{1}{4}\left(\tfrac{3}{4}\right)^3 t^4 + \ldots$$

$$= \tfrac{1}{4}t\left(1 + \tfrac{3}{4}t + \left(\tfrac{3}{4}t\right)^2 + \left(\tfrac{3}{4}t\right)^3 + \ldots\right)$$

The bracketed expression is the sum to infinity of a geometric series with first term 1 and common ratio $\frac{3}{4}t$.

← Pure Year 2, Section 3.5

$$= \tfrac{1}{4}t\left(\frac{1}{1 - \frac{3}{4}t}\right)$$

$$= \frac{\frac{1}{4}t}{1 - \frac{3}{4}t}$$

This could be simplified to $\frac{t}{4 - 3t}$ but in the form given it is clear to see the relationship between G and p.

- **If X is a geometrically distributed discrete random variable X with probability of success in any one trial p, the probability generating function for X is given by**

  $$G_X(t) = \frac{pt}{1 - (1 - p)t}$$

Example 6

The discrete random variable $X \sim$ Negative B(r, p). Prove, from first principles, that the probability generating function of X can be written as

$$G_X(t) = \left(\frac{pt}{1 - (1 - p)t}\right)^r$$

You may quote the following result without proof:

$$\sum_{x=r}^{\infty} \binom{x-1}{r-1} q^{x-r} = (1 - q)^{-r} \text{ where } q = 1 - p.$$

$$P(X = x) = \binom{x-1}{r-1}p^r q^{x-r}, x = r, r+1, r+2, \ldots$$

Write down the probability distribution of X. Use $1 - p = q$.

$$G_X(t) = \sum \binom{x-1}{r-1}p^r q^{x-r}t^x$$

$$= \sum \binom{x-1}{r-1}p^r q^{x-r}t^{x-r}t^r$$

Split the t^x term into two parts which match the powers of p and q.

$$= \sum \binom{x-1}{r-1}(pt)^r(qt)^{x-r}$$

$$= (pt)^r \sum \binom{x-1}{r-1}(qt)^{x-r}$$

$(pt)^r$ can be taken out of the summation since it is independent of x.

$$= (pt)^r(1 - qt)^{-r}$$

$$= \left(\frac{pt}{1-qt}\right)^r = \left(\frac{pt}{1-(1-p)t}\right)^r \text{ as required.}$$

Problem-solving

Since X is an infinite discrete random variable you are required to find the infinite sum of the binomial coefficients. Use the result that $\sum_{x=r}^{\infty}\binom{x-1}{r-1}(q)^{x-r} = (1-q)^{-r}$ as given in the question.

- **If X is a discrete random variable with negative binomial distribution, the probability generating function for X is given by**

$$G_X(t) = \left(\frac{pt}{1-(1-p)t}\right)^r$$

Exercise 7B

1 Write down the probability generating functions for the following distributions:

a $X \sim B(4, 0.5)$ **b** $Y \sim B(6, 0.2)$

c $X \sim B(5, 0.9)$ **d** $X \sim Po(3)$

e $X \sim Po(1.7)$ **f** $Y \sim Po(0.2)$

Hint If a question asks you to 'write down' or 'find' a probability generating function for a standard distribution, then you can quote the standard formulae without proof.

2 Write down the probability generating functions for the following distributions:

a $X \sim Geo(0.3)$ **b** $Y \sim Geo(0.8)$

c $X \sim$ Negative $B(3, 0.4)$ **d** $Y \sim$ Negative $B(5, 0.9)$

3 A dice is biased so that $P(6) = 0.2$. Find the probability generating function of each of the following random variables:

a the number of sixes obtained when the dice is rolled 5 times

b the number of times the dice must be thrown until it shows a six for the first time

c the number of times the dice must be thrown until it shows a six for the second time

E **4** A sail-maker notices that the flaws in a roll of sailcloth occur at an average rate of 0.3 per metre.

a Suggest a suitable model for the random variable X, the number of flaws in a metre of cloth. **(1 mark)**

b Find $P(X = 1)$. **(1 mark)**

c Write down the probability generating function for X. **(2 marks)**

A E **5** Bernice is playing darts and she finds that the probability of hitting a treble score in any one throw is 0.35.

a Suggest a suitable model for the random variable X, the number of throws it takes her to hit a treble. **(1 mark)**

b Find $\text{P}(X = 6)$. **(1 mark)**

c Write down the probability generating function for X. **(2 marks)**

E/P **6** $X \sim \text{B}(4, 0.8)$. Show, from first principles, that the probability generating function for X is

$$\text{G}_X(t) = (0.2 + 0.8t)^4$$ **(5 marks)**

E/P **7** Calls come in to a call centre at a rate of 3.5 calls per five-minute interval. Given that the random variable X is the number of calls that come in during a random five-minute interval and that the calls are independent and random, show, from first principles, that the probability generating function for X is

$$\text{G}_X(t) = \text{e}^{3.5(t - 1)}$$ **(5 marks)**

E/P **8** $Y \sim \text{Geo}(0.7)$. Show, from first principles, that the probability generating function for Y is

$$\text{G}_Y(t) = \frac{0.7t}{1 - 0.3t}$$ **(5 marks)**

Problem-solving

Use the formula for the sum of an infinite convergent geometric series.

E/P **9** The random variable $X \sim \text{B}(n, p)$. Prove, from first principles, that the probability generating function of X is given by

$$\text{G}_X(t) = (1 - p + pt)^n$$ **(8 marks)**

E/P **10** The random variable $X \sim \text{Po}(\lambda)$. Prove, from first principles, that the probability generating function of X is given by

$$\text{G}_X(t) = \text{e}^{\lambda(t - 1)}$$ **(8 marks)**

E/P **11** The random variable $Y \sim \text{Geo}(p)$. Prove, from first principles, that the probability generating function of Y is given by

$$\text{G}_Y(t) = \frac{pt}{1 - (1 - p)t}$$ **(8 marks)**

7.3 Mean and variance of a distribution

You can find the mean and variance of a probability distribution by differentiating the probability generating function.

$$\text{G}_X(t) = \text{E}(t^X) = \sum t^x \text{P}(X = x)$$

$$\Rightarrow \text{G}'_X(t) = \sum x t^{x - 1} \text{P}(X = x) = \text{E}(X t^{X - 1})$$

$$\Rightarrow \text{G}'_X(1) = \sum x \text{P}(X = x) = \text{E}(X)$$

- **If X is a discrete random variable with probability generating function $\text{G}_X(t)$,**
  $\text{E}(X) = \text{G}'_X(1)$

A

$G'_X(t) = \sum xt^{x-1}P(X = x) = E(Xt^{X-1})$
$\Rightarrow G''_X(t) = \sum x(x-1)t^{x-2}P(X = x) = E(X(X-1)t^{X-2})$
$\Rightarrow G''_X(1) = \sum x(x-1)P(X = x) = E(X(X-1)) = E(X^2) - E(X)$
$\Rightarrow E(X^2) = G''_X(1) + E(X) = G''_X(1) + G'_X(1)$

You know that $\text{Var}(X) = E(X^2) - (E(X))^2$. Hence:

- **If X is a discrete random variable with probability generating function $G_X(t)$, $\text{Var}(X) = G''_X(1) + G'_X(1) - (G'_X(1))^2$**

You may be asked to prove these two standard results in your exam.

You can see the above results more clearly by differentiating $G_X(t)$ term-by-term:

If $\quad G_X(t) = P(X = 0) + P(X = 1)t + P(X = 2)t^2 + \ldots + P(X = n)t^n + \ldots$

Then $\quad G'_X(t) = P(X = 1) + 2P(X = 2)t + 3P(X = 3)t^2 \ldots + nP(X = n)t^{n-1} + \ldots$

So $\quad G'_X(1) = \sum xP(X = x)$

And $\quad G_X''(t) = 2P(X = 2) + 6P(X = 3)t + 12P(X = 4)t^2 + \ldots + n(n-1)P(X = n)t^{n-2} + \ldots$

So $\quad G''_X(1) = \sum x(x-1)P(X = x)$

Hint You can see from these expressions that:
$G(0) = P(X = 0)$
$G'(0) = P(X = 1)$
$G''(0) = P(X = 2)$
In general, $G_X^{(n)}(0) = P(X = n)$, where $G_X^{(n)}$ is the nth derivative of G_X.

Example 7

The discrete random variable X has a probability generating function given by

$$G_X(t) = \frac{1}{100\,000}(9 + t^2)^5$$

Find: **a** $E(X)$ **b** $\text{Var}(X)$

a $G'_X(t) = \frac{t}{10\,000}(9 + t^2)^4$ — Use the chain rule to find $G'_X(t)$.

$G'_X(1) = \frac{1}{10\,000}(9 + 1)^4 = 1$

Hence $E(X) = 1$ — State the value of $E(X)$ clearly.

b $G''_X(t) = \frac{9(1 + t^2)(9 + t^2)^3}{10\,000}$ — Find the second derivative of $G_X(t)$.

$\text{Var}(X) = G''_X(1) + G'_X(1) - (G'_X(1))^2$ — Write down the formula for variance.

$\text{Var}(X) = \frac{9(1 + 1)(9 + 1)^3}{10\,000} + 1 - 1^2 = \frac{18\,000}{10\,000} = \frac{9}{5}$ — Use the value of $G'_X(1)$ calculated above.

Example 8

A discrete random variable X has a probability generating function given by $G_X(t) = a + bt + ct^2$ where a, b and c are constants. Given that the expected value of X is $\frac{7}{6}$ and the variance of X is $\frac{29}{36}$, find the values of a, b and c.

$G_X(1) = 1 \Rightarrow a + b + c = 1 \quad (1)$ — $\sum P(X = x) = 1$

$G'_X(t) = b + 2ct$

$\Rightarrow G'_X(1) = b + 2c = \frac{7}{6} \quad (2)$ — Use the formula for the expected value to write an equation in terms of b and c.

$\text{Var}(X) = G''_X(1) + G'_X(1) - (G'_X(1))^2 = \frac{29}{36}$

$\Rightarrow G''_X(1) + \frac{7}{6} - \left(\frac{7}{6}\right)^2 = \frac{29}{36} \Rightarrow G''_X(1) = 1$

$G''_X(t) = 2c = 1 \Rightarrow c = \frac{1}{2}$ — Use the formula for the variance to find c.

Substitute into (2): $b + 1 = \frac{7}{6} \Rightarrow b = \frac{1}{6}$

Substitute into (1): $a + \frac{1}{6} + \frac{1}{2} = 1 \Rightarrow a = \frac{1}{3}$

Exercise 7C

A

1 A discrete random variable X has probability generating function $G_X(t) = \frac{1}{2} + \frac{1}{4}t + \frac{1}{4}t^2$. Use this to find the expected value and variance of X.

2 A discrete random variable X has probability generating function $G_X(t) = \frac{1}{6} + \frac{1}{6}t + \frac{1}{3}t^2 + \frac{1}{3}t^3$. Use this to find the mean and standard deviation of X.

3 Three unbiased coins are spun and the number of tails, X, is noted.

a Write down the probability generating function for X.

b Use your probability generating function to calculate the mean and variance of X.

4 A biased coin is spun four times. If P(head) = 0.6,

a write down the probability generating function of X, the discrete random variable representing the number of heads spun.

b Using your probability generating function from **a**, find:

i the mean of X

ii the standard deviation of X.

E/P **5** A discrete random variable X has probability generating function $G_X(t) = \dfrac{t^2(2 + t)^4}{81}$

Find the mean and variance of X. **(8 marks)**

E/P **6** A discrete random variable X has probability generating function $G_X(t) = \dfrac{9}{(4 - t)^2}$

a Find the mean and standard deviation of X. **(6 marks)**

b Find: **i** $P(X = 0)$ **ii** $P(X = 1)$ **(2 marks)**

Hint Consider the series expansion of $G_X(t)$.

E/P **7** A discrete random variable Y has probability generating function $G_Y(t) = e^{t^2 - 1}$.

a Find the mean and variance of Y. **(6 marks)**

b Find: **i** $P(Y = 0)$ **ii** $P(Y = 2)$ **iii** $P(Y = 3)$ **iv** $P(Y = 4)$ **(4 marks)**

E **8** A bag contains six counters, five red and one yellow. Counters are drawn out and the colour noted before being replaced. Let X represent the number of withdrawals until the yellow counter is drawn.

a State the distribution of X. **(1 mark)**

b Write down the probability generating function of X. **(1 mark)**

c Using your answer to part **b**, find:

i the mean of X **ii** the variance of X. **(7 marks)**

A P **9** The discrete random variable $X \sim B(n, p)$. Use the probability generating function of X to show that $E(X) = np$ and $Var(X) = np(1 - p)$.

P **10** The discrete random variable $X \sim P(\lambda)$. Use the probability generating function of X to show that $E(X) = Var(X) = \lambda$.

E **11** A call centre receives incoming calls at an average rate of four per 15-minute period. Assuming that the incoming calls are independent and random,

a state the distribution of X, the number of calls received in a single 15-minute period **(1 mark)**

b write down the probability generating function of X. **(1 mark)**

c Using your answer to part **b**, show that:

i the mean of X is 4 **(3 marks)**

ii the standard deviation of X is 2. **(4 marks)**

P **12** A discrete random variable X has probability generating function $G_X(t) = a + bt + ct^2$ where a, b and c are constants. Given that $E(X) = \frac{5}{4}$ and $Var(X) = \frac{7}{16}$, find the values of a, b and c.

E/P **13** The discrete random variable X has a probability generating function given by

$$G_X(t) = \frac{at}{b - t^2}, \text{ where } a \text{ and } b \text{ are positive constants.}$$

Given that the mean of X is 1.5, find the values of a and b. **(6 marks)**

E/P **14** A discrete random variable X has probability generating function $G_X(t) = k(1 + t)^4$ where k is a constant.

a Find the value of k. **(1 mark)**

b Write down the probability distribution represented by G. **(1 mark)**

c By explicitly using the probability generating function and showing all steps in your working, show that $E(X) = 2$ and $Var(X) = 1$. **(6 marks)**

E/P **15** Two fair dice are thrown and the random variable X, the smaller of the two numbers, is recorded.

a Find the probability generating function of X. **(3 marks)**

b Use your answer to part **a** to find:

i the mean of X **(3 marks)**

ii the standard deviation of X. **(4 marks)**

Challenge

The random variable X has probability generating function given by $G_X(t) = \frac{t}{2 - t^k}$, where k is a positive integer.

Find:

a $P(X = 0)$ **b** $E(X)$ in terms of k **c** $P(X = 1)$

7.4 Sums of independent random variables

Consider two dice. Dice A has three faces showing the number 1 and three faces showing the number 2. Dice B has two faces showing the number 0 and four faces showing the number 1.

The random variables X and Y represent the scores on dice A and on dice B respectively. The probability distributions of the two dice are shown below:

Dice A:

x	1	2
P($X = x$)	$\frac{1}{2}$	$\frac{1}{2}$

Dice B:

y	0	1
P($Y = y$)	$\frac{1}{3}$	$\frac{2}{3}$

The probability generating functions of X and Y are:

$$\mathrm{G}_X(t) = \tfrac{1}{2}t + \tfrac{1}{2}t^2 \text{ and } \mathrm{G}_Y(t) = \tfrac{1}{3} + \tfrac{2}{3}t$$

Given that the outcomes on the two dice are independent, the distribution of Z, the random variable representing the sum of the scores on the two dice, $Z = X + Y$, can be worked out:

z	1	2	3
P($Z = z$)	$\frac{1}{6}$	$\frac{1}{2}$	$\frac{1}{3}$

$Z = 2$ can occur in two ways: $X = 2$ and $Y = 0$, or $X = 1$ and $Y = 1$. So $\mathrm{P}(Z = 2) = \frac{1}{2} \times \frac{1}{3} + \frac{1}{2} \times \frac{2}{3} = \frac{1}{2}$

The probability generating function of Z is:

$$\mathrm{G}_Z(t) = \tfrac{1}{6}t + \tfrac{1}{2}t^2 + \tfrac{1}{3}t^3$$

If you find the product of the probability generating functions of X and Y, you will find that the resulting function is the same as the probability generating function of Z:

$$\begin{aligned}\mathrm{G}_X(t) \times \mathrm{G}_Y(t) &= \left(\tfrac{1}{2}t + \tfrac{1}{2}t^2\right)\left(\tfrac{1}{3} + \tfrac{2}{3}t\right)\\ &= \tfrac{1}{6}t + \tfrac{1}{3}t^2 + \tfrac{1}{6}t^2 + \tfrac{1}{3}t^3\\ &= \tfrac{1}{6}t + \tfrac{1}{2}t^2 + \tfrac{1}{3}t^3\end{aligned}$$

- **If X and Y are two independent random variables with probability generating function $\mathrm{G}_X(t)$ and $\mathrm{G}_Y(t)$, the probability generating function of $Z = X + Y$ is given by**

  $$\mathrm{G}_Z(t) = \mathrm{G}_X(t) \times \mathrm{G}_Y(t)$$

You will not need to be able to prove this result in your exam.

Example 9

Two independent discrete random variables X and Y have probability generating functions $\mathrm{G}_X(t) = \frac{1}{2} + \frac{1}{2}t$ and $\mathrm{G}_Y(t) = \frac{1}{3} + \frac{1}{2}t + \frac{1}{6}t^2$.

a Find the probability generating function of the random variable $Z = X + Y$.

b Use probability generating functions to show that $\mathrm{E}(Z) = \mathrm{E}(X) + \mathrm{E}(Y)$.

A

a $G_Z(t) = G_X(t) \times G_Y(t)$

$= \left(\frac{1}{2} + \frac{1}{2}t\right)\left(\frac{1}{3} + \frac{1}{2}t + \frac{1}{6}t^2\right)$

$= \frac{1}{6} + \frac{1}{4}t + \frac{1}{12}t^2 + \frac{1}{6}t + \frac{1}{4}t^2 + \frac{1}{12}t^3$

$= \frac{1}{6} + \frac{5}{12}t + \frac{1}{3}t^2 + \frac{1}{12}t^3$

Expand and simplify.

b $G'_X(1) = E(X)$

$G'_X(t) = \frac{1}{2} \Rightarrow E(X) = \frac{1}{2}$

$G'_Y(1) = E(Y)$

$G'_Y(t) = \frac{1}{2} + \frac{1}{3}t \Rightarrow E(Y) = \frac{1}{2} + \frac{1}{3} = \frac{5}{6}$

$G'_Z(1) = E(Z)$

$G'_Z(t) = \frac{5}{12} + \frac{2}{3}t + \frac{1}{4}t^2 \Rightarrow E(Z) = \frac{5}{12} + \frac{2}{3} + \frac{1}{4} = \frac{4}{3}$

Differentiate the expression from part **a**.

Use $E(X) + E(Y)$ from part **b**.

$\frac{1}{2} + \frac{5}{6} = \frac{8}{6} = \frac{4}{3}$ as required.

Example 10

A random variable X has a probability generating function $G_X(t) = \frac{2}{3} + \frac{1}{3}t$.

a Write down the probability distribution of X.

b $Y = 2X + 1$. Write down the probability distribution of Y and hence find the probability generating function of Y.

c Verify that $G_Y(t) = tG_X(t^2)$.

a $P(X = 0) = \frac{2}{3}$; $P(X = 1) = \frac{1}{3}$

Use the p.g.f. to write down the probabilities of the possible outcomes.

b $P(Y = 1) = \frac{2}{3}$; $P(Y = 3) = \frac{1}{3}$

The probability generating function of Y is

$G_Y(t) = \frac{2}{3}t + \frac{1}{3}t^3$

c $G_Y(t) = t\left(\frac{2}{3} + \frac{1}{3}t^2\right) = tG_X(t^2)$ as required.

You can generalise the example above to find the probability generating function for a linear transformation of a random variable X.

- **If the discrete random variable X has probability generating function $G_X(t)$, then the probability generating function of the discrete random variable $Y = aX + b$, where a and b are positive integers, is given by**

  $$G_Y(t) = t^b G_X(t^a)$$

Watch out This result is not given in the formulae booklet, so you need to learn it.

Exercise 7D

A

1 Two independent random variables X and Y have probability generating functions given by $G_X(t) = \frac{1}{4}t + \frac{1}{4}t^2 + \frac{1}{2}t^3$ and $G_Y(t) = \frac{5}{6} + \frac{1}{6}t$.

a Find the probability generating function for the random variable $Z = X + Y$.

b Show that $E(Z) = E(X) + E(Y)$.

E

2 Two independent random variables X and Y have probability generating functions given by $G_X(t) = \frac{1}{4}(1 + t)^2$ and $G_Y(t) = \frac{1}{25}(2 + 3t)^2$.

a Write down the probability distributions of X and Y. **(2 marks)**

b Show that the probability generating function of $Z = X + Y$ can be written in the form $G_Z(t) = a + bt + ct^2 + dt^3 + et^4$ where a, b, c, d and e are constants to be found. **(3 marks)**

c Verify that $E(Z) = E(X) + E(Y)$. **(4 marks)**

E

3 $X \sim \text{Po}(1.3)$ and $Y \sim \text{Po}(2.4)$.

a Write down the probability generating functions for X and Y. **(2 marks)**

b Given that X and Y are independent, write down the probability generating function for $Z = X + Y$. **(1 mark)**

c Use your answers to parts **a** and **b** to show that $E(Z) = E(X) + E(Y)$. **(4 marks)**

E/P

4 Jacintha is rolling a fair six-sided dice until a five appears.

a Show from first principles that the probability generating function for the number of rolls required is given by $G(t) = \frac{t}{6 - 5t}$ **(4 marks)**

Henry rolls a fair ten-sided dice until two fives have appeared.

b Write down the probability generating function for the number of rolls required to obtain two fives. **(1 mark)**

The random variable Z represents the total number of rolls made on both dice.

c Find the probability generating function of Z. **(2 marks)**

d Show that $E(Z) = 26$. **(4 marks)**

E

5 A random variable X has a probability generating function $G_X(t) = k(1 + 2t)^3$.

a Find the value of k. **(2 marks)**

b Find $P(X = 2)$. **(2 marks)**

c Use the probability generating function to show that $E(X) = 2$ and $\text{Var}(X) = \frac{2}{3}$ **(6 marks)**

A second random variable Y has a probability generating function $G_Y(t) = \frac{1}{4} + \frac{3}{4}t$.

Given that X and Y are independent,

d find $E(Y)$ and write down the value of $E(X + Y)$. **(3 marks)**

6 A random variable X has a probability generating function $G_X(t) = \frac{4}{(3-t)^2}$

A second random variable Y has a probability generating function $G_Y(t) = \frac{t}{(3-2t)^3}$

Given that X and Y are independent,

a write down the probability generating function for the random variable $Z = X + Y$ **(1 mark)**

b find $E(Z)$ **(4 marks)**

c show that $\text{Var}(Z) = \frac{39}{2}$ **(6 marks)**

7 Aidan and Chloe each buy 5 scratchcards for different lottery games. The probability of winning a prize on each of Aidan's scratchcards is 0.3. The probability of winning a prize on each of Chloe's scratchcards is 0.4.

The random variable X represents the total number of prize-winning scratchcards.

a Find an expression for the probability generating function of X. **(4 marks)**

b Show that the mean of X is 3.5. **(3 marks)**

8 A random variable X has a probability generating function $G_X(t) = \frac{1}{2}t + \frac{1}{2}t^3$.

Find the probability generating functions for the following random variables:

a $Y = 3X$ **b** $Y = 2X + 3$ **c** $Y = 4X - 5$

9 A random variable X has probability generating function $G_X(t) = \frac{1}{6}t + \frac{1}{6}t^2 + kt^3$, where k is a constant.

a Write down the value of k. **(1 mark)**

b Find $E(X)$. **(3 marks)**

A random variable $Y = 2X - 1$.

c Find the probability generating function of Y. **(2 marks)**

d Verify, using your answers to parts **b** and **c**, that $E(Y) = 2E(X) - 1$. **(3 marks)**

Challenge

1 A discrete random variable $Y = aX + b$. Given that $G_Y(t) = t^bG_X(t^a)$, show that in general $E(Y) = aE(X) + b$.

2 Holly is an archer who hits the bullseye with constant probability 0.6.

a Find the probability generating function $G(t)$ of the number of shots she needs until she hits her first bullseye.

b Show that the probability generating function of the number of shots she needs until she hits her second bullseye is $(G(t))^2$.

c Find, in terms of $G(t)$, the probability generating function of the number of shots she needs until she hits four bullseyes.

Mixed exercise 7

1 The probability generating function of a discrete random variable Y is given by

$$G_Y(t) = k(2 + 2t + 3t^2)$$

a Find the value of k. **(2 marks)**

b Find $P(Y = 1)$. **(2 marks)**

A second random variable, X, has probability generating function

$$G_X(t) = \frac{1}{16}(1 + t + 2t^2)^2$$

Given that X and Y are independent, find:

c the probability generating function of $Z = X + Y$ **(2 marks)**

d $P(Z = 2)$ **(4 marks)**

2 The discrete random variable $X \sim \text{Geo}(p)$. Use the probability generating function of X to show that $E(X) = \frac{1}{p}$ and $\text{Var}(X) = \frac{1-p}{p^2}$ **(5 marks)**

3 $X \sim B(5, 0.4)$. Show, from first principles, that the probability generating function for X is:

$$G_X(t) = (0.4 + 0.6t)^5$$ **(5 marks)**

4 A box of cat treats contains 15 treats, 11 meaty and 4 fishy. Fluffy the cat selects a treat at random. If the treat is meaty, he spits it back in to the box. Let X represent the number of selections until Fluffy selects a fishy treat.

a State the distribution of X. **(1 mark)**

b Write down the probability generating function of X. **(2 marks)**

c Using your answer to part **b**, find:

i the mean of X **(3 marks)**

ii the variance of X. **(4 marks)**

Once Fluffy has selected a fishy treat from the first box, he repeats the process with a second box. The second box contains 12 treats, 7 meaty and 5 fishy. The random variable Z represents the total number of selections needed by Fluffy to get a fishy treat from both boxes.

d Write down the probability generating function of Z. **(2 marks)**

e Find the mean and standard deviation of Z. **(6 marks)**

5 A car insurance company models the number of claims, X, a particular person will make in one year, using a Poisson distribution with mean 0.5.

a Find, in terms of e, the probability that this person will make:

i no claims

ii at least three claims

in a given year. **(5 marks)**

The policy is adjusted so a maximum of 3 claims can be made in any one year. The random variable Y represents the number of claims made.

b Show that the probability generating function of Y is given by

$$G_Y(t) = t^3 + e^{-0.5}\left(1 + \tfrac{1}{2}t + \tfrac{1}{8}t^2 - \tfrac{13}{8}t^3\right)$$ **(3 marks)**

c Use your probability generating function to find, correct to 3 decimal places, the values of E(Y) and Var(Y). **(8 marks)**

E/P **6** The discrete random variable X has probability generating function $G_X(t) = \frac{t^2}{(2-t)^5}$

a Find the mean and standard deviation of X. **(8 marks)**

b Find:

i $P(X = 0)$ **ii** $P(X = 2)$ **(2 marks)**

A second discrete random variable Y has probability generating function $G_Y(t) = \frac{t}{(4-3t)^2}$

c Given that X and Y are independent, find the probability generating functions for:

i $2Y - 1$ **ii** $Z = X + Y$ **(3 marks)**

d Show that $E(Z) = 14$. **(4 marks)**

E **7** The probability generating function of a discrete random variable X is given by

$$G_X(t) = k(1 + 2t^2 + 3t^3)^2$$

a Show that $k = \frac{1}{36}$ **(2 marks)**

b Find $P(X = 4)$. **(2 marks)**

c Show that $E(X) = \frac{13}{3}$ and find Var(X). **(6 marks)**

d Find the probability generating function of $3X - 2$. **(2 marks)**

Challenge

A discrete random variable X has probability generating function

$$G_X(t) = \tan\left(\frac{\pi t}{4}\right)$$

a Show that:

i $P(X = 0) = 0$ **ii** $P(X = 1) = \frac{\pi}{4}$ **iii** $P(X = 2) = 0$

b Show that $E(X) = \text{Var}(X) = \frac{\pi}{2}$

c Find $P(X = 3)$.

Summary of key points

A

1 If a discrete random variable X has probability mass function $\mathrm{P}(X = x)$, then the probability generating function of X is given by $\mathrm{G}_X(t) = \sum \mathrm{P}(X = x)t^x$ where t is a dummy variable.

2 For any probability generating function $\mathrm{G}_X(1) = 1$.

3 The probability generating function of a discrete random variable X is given by

$$\mathrm{G}_X(t) = \mathrm{E}(t^X)$$

4 The probability generating functions for the following standard distributions are given in the table below.

Distribution of X	**$\mathrm{P}(X = x)$**	**P.G.F.**
Binomial $\mathrm{B}(n, p)$	$\binom{n}{x}p^x(1-p)^{n-x}$	$\mathrm{G}_X(t) = (1 - p + pt)^n$
Poisson $\mathrm{Po}(\lambda)$	$\mathrm{e}^{-\lambda}\dfrac{\lambda^x}{x!}$	$\mathrm{G}_X(t) = \mathrm{e}^{\lambda(t-1)}$
Geometric $\mathrm{Geo}(p)$	$p(1-p)^{x-1}$	$\mathrm{G}_X(t) = \dfrac{pt}{1-(1-p)t}$
Negative binomial Negative $\mathrm{B}(r, p)$	$\binom{x-1}{r-1}p^r(1-p)^{x-r}$	$\mathrm{G}_X(t) = \left(\dfrac{pt}{1-(1-p)t}\right)^r$

5 If X is a discrete random variable with probability generating function $\mathrm{G}_X(t)$,

- $\mathrm{E}(X) = \mathrm{G}'_X(1)$
- $\mathrm{Var}(X) = \mathrm{G}''_X(1) + \mathrm{G}'_X(1) - (\mathrm{G}'_X(1))^2$

6 If X and Y are two independent random variables with probability generating function $\mathrm{G}_X(t)$ and $\mathrm{G}_Y(t)$, the probability generating function of $Z = X + Y$ is given by $\mathrm{G}_Z(t) = \mathrm{G}_X(t) \times \mathrm{G}_Y(t)$.

7 If the discrete random variable X has probability generating function $\mathrm{G}_X(t)$, then the probability generating function of the discrete random variable $Y = aX + b$, where a and b are positive integers, is given by

$$\mathrm{G}_Y(t) = t^b\mathrm{G}_X(t^a)$$

8 Quality of tests

Objectives

After completing this chapter you should be able to:

- Know about Type I and Type II errors → **pages 147–153**
- Find Type I and Type II errors using the normal distribution → **pages 153–157**
- Calculate the size and power of a test → **pages 157–162**
- Draw a graph of the power function for a test → **pages 162–167**

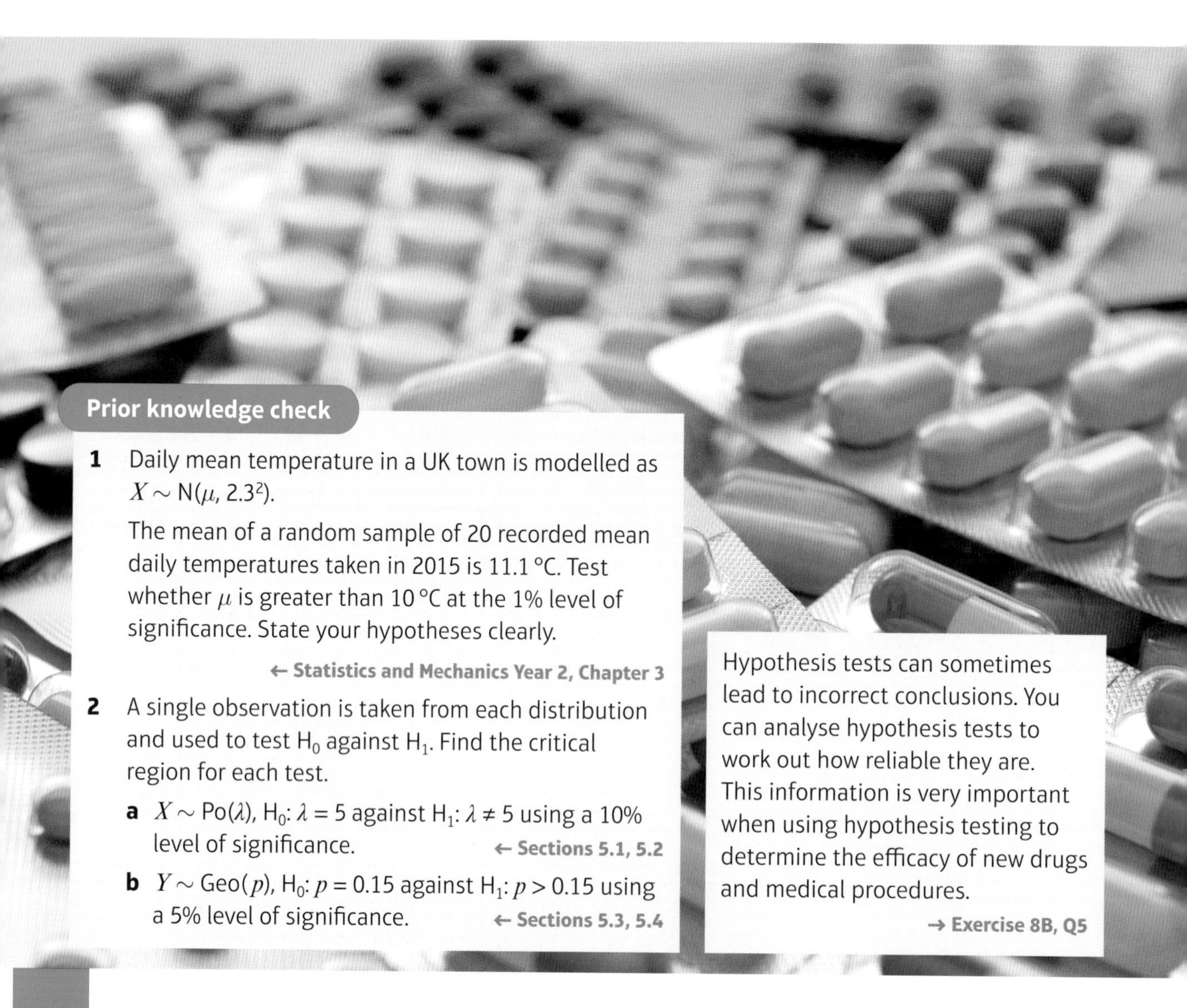

Prior knowledge check

1 Daily mean temperature in a UK town is modelled as $X \sim N(\mu, 2.3^2)$.

The mean of a random sample of 20 recorded mean daily temperatures taken in 2015 is 11.1 °C. Test whether μ is greater than 10 °C at the 1% level of significance. State your hypotheses clearly.

← **Statistics and Mechanics Year 2, Chapter 3**

2 A single observation is taken from each distribution and used to test H_0 against H_1. Find the critical region for each test.

a $X \sim Po(\lambda)$, H_0: $\lambda = 5$ against H_1: $\lambda \neq 5$ using a 10% level of significance. ← **Sections 5.1, 5.2**

b $Y \sim Geo(p)$, H_0: $p = 0.15$ against H_1: $p > 0.15$ using a 5% level of significance. ← **Sections 5.3, 5.4**

Hypothesis tests can sometimes lead to incorrect conclusions. You can analyse hypothesis tests to work out how reliable they are. This information is very important when using hypothesis testing to determine the efficacy of new drugs and medical procedures.

→ **Exercise 8B, Q5**

8.1 Type I and Type II errors

When you carry out a hypothesis test, you make an assumption about the distribution of a test statistic. You then compare the probability of the observed result occurring with the significance level of the test, and decide whether to accept or reject this assumption. This example illustrates a hypothesis test based on the parameter, p, of a binomial distribution.

Example 1

One rainy day during the summer holidays, a family of four were playing a simple game of cards. The game was one of chance so the probability of any particular person winning should have been $\frac{1}{4}$. After playing a number of games, Robert complained that his younger sister Sarah must have been cheating as she kept winning. Their parents quickly intervened and decided to carry out a proper investigation and carefully watched the next 20 games.

Find the critical region for a one-tailed test using a 5% level of significance.

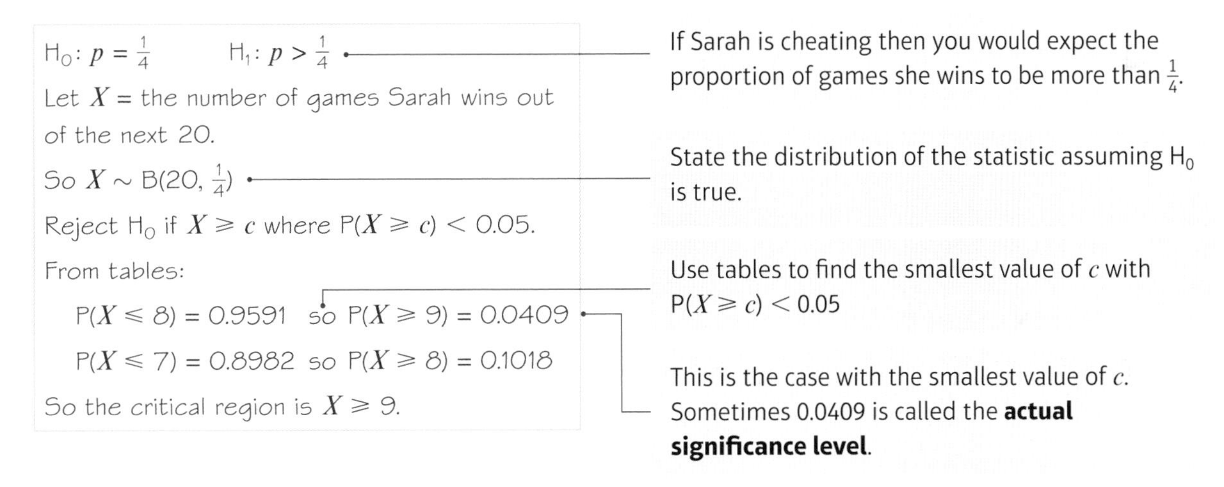

$H_0: p = \frac{1}{4}$ $\quad H_1: p > \frac{1}{4}$

Let X = the number of games Sarah wins out of the next 20.

So $X \sim B(20, \frac{1}{4})$

Reject H_0 if $X \geqslant c$ where $P(X \geqslant c) < 0.05$.

From tables:

$P(X \leqslant 8) = 0.9591$ so $P(X \geqslant 9) = 0.0409$

$P(X \leqslant 7) = 0.8982$ so $P(X \geqslant 8) = 0.1018$

So the critical region is $X \geqslant 9$.

If Sarah is cheating then you would expect the proportion of games she wins to be more than $\frac{1}{4}$.

State the distribution of the statistic assuming H_0 is true.

Use tables to find the smallest value of c with $P(X \geqslant c) < 0.05$

This is the case with the smallest value of c. Sometimes 0.0409 is called the **actual significance level**.

In the example above, if Sarah wins 9 or more games, her parents will reject the null hypothesis, and conclude that $p > \frac{1}{4}$ (or in other words, that Sarah was cheating). It is possible that this conclusion will be incorrect. If $p = \frac{1}{4}$, Sarah might still win 9 or more games by chance. The probability of this occurring is 0.0409, or the **actual significance level** of the test. This is called a **Type I error**.

- **A Type I error is when you reject H_0, but H_0 is in fact true. The probability of a Type I error is the same as the actual significance level of the hypothesis test.**

It is also possible that Sarah *was* cheating, but that she still only wins 8 or fewer games. In this case her parents would accept the null hypothesis, and conclude incorrectly that $p = \frac{1}{4}$. This is called a **Type II error**.

- **A Type II error is when you accept H_0, but H_0 is in fact false.**

Watch out In order to calculate the probability of a Type II error you would need to know the actual value of the parameter p. Because H_0 is false, you usually don't have this information.

A This table summarises the types of error that can occur in a hypothesis test:

		Truth	
		H_0 *is* true	H_0 *is* false
Conclusion of test	Accept H_0	OK	Type II error
	Reject H_0	Type I error	OK

Example 

Use the situation in Example 1.

a Find the probability of a Type I error.

b If in fact Sarah was cheating and $p = 0.35$, find the probability of a Type II error.

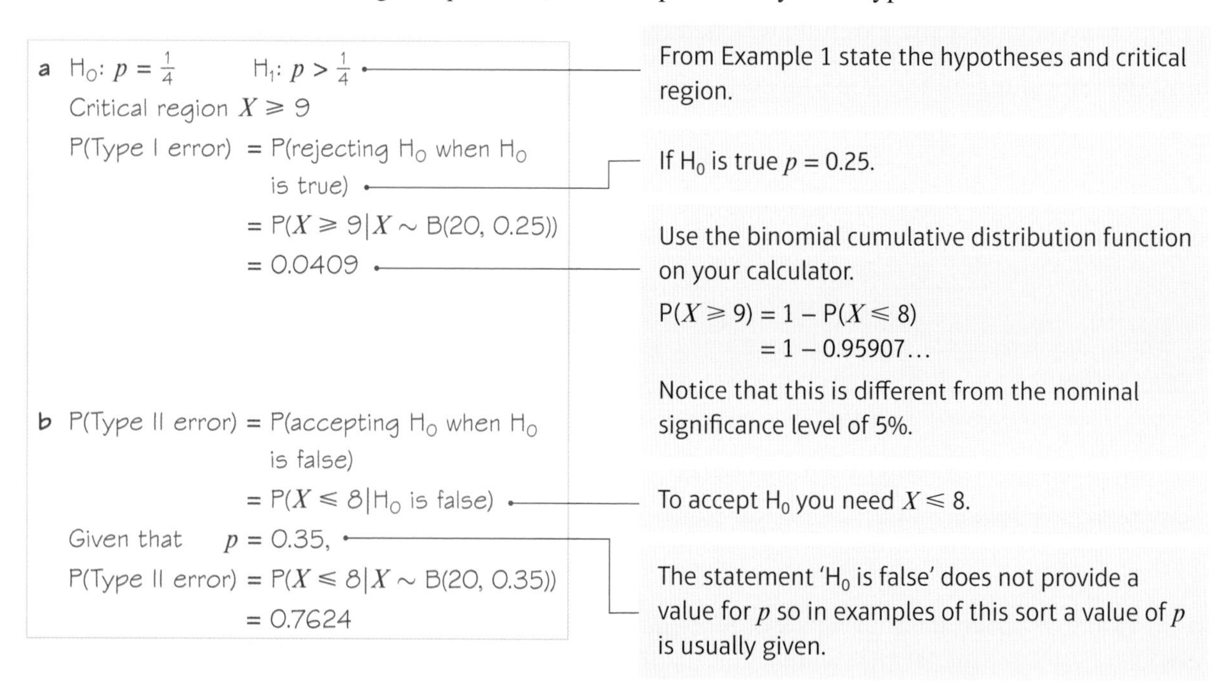

Example 3

Accidents occurred on a stretch of motorway at an average rate of 6 per month. Many of the accidents that occurred involved vehicles skidding into the back of other vehicles. By way of a trial, a new type of road surface that is said to reduce the risk of vehicles skidding is laid on this stretch of road, and during the first month of operation 4 accidents occurred.

a Test this result to see if it gives evidence that there has been an improvement at the 5% level of significance.

b Calculate P(Type I error) for this test.

c If the true average rate of accidents occurring with the new type of road surface was 3.5, calculate the probability of a Type II error.

a You are dealing with a Poisson distribution.

Let λ = the average number of accidents in a month, and X = the number of accidents in any given month, then the hypotheses are

H_0: $\lambda = 6$ (i.e. no change)

H_1: $\lambda < 6$ (i.e. fewer accidents)

Part **a** is a hypothesis test for the mean of a Poisson distribution. ← **Section 5.1**

From tables $P(X \leqslant 4 | \lambda = 6) = 0.2851$.

This is more than 5% so you do not have enough evidence to reject H_0.

The average number of accidents per month has not decreased.

Since it is a one-tailed test the conclusion should be clearly one-tailed.

b In order to reject H_0 you require a value c such that

$P(X \leqslant c | \lambda = 6) < 0.05$

You could have specified as close as possible to 5%.

From the table on page 191, with $\lambda = 6$:

$P(X \leqslant 2) = 0.0620$

and $P(X \leqslant 1) = 0.0174$.

So the critical value c is 1, and the critical region for this test is $X \leqslant 1$.

A Type I error occurs when you reject H_0 when it is true, and the probability of this happening is $P(X \leqslant 1) = 0.0174$.

This is again smaller than the 5% you were aiming for.

c A Type II error occurs when you do not have sufficient evidence to reject H_0 when H_1 is true.

If $\lambda = 3.5$ then H_0 is not true. You do not have sufficient evidence to reject H_0 if $X \geqslant 2$ so

$$\begin{aligned}P(\text{Type II error} | \lambda = 3.5) &= P(X \geqslant 2 | \lambda = 3.5)\\ &= 1 - P(X \leqslant 1 | \lambda = 3.5)\\ &= 1 - 0.1359\\ &= 0.8641\end{aligned}$$

You can also calculate the probabilities of errors from a two-tailed hypothesis test.

Example 4

A coin is spun 20 times and a head is obtained on 7 occasions.

a Test to see whether or not the coin is biased.

b Calculate the probability of a Type I error for this test.

c Given that the coin is biased and that this bias causes the tail to appear 3 times for each head that appears, calculate the probability of a Type II error for the test.

A

a The hypotheses are

$H_0 : p = 0.5 \qquad H_1 : p \neq 0.5$

This is a test for the proportion of a binomial distribution, and since you are testing to see if the coin is biased in either direction, a two-tailed test has to be used.

Let X = the number of heads in 20 spins of the coin.

Assuming H_0 is true then $X \sim B(20, 0.5)$.

For a two-tailed test, at the 5% significance level, you require values c_1 and c_2 so that $P(X \leqslant c_1) \leqslant 0.025$ and $P(X \geqslant c_2) \leqslant 0.025$ (or $P(X \leqslant c_2 - 1) \geqslant 0.975$).

The critical region will be in two parts.

From tables: $P(X \leqslant 6) = 0.0577$

and $P(X \leqslant 5) = 0.0207$

so the value of $c_1 = 5$.

Also: $P(X \geqslant 14) = 1 - P(X \leqslant 13)$
$= 1 - 0.9423$
$= 0.0577$

Alternatively $P(X \leqslant 13) = 0.9423$ and $P(X \leqslant 14) = 0.9793$ so $c_2 - 1 = 14$ and $c_2 = 15$.

$P(X \geqslant 15) = 1 - P(X \leqslant 14)$
$= 1 - 0.9793$
$= 0.0207$

so the value of $c_2 = 15$.

Thus the critical region for X is $X \leqslant 5$ or $X \geqslant 15$.

As 7 falls between 5 and 15 there is insufficient evidence to reject H_0.
The coin is not biased.

Problem-solving

Notice that since $p = 0.5$ the two tails are symmetrical about the mean of 10 and the value of c_2 could have been inferred from that of c_1 in this case.

b A Type I error occurs when you reject H_0 but H_0 is true, and this occurs when $X \leqslant 5$ or $X \geqslant 15$.

In this case there are two probabilities to be found and added.

$$\text{P(Type I error)} = P(X \leqslant 5 | p = 0.5) + P(X \geqslant 15 | p = 0.5)$$
$$= 0.0207 + 0.0207$$
$$= 0.0414$$

c A Type II error occurs when you do not have sufficient evidence to reject H_0 when H_1 is true. You do not have evidence to reject H_0 if $X \geqslant 6$ and $X \leqslant 14$ i.e $6 \leqslant X \leqslant 14$.

$$\text{P(Type II error)} = P(6 \leqslant X \leqslant 14 | p = 0.25)$$
$$= P(X \leqslant 14 | p = 0.25) - P(X \leqslant 5 | p = 0.25)$$
$$= 1.000 - 0.6172$$
$$= 0.3828$$

Remember that
X = the number of heads and
p = the probability of getting a head.
In this case $p = 0.25$.

Example 5

A

Jane knows from experience that 10% of the emails she receives are spam. After her email service upgraded the spam filters, she recorded the number of emails sent up to and including the first spam email. She wants to test, at the 5% significance level, whether this upgrade improved the spam filter.

a Find the critical region for her test.

b Calculate the probability of a Type I error for this test.

c Given that after the upgrade the probability of an email she receives being spam is now 1 in a 100, calculate the probability of a Type II error for the test.

a Let X = number of emails sent up to and including first spam email

$X \sim \text{Geo}(p)$

$H_0: p = 0.1 \quad H_1: p < 0.1$

This is a hypothesis test for the parameter of a geometric distribution. Start by defining your test statistic and stating your null and alternative hypotheses. ← **Section 5.4**

Assume H_0, so that $X \sim \text{Geo}(0.1)$.

For a one-tailed test you need to find a value c so that $P(X \geqslant c) \leqslant 0.05$.

Since $P(X \geqslant c) = (1 - 0.1)^{c-1} = 0.9^{c-1}$, you need an integer c such that

For a geometric distribution $X \sim \text{Geo}(p)$
$P(X \geqslant n) = (1 - p)^{n-1}$ ← **Section 3.1**

$0.9^{c-1} \leqslant 0.05$

$\log 0.9^{c-1} \leqslant \log 0.05$

Take logs of both sides.

$(c - 1)\log 0.9 \leqslant \log 0.05$

$c \log 0.9 \leqslant \log 0.05 + \log 0.9$

$c \geqslant \dfrac{\log 0.05 + \log 0.9}{\log 0.9}$

$c \geqslant 29.4$ (3 s.f.)

Watch out log 0.9 is negative, so change the direction of the inequality when you divide.

So the critical value is $c = 30$, and the critical region is $X \geqslant 30$.

Choose the next integer value above 29.4.

b A Type I error occurs when you reject H_0 but H_0 is true, and this happens when $X \geqslant 30$.

$P(\text{Type I error}) = P(X \geqslant 30 \mid p = 0.1)$
$= (1 - 0.1)^{30-1}$
$= 0.9^{29}$
$= 0.0471$ (4 d.p.)

c A Type II error occurs when you don't have enough evidence to reject H_0 and H_1 is true. You do not have enough evidence to reject H_0 when $X \leqslant 29$.

$P(\text{Type II error}) = P(X \leqslant 29 \mid p = 0.01)$
$= 1 - (1 - 0.01)^{29}$
$= 1 - 0.99^{29}$
$= 0.2528$ (4 d.p.)

$P(X \leqslant n) = 1 - (1 - p)^n$ ← **Section 3.1**

Exercise 8A

A

1 The random variable X is binomially distributed. A sample of 10 is taken, and it is desired to test $H_0: p = 0.25$ against $H_1: p > 0.25$, using a 5% level of significance.

a Calculate the critical region for this test.

b State the probability of a Type I error for this test and, given that the true value of p was later found to be 0.30, calculate the probability of a Type II error.

2 The random variable X is binomially distributed. A sample of 20 is taken, and it is desired to test $H_0: p = 0.30$ against $H_1: p < 0.30$, using a 1% level of significance.

a Calculate the critical region for this test.

b State the probability of a Type I error for this test and, given that the true probability was later found to be 0.25, calculate the probability of a Type II error.

3 The random variable X is binomially distributed. A sample of 10 is taken, and it is desired to test $H_0: p = 0.45$ against $H_1: p \neq 0.45$, using a 5% level of significance.

a Calculate the critical region for this test.

b State the probability of a Type I error for this test and, given that the true probability was later found to be 0.40, calculate the probability of a Type II error.

4 The random variable X has a Poisson distribution. A sample is taken, and it is desired to test $H_0: \lambda = 6$ against $H_1: \lambda > 6$, using a 5% level of significance.

a Find the critical region for this test.

b Calculate the probability of a Type I error and, given that the true value of λ was later found to be 7, calculate the probability of a Type II error.

5 The random variable X has a Poisson distribution. A sample is taken, and it is desired to test $H_0: \lambda = 4.5$ against $H_1: \lambda < 4.5$, using a 5% level of significance.

a Find the critical region for this test.

b Calculate the probability of a Type I error and, given that the true value of λ was later found to be 3.5, calculate the probability of a Type II error.

6 The random variable X has a Poisson distribution. A sample is taken, and it is desired to test $H_0: \lambda = 9$ against $H_1: \lambda \neq 9$, using a 5% level of significance.

a Find the critical region for this test.

b Calculate the probability of a Type I error and, given that the true value of λ was later found to be 8, calculate the probability of a Type II error.

7 The random variable X is geometrically distributed, and it is desired to test $H_0: p = 0.2$ against $H_1: p < 0.2$, using a 5% level of significance.

a Calculate the critical region for this test.

b State the probability of a Type I error for this test and, given that the true probability was found to be $p = 0.05$, calculate the probability of a Type II error.

A **8** The random variable X is geometrically distributed, and it is desired to test $H_0: p = 0.02$ against $H_1: p < 0.02$, using a 1% level of significance.

a Calculate the critical region for this test.

b State the probability of a Type I error for this test and, given that the true probability was found to be $p = 0.01$, calculate the probability of a Type II error.

9 The random variable X is geometrically distributed, and it is desired to test $H_0: p = 0.01$ against $H_1: p \neq 0.01$, using a 5% level of significance.

a Calculate the critical region for this test.

b State the probability of a Type I error for this test and, given that the true probability was found to be $p = 0.1$, calculate the probability of a Type II error.

E/P **10** **a** Define:

i a Type I error **(1 mark)**

ii a Type II error. **(1 mark)**

The discrete random variable $X \sim \text{Geo}(p)$. You wish to test $H_0: p = 0.004$ against $H_1: p \neq 0.004$, using a 10% significance level. The probability in each tail should be as close to 0.05 as possible.

b Find the critical region for this test. **(7 marks)**

c State the probability of a Type I error occurring for this test. **(1 mark)**

E/P **11** Michael has bought a dice with 20 sides, and his friend David suspects that it is landing on 17 more often than it is landing on the other values. They both decide to test this in two different ways, using a 5% significance level. Michael throws the dice 40 times and records the number of times the dice lands on the 17.

a Find the critical region for Michael's test. **(4 marks)**

b State the probability of a Type I error occurring for Michael's test. **(1 mark)**

David decides to throw the dice until the first time it lands on 17.

c Find the critical region for David's test. **(4 marks)**

d State the probability of a Type I error occurring. **(1 mark)**

The actual probability of the dice landing on 17 is 0.0588.

e Calculate the probability of a Type II error occurring in David's test. **(2 marks)**

f Calculate the probability of a Type II error occurring in Michael's test. **(2 marks)**

8.2 Finding Type I and Type II errors using the normal distribution

You need to be able to find Type I and Type II errors using the normal distribution.

Links If you are carrying out a hypothesis test for the mean of a normal distribution, you will be given the value for the population standard deviation, σ or variance, σ^2. The **sample variance** for a sample of size n will be $\frac{\sigma^2}{n}$.

← **Statistics and Mechanics Year 2, Section 3.7**

In the examples in the previous section P(Type I error), which gives the actual significance level, was not equal to the target significance level. This was due to the discrete nature of the distributions used.

- **When a continuous distribution such as the normal distribution is used then P(Type I error) is equal to the significance level of the test.**

Example 6

Bags of sugar having a nominal weight of 1 kg are filled by a machine. From past experience it is known that the weight, X kg, of sugar in the bags is normally distributed with a standard deviation of 0.04 kg. At the beginning of each week a random sample of 10 bags is taken in order to see if the machine needs to be reset. A test is then done at the 5% significance level with $H_0: \mu = 1.00$ kg and $H_1: \mu \neq 1.00$ kg. Find:

a the critical region for this test

b P(Type I error) for this test.

Assuming that the mean weight has in fact changed to 1.02 kg,

c find P(Type II error) for this test.

Online Explore probabilities of Type I and Type II errors in a normal distribution using GeoGebra.

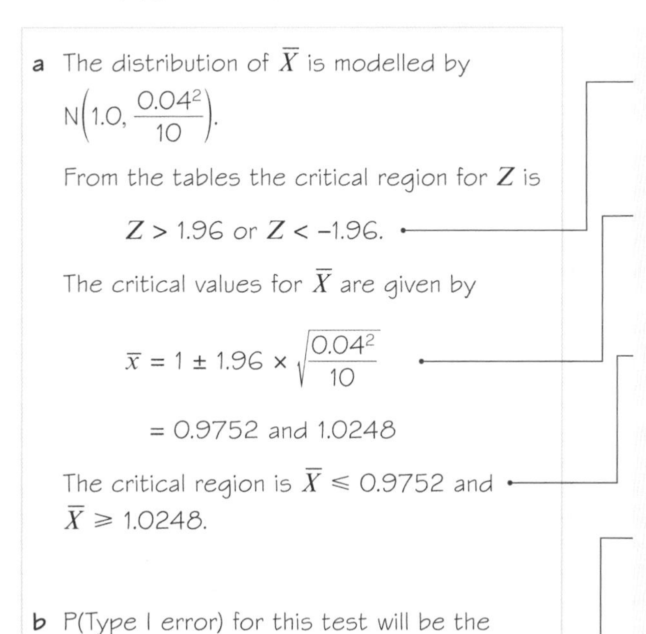

a The distribution of $\overline{X}$ is modelled by $N\left(1.0, \frac{0.04^2}{10}\right)$.

From the tables the critical region for Z is

$Z > 1.96$ or $Z < -1.96$.

The critical values for $\overline{X}$ are given by

$\bar{x} = 1 \pm 1.96 \times \sqrt{\frac{0.04^2}{10}}$

$= 0.9752$ and 1.0248

The critical region is $\overline{X} \leqslant 0.9752$ and $\overline{X} \geqslant 1.0248$.

Since this is a two-tailed test you allow 2.5% at each tail.

The critical region is found by rearranging $\left|\frac{\bar{x} - \mu}{\frac{\sigma}{\sqrt{n}}}\right| > 1.96$ for $\mu = 1.0$, $\sigma = 0.04$ and $n = 10$.

Notice once again that the critical region is in two parts.

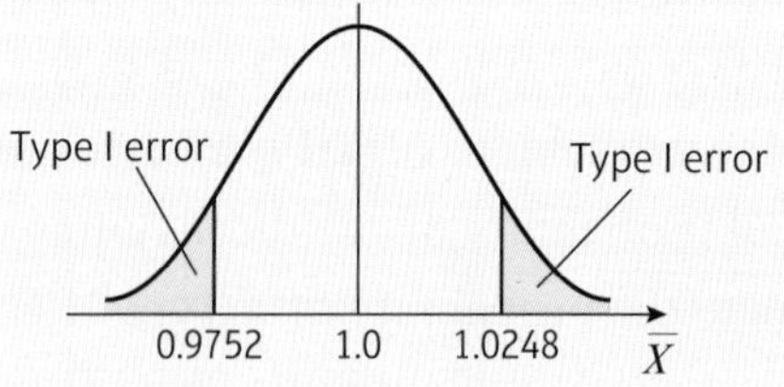

b P(Type I error) for this test will be the same as the significance level = 0.05.

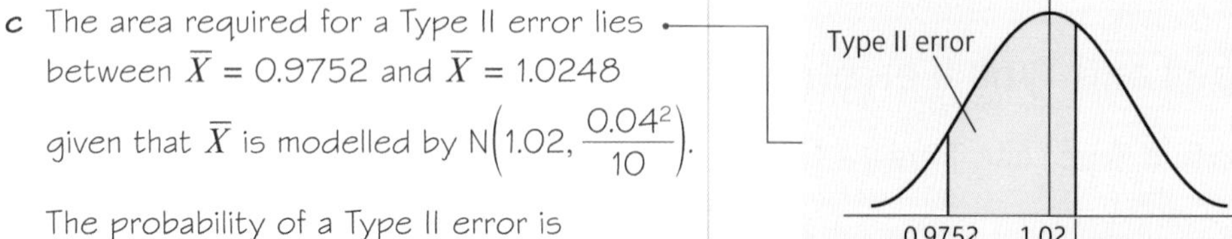

c The area required for a Type II error lies between $\overline{X} = 0.9752$ and $\overline{X} = 1.0248$ given that $\overline{X}$ is modelled by $N\left(1.02, \frac{0.04^2}{10}\right)$.

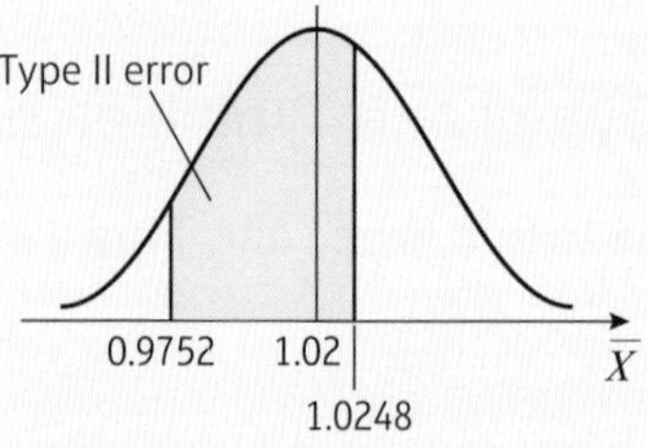

The probability of a Type II error is given by

$P(0.9752 < \overline{X} < 1.0248) = 0.6476$

Use the normal cumulative distribution function on your calculator, with $\sigma = \sqrt{\frac{0.04^2}{10}} = 0.01249\ldots$

A When carrying out hypothesis tests, you want to keep P(Type I error) and P(Type II error) as low as possible. The following example illustrates the relationship between Type I and Type II errors.

Example 7

The weight of jam in a jar, measured in grams, is distributed normally with a mean of 150 g and a standard deviation of 6 g. The production process occasionally leads to a change in the mean weight of jam per jar but the standard deviation remains unaltered.

The manager monitors the production process and for every new batch takes a random sample of 25 jars and weighs their contents to see if there has been any reduction in the mean weight of jam per jar.

Find the critical values for the test statistic $\overline{X}$, the mean weight of jam in a sample of 25 jars, using:

a a 5% level of significance

b a 1% level of significance.

Given that the true value of μ for the new batch is in fact 147,

c find the probability of a Type II error for each of the above critical regions.

a $H_0: \mu = 150$

$H_1: \mu < 150$ (i.e. a one-tailed test)

$\overline{X} \sim N\left(150, \frac{6^2}{25}\right)$, $n = 25$ and $\sigma = 6$

State the hypotheses to define the test. You are looking for a 'reduction' in the mean so a one-tailed test is needed.

The 5% critical region for Z is $Z < -1.6449$ so reject H_0 if

The critical value for Z is found from tables.

$$\frac{\overline{X} - 150}{\frac{6}{\sqrt{25}}} \leqslant -1.6449$$

Note that P(Type I error) in each case is the same as the significance level for the test.

That is, the critical region for $\overline{X}$ is

$$\overline{X} \leqslant \frac{6}{\sqrt{25}} \times (-1.6449) + 150$$

so $\overline{X} \leqslant 148.02612$.

b The 1% critical region for Z is $Z < -2.3263$ so reject H_0 if

$$\frac{\overline{X} - 150}{\frac{6}{\sqrt{25}}} \leqslant -2.3263$$

That is, the critical region for $\overline{X}$ is

$$\overline{X} \leqslant \frac{6}{\sqrt{25}} \times (-2.3263) + 150$$

so $\overline{X} \leqslant 147.20844$.

c 5% test P(Type II error)

$= P(\overline{X} > 148.026... \mid \mu = 147)$

$= 0.1963$ (4 d.p.)

Use your calculator with $\sigma = \sqrt{\frac{6^2}{25}} = 1.2$.

1% test P(Type II error)

$= P(\overline{X} > 147.2084 \mid \mu = 147)$

$= 0.4311$ (4 d.p.)

A Notice how in this example if we try to *reduce* P(Type I error) from 5% to 1% then P(Type II error) *increases* from 0.1963 to 0.4311. A more detailed study of the interplay between these two probabilities follows later in this chapter. However, you should be aware of this phenomenon and appreciate one of the reasons why we do not always use a significance level that is very small. The value of 5% is a commonly used level and, in a situation where a particular significance level is not given, this value is recommended.

This does not mean that other significance levels are never used. When, for example, the results of the research are highly important and making a Type I error could be very serious, a 1% significance level might be used. In other cases a significance level of 10% might be used. An alternative method of reducing the probability of a Type II error is to increase the sample size but this can increase the cost or duration of a survey or experiment.

The relationship between the probabilities of Type I and Type II errors can be illustrated by imagining pushing down on one side of a balloon.

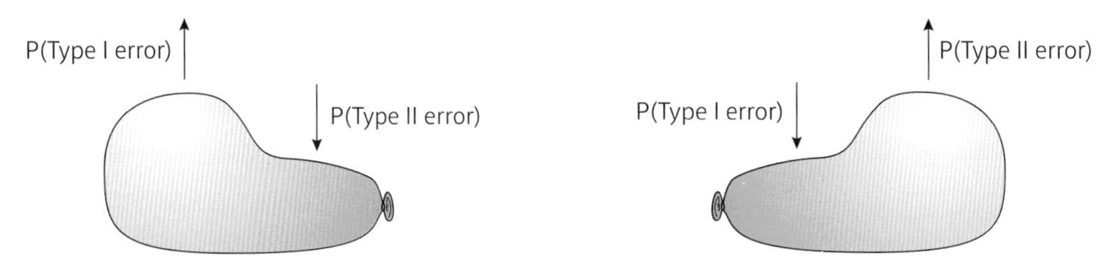

The only way to push down on both sides at once (and reduce the overall thickness) is to allow the air to move sideways. Using a larger balloon would allow you to reduce the overall thickness (this is equivalent to increasing the size of the sample n).

Exercise 8B

1 The random variable $X \sim N(\mu, 3^2)$. A random sample of 20 observations of X is taken, and the sample mean $\bar{x}$ is taken to be the test statistic. It is desired to test $H_0: \mu = 50$ against $H_1: \mu > 50$, using a 1% level of significance.

a Find the critical region for this test.

b State the probability of a Type I error for this test.

Given that the true mean was later found to be 53,

c find the probability of a Type II error.

2 The random variable $X \sim N(\mu, 2^2)$. A random sample of 16 observations of X is taken, and the sample mean $\bar{x}$ is taken to be the test statistic. It is desired to test $H_0: \mu = 30$ against $H_1: \mu < 30$, using a 5% level of significance.

a Find the critical region for this test.

b State the probability of a Type I error for this test.

Given that the true mean was later found to be 28.5,

c find the probability of a Type II error.

A **3** The random variable $X \sim N(\mu, 4^2)$. A random sample of 25 observations of X is taken, and the sample mean $\bar{x}$ is taken to be the test statistic. It is desired to test H_0: $\mu = 40$ against $H_1 : \mu \neq 40$, using a 1% level of significance.

a Find the critical region for this test.

b State the probability of a Type I error.

Given that the true mean was later found to be 42,

c find the probability of a Type II error.

E **4** A manufacturer claims that the average outside diameter of a particular washer produced by his factory is 15 mm. The diameter is assumed to be normally distributed with a standard deviation of 1 mm. The manufacturer decides to take a random sample of 25 washers from each day's production in order to monitor any changes in the mean diameter.

a Using a significance level of 5%, find the critical region to be used for this test. **(4 marks)**

Given that the average diameter had in fact increased to 15.6 mm,

b find the probability that the day's production would be wrongly accepted. **(2 marks)**

E/P **5** The number of patients that a medic can inoculate with a vaccine in one day can be modelled by a normal distribution with mean 40 and standard deviation 8. The manufacturer of the vaccine claims that a new method of inoculation will speed up the rate at which the medic works.

A random sample of 30 medics tried out the new method of inoculation and the average number of patients they dealt with per day $\bar{X}$ was recorded.

a Using a 5% significance level, find the critical value of $\bar{X}$. **(4 marks)**

The average number of patients dealt with per day using the new method of inoculation was in fact 42.

b Find the probability of making a Type II error. **(2 marks)**

The manufacturer of the vaccine would like to lessen the probability of a Type II error being made and recommends that the significance level be changed.

c State, giving a reason, what recommendation you would make. **(1 mark)**

8.3 Calculate the size and power of a test

You need to be able to calculate the size and power of a test.

You have already seen that a Type I error occurs when the null hypothesis is rejected when it is in fact true. The probability of a Type I error will be written as α and is often known as **the size of the test**.

- **The size of a test is the probability of rejecting the null hypothesis when it is in fact true and this is equal to the probability of a Type I error.**

The size of a test, as you have seen, is the actual significance level of the test and this is usually chosen before the test is carried out.

When conducting a hypothesis test you should also be interested in the probability of rejecting the null hypothesis when it is in fact untrue, as this is clearly a desirable feature of a test. The probability of rejecting the null hypothesis H_0 when it is untrue, is known as **the power of the test**.

A ■ **The power of a test is the probability of rejecting the null hypothesis when it is not true.**

Power = 1 − P(Type II error)
= P(being in the critical region when H_0 is false)

The greater the power of a test, the greater the probability of rejecting H_0 when H_0 is false. It follows that the higher the power, the better the test.

The table on page 148 can now be rewritten to show the probabilities for the different situations.

		Truth	
		H_0 *is* true	H_0 *is* false
Conclusion of test	Accept H_0	OK	P(Type II error)
	Reject H_0	Size = P(Type I error)	Power = 1 − (Type II error)

The size and power both relate to rejecting H_0.

The size relates to when H_0 is true and a Type I error has been made.

The power relates to when H_0 is false and a correct decision was made.

If the power is greater than 0.5, the probability of coming to the correct conclusion (rejecting H_0 when H_0 is false) is greater than the probability of coming to the wrong conclusion (accepting H_0 when H_0 is false).

On page 156 you were told that, generally, if you increase the sample size the probability of a Type II error decreases. It follows that the larger the sample size, the greater the power of the test. Increasing the sample size is preferable to increasing the significance level as a way of increasing the power of a test.

Example 8

The random variable X has a binomial distribution. A random sample of size 25 was taken to test H_0: $p = 0.30$ against H_1: $p < 0.30$ using a 10% level of significance.

a Find the critical region for this test.

b Find the size of this test.

Given that $p = 0.20$,

c calculate the power of this test.

a $X \sim B(25, p)$
H_0: $p = 0.30$ $\quad$ H_1: $p < 0.30$
Assume H_0 so that $X \sim (25, 0.30)$.
H_0 is rejected when $X \leqslant c$ where
$P(X \leqslant c) \leqslant 0.10$.
From tables:
$P(X \leqslant 4) = 0.0905$ — Use tables of B(25, 0.30).
$P(X \leqslant 5) = 0.1935$
So the critical region is $X \leqslant 4$.

b Size = P(Type I error)
$= P(X \leqslant 4 \mid p = 0.30)$
$= 0.0905$

The size is the actual significance level of the test. Use your calculator to find $P(X \leqslant 4 \mid p = 0.30)$.

c If $p = 0.20$ then H_0 is false.
Power = P(rejecting $H_0 \mid H_0$ is false, i.e. $p = 0.2$)
$= P(X \leqslant 4 \mid p = 0.20)$
$= 0.4207$

Watch out Calculate the power directly. There is no need to calculate P(Type II error) first. Remember to change the p-value in your calculator from 0.30 to 0.20.

Example 9

Jam is sold in jars. The amount of jam, in grams, in a jar is normally distributed with mean μ and standard deviation 5. The manufacturer claims that μ is 106 and quality control officers will take action against the manufacturer if $\mu < 106$. A random sample of 30 jars is examined and a 5% level of significance is used.

a Find the critical region for the sample mean using this test.

Given that in fact $\mu = 102$,

b find the power of this test.

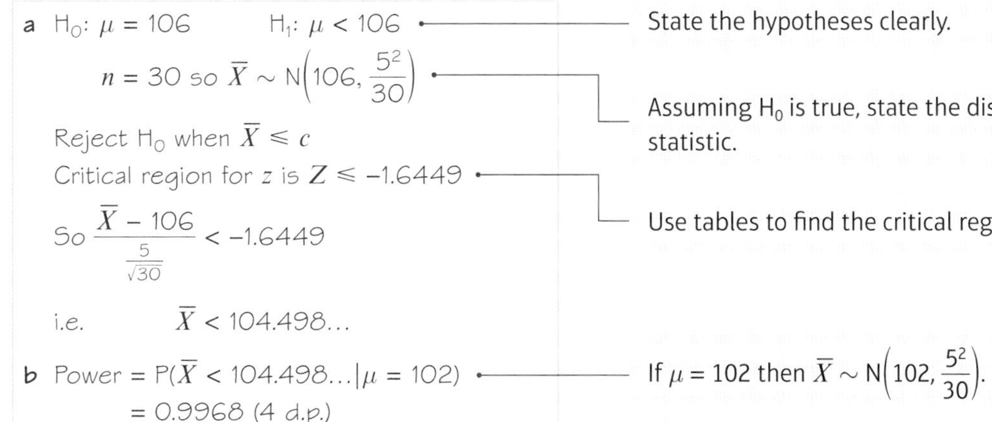

a $H_0: \mu = 106 \qquad H_1: \mu < 106$

State the hypotheses clearly.

$n = 30$ so $\bar{X} \sim N\left(106, \frac{5^2}{30}\right)$

Assuming H_0 is true, state the distribution of the statistic.

Reject H_0 when $\bar{X} \leqslant c$

Critical region for z is $Z \leqslant -1.6449$

Use tables to find the critical region for Z.

So $\frac{\bar{X} - 106}{\frac{5}{\sqrt{30}}} < -1.6449$

i.e. $\bar{X} < 104.498...$

b Power $= P(\bar{X} < 104.498... \mid \mu = 102)$
$= 0.9968$ (4 d.p.)

If $\mu = 102$ then $\bar{X} \sim N\left(102, \frac{5^2}{30}\right)$.

Example 10

A particular mobile-phone provider fails to deliver text messages with probability p.

Brooke wants to investigate whether $p > 0.02$.

Using $H_0: p = 0.02$ and $H_1: p > 0.02$, Brooke notes the number of text messages she is able to send successfully up until the first failure. If this value is less than or equal to 5 she rejects H_0. If it is more than 100 she accepts H_0. If it is more than 5 but less than or equal to 100 she notes the number of additional text messages she is able to send successfully up until the next failure. She rejects H_0 if this is less than or equal to 5 and accepts it otherwise.

a Find the size of this test.

b Calculate the power of this test when $p = 0.015$.

A

a Let X = number of messages sent up to and including first failure

Then $X \sim Geo(p)$

Assume H_0 is true, so that $X \sim Geo(0.02)$.

You need to calculate the probability that H_0 is rejected, assuming that it is true. You are given the critical region, so find $P(H_0 \text{ rejected} \mid p = 0.02)$.

$P(X \leqslant 5) = 1 - (1 - 0.02)^5$
$= 0.09607\ldots$

$P(5 < X \leqslant 100)$
$= P(X \leqslant 100) - P(X \leqslant 5)$
$= 1 - (1 - 0.02)^{100} - 0.09607\ldots$
$= 0.77130\ldots$

$P(H_0 \text{ rejected} \mid p = 0.02)$
$= P(X \leqslant 5) + P(5 < X \leqslant 100) \times P(X \leqslant 5)$
$= 0.09607\ldots + 0.77130\ldots \times 0.09607\ldots$
$= 0.17018\ldots$

The size of the test is 0.1702 (4 d.p.).

Problem-solving

You can use Geo(0.02) to model the number of text messages up to and including the first failure. After the first failure, the number of text messages up to and including the next failure also has distribution Geo(0.02).

b Assume $p = 0.015$ so that $Y \sim Geo(0.015)$.

The power of the test is $P(H_0 \text{ is rejected} \mid p = 0.015)$. Repeat your calculation using a different assumed value of p.

$P(Y \leqslant 5) = 1 - (1 - 0.015)^5 = 0.07278\ldots$

$P(5 < Y \leqslant 100)$
$= P(Y \leqslant 100) - P(Y \leqslant 5)$
$= 1 - (1 - 0.015)^{100} - 0.07278\ldots$
$= 0.70660\ldots$

$P(H_0 \text{ rejected} \mid p = 0.015)$
$= P(Y \leqslant 5) + P(5 < Y \leqslant 100) \times P(Y \leqslant 5)$
$= 0.12421\ldots$

The power of the test when $p = 0.015$ is 0.1242 (4 d.p.).

This is quite a small value for the power of the test. This suggests that the test is not very useful when $p = 0.15$.

Exercise 8C

1 The random variable $X \sim N(\mu, 3^2)$. A random sample of 25 observations of X is taken and the sample mean $\bar{x}$ is taken as the test statistic. It is desired to test $H_0: \mu = 20$ against $H_1: \mu > 20$ using a 5% significance level.

a Find the critical region for this test.

b Given that $\mu = 20.8$, find the power of this test.

2 The random variable X has a binomial distribution. A sample of 20 is taken from it. It is desired to test $H_0: p = 0.35$ against $H_1: p > 0.35$ using a 5% significance level.

a Calculate the size of this test.

b Given that $p = 0.36$, calculate the power of this test.

3 The random variable X has a Poisson distribution. A sample is taken and it is desired to test $H_0: \lambda = 4.5$ against $H_1: \lambda < 4.5$ using a 5% significance level.

a Find the size of this test.

b Given that $\lambda = 4.1$, find the power of this test.

4 A manufacturer claims that a particular rivet produced in his factory has a diameter of 2 mm, and that the diameter is normally distributed with a variance of 0.004 mm^2.
A random sample of 25 rivets is taken from a day's production to test whether the mean diameter had altered, up or down, from the stated figure. A 5% significance level is to be used for this test.

If the mean diameter had in fact altered to 2.02 mm, calculate the power of this test. **(5 marks)**

5 In a binomial experiment consisting of 10 trials the random variable X represents the number of successes, and p is the probability of a success.

In a test of $H_0: p = 0.3$ against $H_1: p > 0.3$, a critical region of $x \geqslant 7$ is used.

Find the power of this test when

a $p = 0.4$ **(3 marks)**

b $p = 0.8$. **(3 marks)**

c Comment on your results. **(1 mark)**

6 Explain briefly what you understand by

a a Type I error **(1 mark)**

b the size of a significance test. **(1 mark)**

A single observation is made on a random variable X, where $X \sim N(\mu, 10)$.
The observation, x, is to be used to test $H_0: \mu = 20$ against $H_1: \mu > 20$. The critical region is chosen to be $x \geqslant 25$.

c Find the size of the test. **(2 marks)**

7 The random variable X has a geometric distribution. It is desired to test $H_0: p = 0.01$ against $H_1: p > 0.01$ using a 5% significance level.

a Find the critical region for this test.

b Given that $p = 0.2$, calculate the power of this test.

8 The random variable X has a geometric distribution. It is desired to test $H_0: p = 0.01$ against $H_1: p \neq 0.01$ using a 5% significance level.

a Find the critical region for this test.

b Given that $p = 0.02$, calculate the power of this test.

9 The wallpaper produced by a certain manufacturer has defects that occur randomly at a constant rate of λ per roll. If λ is thought to be greater than 0.8 then action has to be taken.

Using $H_0: \lambda = 0.8$ and $H_1: \lambda > 0.8$, a quality control manager takes a sample of 10 rolls and rejects H_0 if there are 12 or more defects. If there are 9 or fewer defects then H_0 is accepted. If there are 10 or 11 defects, a second sample of 10 rolls is taken and H_0 is rejected if there are 8 or more defects in this second sample, otherwise it is accepted.

a Find the size of this test. **(4 marks)**

b Find the power of this test when $\lambda = 1$. **(3 marks)**

10 A sweet manufacturer makes boxes of jelly beans. The number of jelly beans in each box is assumed to be normally distributed with standard deviation 5.

A consumer group wants to test the manufacturer's claim that the mean number of jelly beans in each box is 80. The group takes repeated samples of 20 boxes and records the mean number of jelly beans per box in each sample.

The random variable X represents the number of samples the group need to take before they obtain a sample with a mean less than 79.

If $X \leqslant 10$ the group rejects the company's claim.

a Find the size of this test. **(5 marks)**

b Given that the actual mean number of jelly beans in each box is 81, find the power of this test. **(5 marks)**

Challenge

A jam factory has an automated system for sealing their jars, and the expected probability of error when the machines are well calibrated is 8%. The jars are sealed and placed into boxes of 60. To see whether the machine that sealed all the jars in a specific box needs recalibrating, a series of tests is performed. The first box is inspected by taking a sample of 20 jars and performing a test, with 5% significance level, to see whether the probability of a defective seal is greater than 8%. If the first box fails the test they conclude that the machine needs recalibrating, but if it passes the test they move on to the second box, and perform the same test. Once again, if the box fails the test they conclude that the machine needs recalibrating, but if it passes the test they move on to the next box, and so on until a box fails the test.

a What is the maximum number of boxes that can be inspected such that the probability of a Type I error is smaller than 10%?

The factory decides to conclude that the machine does not need recalibrating if the first four boxes all pass the test.

b Given that after the second box the machine is decalibrated, increasing the probability of a defective seal to 20%, find the power of the test, knowing that only 4 boxes were inspected. You may assume that the probability of a defective seal for each jar in the first two boxes is 8%.

8.4 The power function

So far you have calculated the probability of a Type II error or the power only when you have been given a particular value of the population parameter of interest. Population parameters are seldom known, and if they were known there would be little point in doing the test anyway. Sometimes past experience can give you some idea of likely values of the parameters but, in general, since you do not know the value of the parameter, you cannot decide the power of the test concerned. It is, however, possible in these cases to calculate the power as a function of the relevant parameter (which we shall generalise as θ). Such a function is known as a **power function**.

■ **The power function of a test is the function of the parameter θ which gives the probability that the test statistic will fall in the critical region of the test if θ is the true value of the parameter.**

A power function enables you to calculate the power of the test for any given value of θ, and thus to plot a graph of power against θ.

Example 11

Past experience has shown that the number of accidents that take place at a road junction has a Poisson distribution with an average of 3.5 accidents per month. A trading estate is built along one of the roads leading away from the junction and the local council is anxious that this may have increased the accident rate. To see if the number of accidents had increased, a test was set up with the null hypothesis H_0: $\lambda = 3.5$ and with the alternative hypothesis being accepted if the number of accidents X within the first month after the alteration was $\geqslant 7$.

a Find the size of the test.

b Find the power function for the test and sketch the graph of the power function.

a Size of test = P(reject H_0 when it is true)

$= P(X \geqslant 7 | X \sim Po(3.5))$

$= 1 - 0.9347 = 0.0653$

You can use conditional probability notation to write your assumptions quickly.

b Power function = P(reject H_0 when it is false)

$= P(X \geqslant 7 | X \sim Po(\lambda))$

$= 1 - P(X \leqslant 6 | X \sim Po(\lambda))$

$= 1 - e^{-\lambda}\left(1 + \lambda + \frac{\lambda^2}{2} + \frac{\lambda^3}{6} + \frac{\lambda^4}{24} + \frac{\lambda^5}{120} + \frac{\lambda^6}{720}\right)$

Problem-solving

You do not know the value of λ. Your power function will be given in terms of this unknown parameter.

Use the finite sum of the probabilities for $X = 1, 2, 3, 4, 5, 6$ to find the power function.

This enables values of the power of the test to be calculated for different values of λ.

$\lambda = 4$ gives power = 0.1107
$\lambda = 5$ gives power = 0.2378
$\lambda = 6$ gives power = 0.3937
$\lambda = 7$ gives power = 0.5503
$\lambda = 8$ gives power = 0.6866
$\lambda = 9$ gives power = 0.7932
$\lambda = 10$ gives power = 0.8699

Often in an examination a partially completed table will be given.

The graph is as shown below.

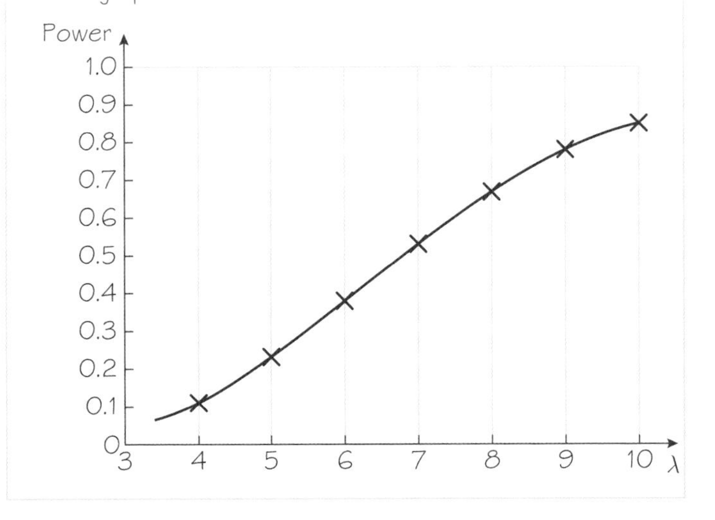

A Power functions are particularly useful when comparing two different tests.

- **When comparing two tests of comparable size, you should recommend the test with the higher power within the likely range of the parameter.**

Example 12

A manufacturer of sweets supplies a mixed assortment of chocolates in a jar. He claims that 40% of the chocolates have a 'hard centre', the remainder being 'soft centred'.

A shopkeeper does not believe the manufacturer's claim and proposes to test it using the following hypotheses.

$H_0: p = 0.4 \qquad H_1: p < 0.4$

where p is the proportion of 'hard centres' in the jar. Two tests are proposed.

In test A he takes a random sample of 10 chocolates from the jar and rejects H_0 if the number of 'hard centres' is less than 2.

a Find the size of test A.

b Show that the power function of test A is given by

$(1 - p)^{10} + 10p(1 - p)^9$.

In test B he takes a random sample of 5 chocolates from the jar and if there are no 'hard centres' he rejects H_0, otherwise he takes a second sample of 5 chocolates and H_0 is rejected if there are no further 'hard centres' on this second occasion.

c Find the size of test B.

d Find an expression for the power function of test B.

The powers for test A and test B for various values of p are given in the table.

p	0.1	0.2	0.25	0.3	0.35
Power for test A	0.74	r	0.24	s	0.09
Power for test B	0.83	0.54	0.42	0.31	0.22

e Calculate values for r and s.

f State, giving a reason, which of the two tests the shopkeeper should use.

a Size of test $A = P(X < 2 | X \sim B(10, 0.4))$
$= 0.0464$ (4 d.p.)

b Power of test $A = P(X < 2 | X \sim B(10, p))$
$= P(X = 0) + P(X = 1)$
$= (1 - p)^{10} + 10p(1 - p)^9$

Write out the probabilities in terms of p. This is already in the desired form, so you don't need to factorise, but you could also write this power function as $(1 - p)^9(1 + 9p)$.

c Size of test $B = P(\text{reject } H_0 | p = 0.4)$
$= P(X = 0) + (1 - P(X = 0)) \times P(X = 0)$
$= 0.6^5 + (1 - 0.6)^5 \times 0.6^5$
$= 0.0786$

d Power of test B = P(0 hard centres in first 5)
+ P(0 hard centres in second 5 and > 0 hard centres in first 5)

$$= P(X = 0 \mid p) + (1 - P(X = 0 \mid p)) \times P(X = 0 \mid p)$$
$$= (1 - p)^5 + (1 - (1 - p)^5)(1 - p)^5$$
$$= (1 - p)^5 (1 + 1 - (1 - p)^5)$$
$$= (1 - p)^5 (2 - (1 - p)^5)$$
$$= 2(1 - p)^5 - (1 - p)^{10}$$

Problem-solving

Consider the conditions that are necessary for H_0 to be rejected:

First sample — Second sample

- $X = 0$ and H_0 rejected
- $X > 0$
  - $X = 0$ and H_0 rejected
  - $X > 0$

e Test A: $p = 0.2$ Power $= (1 - 0.2)^{10} + 10(0.2)(1 - 0.2)^9$
$= 0.38$
so $r = 0.38$

$p = 0.3$ Power $= (1 - 0.3)^{10} + 10(0.3)(1 - 0.3)^9$
so $s = 0.15$

f Power for test B > Power for test A for all the given values of p, so he should use test B.

The reason for the final comment should be based upon the calculations of the power.

Example 13

A local park believes the fox population in the area has decreased. They want to test for the probability, p, that a fox will be observed on any given day. They count the number of days, X, that pass until the first observation of a fox. They test H_0: $p = 0.1$ against H_1: $p < 0.1$ and reject H_0 if $X > 30$.

a Find the size of this test.

b Find the power function for the test.

a Size of test $= P(X > 30 \mid X \sim Geo(0.1))$
$= (1 - 0.1)^{30}$
$= 0.9^{30} = 0.0424$ (4 d.p.)

If $X \sim$ Geo(p), then $P(X > x) = (1 - p)^x$. ← Section 3.1

b Power function $= P(X > 30 \mid X \sim Geo(p))$
$= (1 - p)^{30}$

Exercise 8D

1 A single observation x is taken from a Poisson distribution with parameter λ. This observation is to be used to test H_0: $\lambda = 6.5$ against H_1: $\lambda < 6.5$. The critical region chosen was $x \leqslant 2$.

a Find the size of the test. **(4 marks)**

b Show that the power function of this test is given by

$$e^{-\lambda}\left(1 + \lambda + \tfrac{1}{2}\lambda^2\right)$$ **(3 marks)**

The table gives the value of the power function to two decimal places.

λ	1	2	3	4	5	6
Power	0.92	s	0.42	0.24	t	0.06

c Calculate values for s and t. **(1 mark)**

d Draw a graph of the power function. **(1 mark)**

e Find the values of λ for which the test is more likely than not to come to the correct conclusion. **(1 mark)**

E/P **2** In a binomial experiment consisting of 12 trials, X represents the number of successes and p the probability of a success.

In a test of H_0: $p = 0.45$ against H_1: $p < 0.45$ the null hypothesis is rejected if the number of successes is 2 or less.

a Find the size of this test. **(4 marks)**

b Show that the power function for this test is given by

$$(1 - p)^{12} + 12p(1 - p)^{11} + 66p^2(1 - p)^{10}$$ **(3 marks)**

c Find the power of this test when p is 0.3. **(1 mark)**

3 In a binomial experiment consisting of 10 trials, the random variable X represents the number of successes and p the probability of a success.

In a test of H_0: $p = 0.4$ against H_1: $p > 0.4$, a critical region of $x \geqslant 8$ was used.

Find the power of this test when:

a $p = 0.5$

b $p = 0.8$.

c Comment on your results.

4 A certain gambler always calls heads when a coin is spun. Before he uses a coin he tests it to see whether or not it is fair and uses the following hypotheses:

$$H_0: p = \tfrac{1}{2} \qquad H_1: p < \tfrac{1}{2}$$

where p is the probability that the coin lands heads on a particular spin. Two tests are proposed.

In test A the coin is spun 10 times and H_0 is rejected if the number of heads is 2 or fewer.

a Find the size of test A. **(4 marks)**

b Explain why the power of test A is given by

$$(1 - p)^{10} + 10p(1 - p)^9 + 45p^2(1 - p)^8$$ **(3 marks)**

In test B the coin is first spun 5 times. If no heads result, H_0 is immediately rejected. Otherwise the coin is spun a further 5 times and H_0 is rejected if no heads appear on this second occasion.

c Find the size of test B. **(4 marks)**

d Find an expression for the power of test B in terms of p. **(3 marks)**

The power for test A and the power for test B are given in the table for various values of p.

p	0.1	0.2	0.25	0.3	0.35	0.4
Power for test A	0.9298	0.6778		0.3828		0.1673
Power for test B	0.8323	0.5480	0.4183	0.3079	0.2186	0.1495

e Find the power for test A when p is 0.25 and 0.35. **(2 marks)**

f Giving a reason, advise the gambler about which test he should use. **(1 mark)**

A **5** In an experiment the probability of success in each trial is constant, and the random variable X represents the number of trials needed to get one success. A test of H_0: $p = 0.15$ against H_1: $p < 0.15$ with a 1% significance level is used.

a Find the size of the test.

b Find the power function.

E/P **6** A cyclist uses new tyres every time he does a time trial. He has found that on one specific route he has a probability of 0.9 of not getting a flat tyre. After changing tyre brands he believes that the new tyres are more resistant, and decides to perform a test, with 5% significance level, by doing 10 trials on the route and seeing how many times he would complete it without a flat tyre.

a Find the size of the test. **(4 marks)**

b Find the power function of the test. **(2 marks)**

c Find the power function for the test if, instead of 10 trials, he had done 12 trials. **(5 marks)**

d Given that the probability of completing the trial without a flat tyre with the new brand is 0.95, calculate which number of trials gives a more accurate test result. **(3 marks)**

Mixed exercise 8

E **1** The random variable X is binomially distributed. A sample of 15 observations is taken and it is desired to test H_0: $p = 0.35$ against H_1: $p > 0.35$ using a 5% significance level.

a Find the critical region for this test. **(4 marks)**

b State the probability of making a Type I error for this test. **(2 marks)**

The true value of p was found later to be 0.5.

c Calculate the power of this test. **(2 marks)**

E **2** The random variable X has a Poisson distribution. A sample is taken and it is desired to test H_0: $\lambda = 3.5$ against H_1: $\lambda < 3.5$ using a 5% significance level.

a Find the critical region for this test. **(4 marks)**

b State the probability of committing a Type I error for this test. **(2 marks)**

Given that the true value of λ is 3.0,

c find the power of this test. **(2 marks)**

E **3** The random variable $X \sim N(\mu, 9)$. A random sample of 18 observations is taken, and it is desired to test H_0: $\mu = 8$ against H_1: $\mu \neq 8$, at the 5% significance level. The test statistic to be used is

$$Z = \frac{X - \mu}{\frac{\sigma}{\sqrt{n}}}$$

a Find the critical region for this test. **(4 marks)**

b State the probability of a Type I error for this test. **(2 marks)**

Given that μ was later found to be 7,

c find the probability of making a Type II error. **(2 marks)**

d State the power of this test. **(1 mark)**

E/P **4** A bird observatory wishes to test whether the migration rate of geese has changed from that of 10 per day. First they take note of how many geese are observed flying in a migratory pattern on a specific day. If the number of geese migrating is greater than or equal to 4 and less than or equal to 17, then they conclude that the rate has not changed. If they observe 3 or fewer geese, then on the following day they conduct further observations, and if they observe 2 or fewer geese they conclude that the rate has decreased, otherwise they conclude that it hasn't changed. If on the first day they observe 18 or more geese migrating, then on the following day they also conduct further observations and if they observe 19 or more geese migrating they conclude that the rate has increased, otherwise they conclude that it has not changed.

a Find the size of the test. **(4 marks)**

Given that the migration rate of the geese actually dropped to 5 per day,

b find the power of the test. **(6 marks)**

E **5** A single observation, x, is taken from a Poisson distribution with parameter λ. The observation is used to test H_0: $\lambda = 4.5$ against H_1: $\lambda > 4.5$. The critical region chosen for this test was $x \geqslant 8$.

a Find the size of this test. **(4 marks)**

b The table gives the power of the test for different values of λ.

λ	1	2	3	4	5	6	7	8	9	10
Power	0	0.0011	0.0119	r	0.1334	s	0.4013	0.5470	t	0.7798

i Find values for r, s and t. **(2 marks)**

ii Using graph paper, plot the power function against λ. **(2 marks)**

E **6** In a binomial experiment consisting of 15 trials, X represents the number of successes and p the probability of success.

In a test of H_0: $p = 0.45$ against H_1: $p < 0.45$ the critical region for the test was $X \leqslant 3$.

a Find the size of the test. **(4 marks)**

b Use the binomial cumulative distribution function to complete the table given below. **(3 marks)**

p	0.1	0.2	0.3	0.4	0.5
Power	0.944	s	0.2969	t	0.0176

c Draw the graph of the power function for this test. **(1 mark)**

E **7** A company buys rope from Bindings Ltd and it is known that the number of faults per 100 m of their rope follows a Poisson distribution with mean 2. The company is offered 100 m of rope by Tieup, a newly established rope manufacturer. The company is concerned that the rope from Tieup might be of poor quality.

a Write down the null and alternative hypotheses appropriate for testing that rope from Tieup is in fact as reliable as that from Bindings Ltd. **(1 mark)**

b Derive a critical region to test your null hypothesis with a size of approximately 0.05. **(4 marks)**

c Calculate the power of this test if rope from Tieup contains an average of 4 faults per 100 m. **(3 marks)**

8 The number of faulty garments produced per day by machinists in a clothing factory has a Poisson distribution with mean 2. A new machinist is trained and the number of faulty garments made in one day by the new machinist is counted.

a Write down the appropriate null and alternative hypotheses involved in testing the theory that the new machinist is less reliable than the other machinists. **(1 mark)**

b Derive a critical region, of size approximately 0.05, to test the null hypothesis. **(4 marks)**

c Calculate the power of this test if the new machinist produces an average of 3 faulty garments per day. **(3 marks)**

The number of faulty garments produced by the new machinist over three randomly selected days is counted.

d Derive a critical region, of approximately the same size as in part **b**, to test the null hypothesis. **(2 marks)**

e Calculate the power of this test if the machinist produces an average of 3 faulty garments per day. **(3 marks)**

f Comment briefly on the difference between the two tests. **(1 mark)**

9 A single observation, x, is to be taken from a Poisson distribution with parameter μ. This observation is to be used to test H_0: $\mu = 6$ against H_1: $\mu < 6$. The critical region is chosen to be $x \leqslant 2$.

a Find the size of the critical region. **(1 mark)**

b Show that the power function for this test is given by

$$\frac{1}{2}e^{-\mu}(2 + 2\mu + \mu^2)$$ **(4 marks)**

The table gives the values of the power function to 2 decimal places.

μ	1.0	1.5	2.0	4.0	5.0	6.0	7.0
Power	0.92	0.81	s	0.24	t	0.06	0.03

c Calculate the values of s and t. **(3 marks)**

d Draw a graph of the power function. **(2 marks)**

e Estimate the range of values of μ for which the power of this test is greater than 0.8. **(3 marks)**

E/P **10** A proportion p of the items produced by a laboratory are defective. A technician selects a random sample of 10 items from each batch produced to check whether or not there is evidence that $p > 0.10$. The criterion that the technician uses for rejecting the hypothesis that p is 0.10 is that there are more than 4 defective items in the sample.

a Find the size of the test. **(2 marks)**

The table gives some values, to 2 decimal places, of the power function of this test.

p	0.15	0.20	0.25	0.30	0.35	0.40
Power	0.01	0.03	u	0.15	0.25	0.37

b Find the value of u. **(2 marks)**

A supervisor checks the production by taking a random sample of 5 items from each batch produced. The hypothesis that $p = 0.10$ is rejected if more than 2 defective items are found in the sample.

A

c Find P(Type 1 error) using the supervisor's test. **(2 marks)**

The table gives some values, to 2 decimal places, of the power function for the test in part **c**.

p	0.15	0.20	0.25	0.30	0.35	0.40
Power	0.03	0.06	0.10	0.16	v	0.32

d Find the value of v. **(2 marks)**

e Using the same axes, on graph paper draw the graphs of the power functions of these two tests. **(4 marks)**

f **i** State the value of p where the graphs cross.

ii Explain the significance of p being greater than this value. **(2 marks)**

g Suggest two advantages of using the test with sample size 5. **(2 marks)**

E/P **11** Accidents on a stretch of motorway occur at an average rate of λ per week. A road safety officer takes a random sample of 10 weeks to test whether or not there is evidence that $\lambda > 0.3$. The criterion that the officer uses for rejecting the hypothesis that $\lambda = 0.3$ is that there are more than 5 accidents in the sample.

a Find the size of the test. **(2 marks)**

The table gives some values, to 2 decimal places, of the power function of this test.

λ	0.4	0.5	0.6	0.7	0.8	0.9	1.0
Power	0.21	a	0.55	0.70	0.81	0.88	0.93

b Find the value of a. **(2 marks)**

The road safety manager would like to design a test of whether or not $\lambda > 0.3$, using a larger sample. The manager chooses a random sample of 15 weeks and requires the probability of a Type I error to be less than 5%.

c Find the criterion to reject the hypothesis that $\lambda = 0.3$ which makes the test as powerful as possible. **(2 marks)**

d Hence state the size of this second test. **(1 mark)**

The table gives some values, to 2 decimal places, of the power function for the test in part **c**.

λ	0.4	0.5	0.6	0.7	0.8	0.9	1.0
Power	0.15	0.34	0.54	0.72	0.85	b	0.96

e Find the value of b. **(2 marks)**

f Using the same axes, on graph paper draw the graphs of the power functions of these two tests. **(4 marks)**

g **i** State the value of λ where the graphs cross.

ii Explain the significance of λ being greater than this value. **(2 marks)**

Challenge

A Jane and Emma decide to test a pair of dice from a new board game. They suspect that at least one of them has probability higher than $\frac{1}{6}$ of showing the value one.

Jane decides to throw both dice 12 times. If a pair of ones appears 2 or more times, she concludes that at least one of the dice is biased.

a Find the size of Jane's test.

b Express the power of Jane's test in terms of the parameter p, which represents the probability of obtaining a pair of ones.

Emma decides to throw one dice 6 times. If the value one appears 4 or more times she concludes that the dice is biased. If it appears fewer than 4 times, then she throws the other dice 6 times, and concludes that the second dice is biased if the value one appears 4 or more times.

c Find the size of Emma's test.

Now assume that one of the dice is fair, and let q be the probability of obtaining the value one on the other dice.

d Show that the power of Jane's test is given by the expression

$$1 - \left(1 - \frac{q}{6}\right)^{12} - 2q\left(1 - \frac{q}{6}\right)^{11}$$

e Show that the power of Emma's test is given by the expression

$$0.0087 + 14.8695q^4 - 23.7912q^5 + 9.913q^6$$

Below is a graph of the power function for Emma's test.

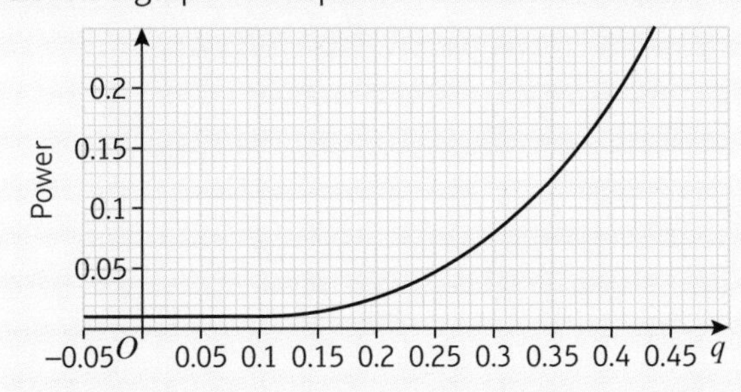

f By using a table of values, draw the graph of the power function for Jane's test.

g Given that the parameter q lies between 0.1 and 0.4, explain, giving your reasons, which test you would recommend.

Summary of key points

A

1 A **Type I error** is when you reject H_0, but H_0 is in fact true. The probability of a Type I error is the same as the actual significance level of the hypothesis test.

2 A **Type II** error is when you accept H_0, but H_0 is in fact false.

3 When a continuous distribution such as the normal distribution is used then P(Type I error) is equal to the significance level of the test.

4 The **size of a test** is the probability of rejecting the null hypothesis when it is in fact true and this is equal to the probability of a Type I error.

5 The **power of a test** is the probability of rejecting the null hypothesis when it is not true.
Power = 1 − P(Type II error) = P(being in the critical region when H_0 is false)

6 The **power function** of a test is the function of the parameter θ which gives the probability that the test statistic will fall in the critical region of the test if θ is the true value of the parameter.

7 When comparing two tests of comparable size, you should recommend the test with the higher power within the likely range of the parameter.

Review exercise

(E/P) **1** A quality control manager regularly samples 20 items from a production line and records the number of defective items x. The results of 100 such samples are given below.

x	0	1	2	3	4	5	6	7 or more
Frequency	17	31	19	14	9	7	3	0

a Estimate the proportion of defective items from the production line. **(1)**

The manager claims that the number of defective items in a sample of 20 can be modelled by a binomial distribution. He uses the answer in part **a** to calculate the expected frequencies given below.

x	0	1	2	3	4	5	6	7 or more
Expected frequency	12.2	27.0	r	19.0	s	3.2	0.9	0.2

b Find the value of r and the value of s giving your answers to 1 decimal place. **(2)**

c Stating your hypotheses clearly, use a 5% level of significance to test the manager's claim. **(6)**

d Explain what the analysis in part **c** tells the manager about the occurrence of defective items from this production line. **(2)**

← Section 6.4

(E/P) **2** Five coins were spun 100 times and the number of heads recorded. The results are shown in the table below.

Number of heads	0	1	2	3	4	5
Frequency	6	18	29	34	10	3

a Suggest a suitable distribution to model the number of heads when five unbiased coins are spun. **(1)**

b Test, at the 10% level of significance, whether or not the five coins are unbiased. State your hypotheses clearly. **(6)**

← Section 6.4

(E/P) **3** Ten cuttings were taken from each of 100 randomly selected garden plants. The number of cuttings that did not grow were recorded.

The results are as follows.

Number which did not grow	0	1	2	3	4	5	6	7	8, 9 or 10
Frequency	11	21	30	20	12	3	2	1	0

a Show that the probability of a randomly selected cutting, from this sample, not growing is 0.223. **(2)**

A gardener believes that a binomial distribution might provide a good model for the number of cuttings, out of 10, that do not grow.

He uses a binomial distribution, with the probability 0.2 of a cutting not growing. The calculated expected frequencies are as follows.

Number which did not grow	0	1	2	3	4	5 or more
Expected frequency	r	26.84	s	20.13	8.81	t

b Find the values of r, s and t. **(3)**

c State clearly the hypotheses required to test whether or not this binomial distribution is a suitable model for these data. **(1)**

The test statistic for the test is 4.17 and the number of degrees of freedom used is 4.

d Explain fully why there are 4 degrees of freedom. **(2)**

e Stating clearly the critical value used, carry out the test using a 5% level of significance. **(4)**

← Section 6.4

E/P **4** The number of times per day a computer fails and has to be restarted is recorded for 200 days. The results are summarised in the table.

Number of restarts	Frequency
0	99
1	65
2	22
3	12
4	2

Test whether or not a Poisson model is suitable to represent the number of restarts per day. Use a 5% level of significance and state your hypothesis clearly. **(6)**

← Section 6.4

E/P **5** The Director of Studies at a large college believes that students' grades in Mathematics are independent of their grades in English. She examined the results of a random group of candidates who had studied both subjects and she recorded the number of candidates in each of the 6 categories shown.

		Maths grade		
		A or B	C or D	E or U
English grade	A or B	25	25	10
	C to U	5	30	15

a Stating your hypotheses clearly, test the Director's belief using a 10% level of significance. You must show each step of your working. **(7)**

The Head of English suggested that the Director was losing accuracy by combining the English grades C to U in one row. He suggested that the Director should split the English grades into two rows, grades C or D and grades E or U as for Mathematics.

b State why this might lead to problems in performing the test. **(2)**

← Section 6.5

E **6** People over the age of 65 are offered an annual flu injection. A health official took a random sample from a list of patients who were over 65. She recorded their gender and whether or not the offer of an annual flu injection was accepted or rejected. The results are summarised below.

		Accepted	Rejected
Gender	Male	170	110
	Female	280	140

Using a 5% significance level, test whether or not there is an association between gender and acceptance or rejection of an annual flu injection. State your hypotheses clearly. **(7)**

← Section 6.5

E **7** Students in a mixed sixth form college are classified as taking courses in either arts, science or humanities. A random sample of students from the college gave the following results.

		Course		
		Arts	Science	Humanities
Gender	Boy	30	50	35
	Girl	40	20	42

Showing your working clearly, test, at the 1% level of significance, whether or not there is an association between gender and the type of course taken. State your hypotheses clearly. **(7)**

← Section 6.5

E **8** A researcher carried out a survey of three treatments for a fruit tree disease.

	No action	Remove diseased branches	Spray with chemicals
Tree died within 1 year	10	5	6
Tree survived for 1–4 years	5	9	7
Tree survived beyond 4 years	5	6	7

Test, at the 5% level of significance, whether or not there is any association between the treatment of the trees and their survival. State your hypotheses and conclusion clearly. **(7)**

← Section 6.5

E **9** A research worker studying colour preference and the age of a random sample of 50 children obtained the results shown below.

Age in years	Red	Blue	Totals
4	12	6	18
8	10	7	17
12	6	9	15
Totals	28	22	50

Using a 5% significance level, carry out a test to decide whether or not there is an association between age and colour preference. State your hypotheses clearly. **(7)**

← Section 6.5

E **10** A celebrity receives fan mail six days a week. She thinks that the deliveries of mail are uniformly distributed throughout the week. The deliveries over a five-week period are as follows:

Day	Mon	Tues	Wed	Thurs	Fri	Sat
Frequency	20	15	18	23	19	25

Test the celebrity's assertion using a 1% level of significance. **(6)**

← Section 6.4

A E **11** Philomena has a large collection of DVDs, but she estimates that she only likes 40% of them. Every evening she picks DVDs from the rack at random until she finds one she likes. Over the course of two months she records the number of DVDs that she has picked each evening before finding one she likes. Her data is shown in the table below.

Number of DVDs	1	2	3	4	5
Frequency	30	18	12	1	1

a Calculate the expected frequencies if the number of DVDs chosen is modelled as a geometric random variable $X \sim \text{Geo}(0.4)$. **(3)**

Philomena wants to test her belief that the proportion of her DVDs that she likes is actually 40%.

b Write down suitable null and alternative hypotheses. **(2)**

c Is Philomena right in her belief? Use a 1% level of significance. **(6)**

d State the effect on Philomena's conclusion if she had used a 2% level of significance. **(1)**

← Section 6.6

E/P **12** The probability generating function of a discrete random variable X is given by

$$G_X(t) = k(3 + 2t + t^2)^2$$

a Find the value of k. **(2)**

b Find $P(X = 1)$. **(2)**

← Section 7.1

E/P **13** Billy is practicing archery. His probability of hitting a 'gold' in any one attempt is 0.24. In his first practice session, he fires arrows at a target until he hits a 'gold'.

a Suggest a suitable model for the random variable X, the number of arrows it takes him to hit a 'gold'. **(1)**

b Find $P(X = 7)$. **(1)**

c Write down the probability generating function for X. **(2)**

A

In his second practice session, he fires 15 arrows at the target.

d Write down the probability generating function for the random variable Y, the number of times he hits a 'gold'. **(2)**

In his third practice session, he continues to fire arrows until he has hit a 'gold' four times.

e Write down the probability generating function for the random variable Z, the number of shots it takes to hit four 'golds'. **(2)**

← Section 7.2

E/P **14** Calls come into a help desk at a rate of 1.7 per two-minute interval. Given that the random variable X is the number of calls that come in during a random two-minute interval and that the calls are independent and random, show, from first principles, that the probability generating function for X is:

$$G_X(t) = e^{1.7(t-1)}$$ **(5)**

← Section 7.2

E/P **15** A discrete random variable X has probability generating function

$$G_X(t) = \frac{4}{(3-t)^2}$$

a Find the mean and standard deviation of X. **(4)**

b Find:

i $P(X = 0)$

ii $P(X = 1)$ **(4)**

← Section 7.3

E/P **16** A random variable X has a probability generating function

$$G_X(t) = kt^2(1 + 3t^2)^2$$

a Find the value of k. **(2)**

b Find $P(X = 4)$. **(2)**

c Use the probability generating function to show $E(X) = 5$ and $Var(X) = \frac{3}{8}$ **(6)**

A

A second random variable Y has a probability generating function

$$G_Y(t) = \left(\frac{1}{4} + \frac{3}{4}t\right)^2$$

Given that X and Y are independent,

d find $E(Y)$ and write down the value of $E(X + Y)$. **(3)**

← Section 7.3, 7.4

E/P **17** The probability generating function of a discrete random variable X is given by:

$$G_X(t) = k(t + 4t^2 + 2t^3)^2$$

a Show that $k = \frac{1}{49}$. **(2)**

b Find $P(X = 3)$. **(2)**

c Show that $E(X) = \frac{30}{7}$ and find $Var(X)$. **(6)**

d Find a probability generating function of $3X - 2$. **(2)**

← Section 7.3, 7.4

E **18** **a** Define

i a Type I error **(1)**

ii a Type II error **(1)**

A small aviary, that leaves the eggs with the parent birds, rears chicks at an average rate of 5 per year. In order to increase the number of chicks reared per year, it is decided to remove the eggs from the aviary as soon as they are laid and put them in an incubator. At the end of the first year of using an incubator 7 chicks had been successfully reared.

b Assuming that the number of chicks reared per year follows a Poisson distribution test, at the 5% significance level, whether or not there is evidence of an increase in the number of chicks reared per year. State your hypotheses clearly. **(4)**

c Calculate the probability of the Type I error for this test. **(2)**

d Given that the true average number of chicks reared per year when the eggs are hatched in an incubator is 8, calculate the probability of a Type II error. **(2)**

← Section 8.1

A E/P **19** A butter-packing machine cuts butter into blocks. The weight of a block of butter is normally distributed with a mean weight of 250 g and a standard deviation of 4 g.

A random sample of 15 blocks is taken to monitor any change in the mean weight of the blocks of butter.

a Find the critical region of a suitable test using a 2% level of significance. **(4)**

b Assuming the mean weight of a block of butter has increased to 254 g, find the probability of a Type II error. **(2)**

← Section 8.2

E/P **20** It is suggested that a Poisson distribution with parameter λ can model the number of currants in a currant bun. A random bun is selected in order to test the hypotheses H_0: $\lambda = 8$ against H_1: $\lambda \neq 8$, using a 10% level of significance.

a Find the critical region for this test, such that the probability in each tail is as close as possible to 5%. **(4)**

b Given that $\lambda = 10$, find:

i the probability of a Type II error **(2)**

ii the power of the test. **(2)**

← Section 8.1, 8.3

E/P **21** A train company claims that the probability p of one of its trains arriving late is 10%. A regular traveller on the company's trains believes that the probability is greater than 10% and decides to test this by randomly selecting 12 trains and recording the number of trains that were late, X.
The traveller sets up the hypotheses H_0: $p = 0.1$ and H_1: $p > 0.1$ and accepts the null hypothesis if $x \leqslant 2$

a Find the size of the test. **(2)**

b Show that the power function of the test is **(4)**

$$1 - (1 - p)^{10} (1 + 10p + 55p^2).$$

c Calculate the power of the test when

i $p = 0.2$ **(2)**

ii $p = 0.6$ **(2)**

d Comment on your results from part **c**. **(2)**

← Section 8.1, 8.3

A E/P **22** **a** Define

i the power of a test **(1)**

ii the size of a test. **(1)**

Jane claims that she can read Alan's mind. To test this claim Alan randomly chooses a card with one of 4 symbols on it. He then concentrates on the symbol. Jane then attempts to read Alan's mind by stating what symbol she thinks is on the card. The experiment is carried out 8 times and the number of times, X, that Jane is correct is recorded.

The probability of Jane stating the correct symbol is denoted by p.

To test the hypothesis H_0: $p = 0.25$ against H_1: $p > 0.25$, a critical region of $X > 6$ is used.

b Find the size of this test. **(2)**

c Show that the power function of this test is $8p^7 - 7p^8$ **(4)**

Given that $p = 0.3$, calculate:

d the power of this test, **(2)**

e the probability of a Type II error. **(2)**

f Suggest two ways in which you might reduce the probability of a Type II error. **(2)**

← Section 8.1, 8.3, 8.4

E/P **23** The number of burglaries per year in a particular county follows a Poisson distribution with mean λ per 1000 households. A police commissioner claims that the mean number of burglaries per year has decreased.

Using H_0: $\lambda = 4$ and H_1: $\lambda < 4$ the commissioner takes a sample of 1000 households and rejects H_0 if there are 2 or fewer burglaries. If there are 5 or more burglaries then H_0 is accepted.

If there are 3 or 4 burglaries, a second sample of 1000 households is taken and H_0 is rejected if there are 2 or fewer burglaries in this second sample, otherwise it is accepted.

a Find the size of this test. **(3)**

A

b Show that the power function for this test is given by

$$e^{-2\lambda}\left(1 + \lambda + \frac{\lambda^2}{2}\right) + e^{\lambda}\left(\frac{\lambda^3}{6} + \frac{\lambda^4}{24}\right) \quad \textbf{(5)}$$

c Find the probability of a Type II error when $\lambda = 3$. **(2)**

← Section 8.1, 8.3, 8.4

E/P **24** A drug is claimed to produce a cure to a certain disease in 35% of people who have the disease. To test this claim a sample of 20 people having this disease is chosen at random and given the drug. If the number of people cured is between 4 and 10 inclusive, the claim will be accepted. Otherwise the claim will not be accepted.

a Write down suitable hypotheses to carry out this test. **(2)**

b Find the probability of making a Type I error. **(2)**

The table below gives the value of the probability of the Type II error, to 4 decimal places, for different values of p where p is the probability of the drug curing a person with the disease.

P(cure)	0.2	0.3	0.4	0.5
P(Type II error)	0.5880	r	0.8565	s

c Calculate the value of r and the value of s. **(2)**

d Calculate the power of the test for $p = 0.2$ and $p = 0.4$ **(2)**

e Comment, giving your reasons, on the suitability of this test procedure. **(2)**

← Section 8.1, 8.3

E/P **25** A manager in a flour mill believes that the machines are working incorrectly and the proportion p of underweight bags of flour is more than 5%. She decides to test this by randomly selecting a sample of 5 bags and recording the number x that are underweight. The manager sets up the hypotheses H_0: $p = 0.05$ and H_1: $p > 0.05$ and rejects the null hypothesis if $x > 1$.

a Find the size of the test. **(2)**

b Show that the power function of the test is $1 - (1 - p)^4(1 + 4p)$ **(3)**

The manager goes on holiday and her assistant checks the production by randomly selecting a sample of 10 bags of flour. The assistant rejects the hypothesis that $p = 0.05$ if more than 2 underweight bags are found in the sample.

c Find the probability of a Type I error using the assistant's test. **(2)**

The table below gives some values, to 2 decimal places, of the power function for the assistant's test.

p	0.10	0.15	0.20	0.25
Power	0.07	0.18	0.32	a

d Find the value of a. **(3)**

e On the same axes, draw the graph of the power function for the manager's and the assistant's tests. **(4)**

f Given that $p > 0.2$, state, with a reason, which test you would recommend. **(2)**

The assistant suggests that they should use his sampling method rather than the manager's.

g Give two reasons why the manager might not agree to this change. **(2)**

← Section 8.1, 8.3, 8.4

Challenge

1 A manufacturer claims that the batteries used in his mobile phones have a mean lifetime of 360 hours and a standard deviation of 20 hours, when the phone is left on standby. To test this claim 100 phones were left on standby until the batteries ran flat. The lifetime t hours of the batteries was recorded. The results are as follows.

t	300–	320–	340–	350–	360–	370–	380–	400–
Frequency	1	9	28	20	16	18	7	1

A researcher believes that a normal distribution might provide a good model for the lifetime of the batteries

She calculated the expected frequencies as follows using the distribution N (360, 20).

t	< 320	320–	340–	350–	360–	370–	380–	400–
Expected frequency	2.28	13.59	24.26	r	s	14.98	13.59	2.28

a Find the values of r and s.

b Stating clearly your hypotheses, test, at the 1% level of significance, whether or not this normal distribution is a suitable model for these data.

← Section 6.4

2 You can use the following rule to find the distribution of a sum of N identical independent random variables, where N is itself an independent random variable:

$X_1, X_2, X_3, \ldots$ are identically distributed independent random variables, each with probability generating function $\mathrm{G}_X(t)$, and N is a random variable with probability generating function $\mathrm{G}_N(t)$.

If $S = \sum_{i=1}^{N} X_1 + X_2 + \ldots + X_N$ then the probability generating function of S is given by $\mathrm{G}_S(t) = \mathrm{G}_N((\mathrm{G}_X(t))$.

a Use the above result to show further that:

i $\mathrm{E}(S) = \mathrm{E}(N)\mathrm{E}(X)$

ii $\mathrm{Var}(S) = \mathrm{E}(N)\mathrm{Var}(X) + (\mathrm{E}(X))^2\,\mathrm{Var}(N)$

A bank models the number of people who use its external cash machine each hour as a Poisson random variable Po(λ). Each person who uses the cash machine makes a balance enquiry with probability p.

b Show that the total number of balance enquiries made at the external cash machine each hour also has a Poisson distribution and determine its parameter.

The bank models the number of people who use its internal cash machine each hour as a binomial random variable B(n, q), where n is the number of customers who visit the bank each hour, and q is the probability that each customer uses the machine. Given that each of these people also makes a balance enquiry with probability p,

c show that the total number of balance enquiries made at the internal cash machine also has a binomial distribution, and determine its parameters.

Given further that $\lambda = 75$, $n = 80$ and $q = 0.25$, and that each person who uses a cash machine withdraws either £0, £10, £20 or £50 with probabilities 0.1, 0.3, 0.4 and 0.2 respectively,

d find the mean and standard deviation of the total amount of money withdrawn each hour at

i the external cash machine

ii the internal cash machine.

← Sections 7.1, 7.2, 7.3, 7.4

Exam-style practice

Further Mathematics
AS Level
Further Statistics 1

Time: 50 minutes
You must have: Mathematical Formulae and Statistical Tables, Calculator

1 The discrete random variable X has probability distribution given by:

x	-2	-1	0	1	2
$P(X = x)$	0.1	a	0.15	0.2	b

The random variable $Y = 2X + 3$. Given that $E(Y) = 4.48$,

a find the values of a and b **(5)**

b calculate the exact value of $Var(X)$ **(3)**

c find $P(Y - 2 > X)$. **(2)**

2 A call centre receives calls about insurance at a rate of 3.2 per ten-minute interval and calls about utility bills at a rate of 4.1 per ten-minute interval. Calls about insurance and calls about utility bills are independent of each other.

a In a ten-minute interval, calculate the probability that the company receives exactly 3 calls of each type. **(2)**

b In a ten-minute interval, calculate the probability that the company receives at least 7 calls in total. **(2)**

c In a one-hour period, calculate the probability that the company receives fewer than 45 calls in total. **(2)**

3 A sports club collects data on the gender of its members and the sport that they play. A random sample of 250 members is taken and the data is recorded in the table below:

	Hockey	Cricket	Squash
Male	61	45	32
Female	66	23	23

A test is to be carried out at a 2.5% level of significance to determine whether or not there is an association between gender and choice of sport.

a Write down suitable null and alternative hypotheses. **(2)**

b Calculate the test statistic for this test. **(4)**

c State the number of degrees of freedom of the test. **(1)**

d State whether or not the null hypothesis is accepted. Give a reason for your answer. **(2)**

e State the effect on your answer to part **d** if the test was carried out at the 5% level of significance. **(1)**

4 A large pottery believes that 0.5% of the bowls that they make contain a defect. A quality control manager takes a random sample of 750 bowls.

a Find the mean and variance of the number of bowls in the sample with a defect. **(2)**

b By using a Poisson approximation, estimate the probability that more than three bowls in the sample have a defect. **(2)**

c Give a reason to support the use of a Poisson approximation. **(1)**

5 Jennifer spins four identical coins 100 times and records the number of heads each time. The results are shown in the table below.

Number of heads	0	1	2	3	4
Frequency	6	18	35	26	15

a Use these results to estimate the probability of any single coin landing on heads. **(2)**

She believes that the binomial distribution is a suitable model for the number of heads. Using your answer to part **a**,

b Carry out a test at the 10% level of significance to check Jennifer's claim. You must state your hypotheses clearly. **(7)**

Exam-style practice

Further Mathematics A Level Further Statistics 1

Time: 1 hour and 30 minutes
You must have: Mathematical Formulae and Statistical Tables, Calculator

1 Johsva works in a call centre. He calls people from a list until the first person answers. Over the course of a week, he records the number of calls he has to make before a person answers. The results are shown in the frequency table below:

Number of calls	1	2	3	4	5	6
Frequency	52	31	12	7	1	1

Joshva believes that the distribution of the random variable 'number of calls made until someone answers' is Geo(0.4).

Test, at the 5% level of significance, whether Joshva's belief is correct. You must state your hypotheses clearly. **(10)**

2 The probability generating function of the discrete random variable X is given by:

$$G_X = k(1 + 2t + 3t^2)^3$$

a Show that $k = \frac{1}{216}$ **(2)**

b Find $P(X = 2)$. **(2)**

c Show that $\text{Var}(X) = \frac{5}{3}$ **(8)**

d Write down a probability generating function for $Y = 2X + 3$. **(2)**

3 Jagdeep is practising darts. He continues to throw darts at the board until he hits the bullseye r times. The random variable Y represents the total number of darts he throws. Given that the mean and variance of Y are 20 and $37\frac{1}{7}$ respectively,

a write down a suitable model for this situation and state two assumptions that must be made for it to be valid. **(3)**

b Find the value of p, the probability of Jagdeep hitting the bullseye with any one dart. **(4)**

c Find the value of r. **(1)**

4 The number of flaws, X, in a ten-metre length of cloth is modelled as $X \sim \text{Po}(2.1)$.
A random sample of 200 ten-metre lengths of the cloth is taken.

a find the probability that the sample mean, $\overline{X}$, is greater than 2.3. **(4)**

A tailor decides to modify the production process to reduce the rate of appearance of flaws. After this modification, he takes a random sample of twenty metres of cloth and finds that there is one flaw in it.

b Test, at the 5% significance level, whether there is evidence that the rate of appearance of flaws has been reduced. **(5)**

5 The discrete random variable X has a probability distribution as shown in the table below.

x	-3	-2	0	2	3	5
$\mathrm{P}(X = x)$	p	q	r	q	q	r

The random variable Y is defined as $Y = 2X + 5$. Given that $\mathrm{E}(Y) = 4.9$ and that $\mathrm{P}(Y < 9) = 0.55$, find:

a the values of p, q and r **(7)**

b $\mathrm{P}(X > 2Y - 3)$. **(2)**

6 A footballer is practising penalty kicks. He takes 500 penalties per day and the probability of a random penalty missing the goal is p.
The probability that the footballer never misses the goal in four consecutive kicks is 0.8853.

a Find the mean and variance of the number of missed penalties each day. **(7)**

b Explain why a Poisson approximation can be used to find the probability that the footballer misses more than r penalties per day. **(2)**

c Use a suitable Poisson approximation to find the probability that the footballer misses more than 17 penalties on a given day. **(2)**

d Comment on the accuracy of the approximation obtained in part **c**. **(2)**

7 Philip is rolling a six-sided dice to test if it is biased against rolling a six.
He rolls the dice 25 times and records the number of times a six appears.

a Using a 10% significance level, find the critical region for Philip's test and write down the size of the test. **(3)**

b Show that the power function for Philip's test is given by

$$(1 - p)^{24}(1 + 24p)$$ **(3)**

Gemma carries out a different experiment such that she continues to roll the dice until a six appears.
Given that the critical region for Gemma's test is 'greater than or equal to 13',

c find the size of Gemma's test and write down the power function for her test. **(3)**

d Give two reasons why you would recommend Philip's test over Gemma's test when $p = 0.09$. **(3)**

BINOMIAL CUMULATIVE DISTRIBUTION FUNCTION

The tabulated value is $P(X \leqslant x)$, where X has a binomial distribution with index n and parameter p.

$p =$	0.05	0.10	0.15	0.20	0.25	0.30	0.35	0.40	0.45	0.50
$n = 5, x = 0$	0.7738	0.5905	0.4437	0.3277	0.2373	0.1681	0.1160	0.0778	0.0503	0.0312
1	0.9774	0.9185	0.8352	0.7373	0.6328	0.5282	0.4284	0.3370	0.2562	0.1875
2	0.9988	0.9914	0.9734	0.9421	0.8965	0.8369	0.7648	0.6826	0.5931	0.5000
3	1.0000	0.9995	0.9978	0.9933	0.9844	0.9692	0.9460	0.9130	0.8688	0.8125
4	1.0000	1.0000	0.9999	0.9997	0.9990	0.9976	0.9947	0.9898	0.9815	0.9688
$n = 6, x = 0$	0.7351	0.5314	0.3771	0.2621	0.1780	0.1176	0.0754	0.0467	0.0277	0.0156
1	0.9672	0.8857	0.7765	0.6554	0.5339	0.4202	0.3191	0.2333	0.1636	0.1094
2	0.9978	0.9842	0.9527	0.9011	0.8306	0.7443	0.6471	0.5443	0.4415	0.3438
3	0.9999	0.9987	0.9941	0.9830	0.9624	0.9295	0.8826	0.8208	0.7447	0.6563
4	1.0000	0.9999	0.9996	0.9984	0.9954	0.9891	0.9777	0.9590	0.9308	0.8906
5	1.0000	1.0000	1.0000	0.9999	0.9998	0.9993	0.9982	0.9959	0.9917	0.9844
$n = 7, x = 0$	0.6983	0.4783	0.3206	0.2097	0.1335	0.0824	0.0490	0.0280	0.0152	0.0078
1	0.9556	0.8503	0.7166	0.5767	0.4449	0.3294	0.2338	0.1586	0.1024	0.0625
2	0.9962	0.9743	0.9262	0.8520	0.7564	0.6471	0.5323	0.4199	0.3164	0.2266
3	0.9998	0.9973	0.9879	0.9667	0.9294	0.8740	0.8002	0.7102	0.6083	0.5000
4	1.0000	0.9998	0.9988	0.9953	0.9871	0.9712	0.9444	0.9037	0.8471	0.7734
5	1.0000	1.0000	0.9999	0.9996	0.9987	0.9962	0.9910	0.9812	0.9643	0.9375
6	1.0000	1.0000	1.0000	1.0000	0.9999	0.9998	0.9994	0.9984	0.9963	0.9922
$n = 8, x = 0$	0.6634	0.4305	0.2725	0.1678	0.1001	0.0576	0.0319	0.0168	0.0084	0.0039
1	0.9428	0.8131	0.6572	0.5033	0.3671	0.2553	0.1691	0.1064	0.0632	0.0352
2	0.9942	0.9619	0.8948	0.7969	0.6785	0.5518	0.4278	0.3154	0.2201	0.1445
3	0.9996	0.9950	0.9786	0.9437	0.8862	0.8059	0.7064	0.5941	0.4770	0.3633
4	1.0000	0.9996	0.9971	0.9896	0.9727	0.9420	0.8939	0.8263	0.7396	0.6367
5	1.0000	1.0000	0.9998	0.9988	0.9958	0.9887	0.9747	0.9502	0.9115	0.8555
6	1.0000	1.0000	1.0000	0.9999	0.9996	0.9987	0.9964	0.9915	0.9819	0.9648
7	1.0000	1.0000	1.0000	1.0000	1.0000	0.9999	0.9998	0.9993	0.9983	0.9961
$n = 9, x = 0$	0.6302	0.3874	0.2316	0.1342	0.0751	0.0404	0.0207	0.0101	0.0046	0.0020
1	0.9288	0.7748	0.5995	0.4362	0.3003	0.1960	0.1211	0.0705	0.0385	0.0195
2	0.9916	0.9470	0.8591	0.7382	0.6007	0.4628	0.3373	0.2318	0.1495	0.0898
3	0.9994	0.9917	0.9661	0.9144	0.8343	0.7297	0.6089	0.4826	0.3614	0.2539
4	1.0000	0.9991	0.9944	0.9804	0.9511	0.9012	0.8283	0.7334	0.6214	0.5000
5	1.0000	0.9999	0.9994	0.9969	0.9900	0.9747	0.9464	0.9006	0.8342	0.7461
6	1.0000	1.0000	1.0000	0.9997	0.9987	0.9957	0.9888	0.9750	0.9502	0.9102
7	1.0000	1.0000	1.0000	1.0000	0.9999	0.9996	0.9986	0.9962	0.9909	0.9805
8	1.0000	1.0000	1.0000	1.0000	1.0000	1.0000	0.9999	0.9997	0.9992	0.9980
$n = 10, x = 0$	0.5987	0.3487	0.1969	0.1074	0.0563	0.0282	0.0135	0.0060	0.0025	0.0010
1	0.9139	0.7361	0.5443	0.3758	0.2440	0.1493	0.0860	0.0464	0.0233	0.0107
2	0.9885	0.9298	0.8202	0.6778	0.5256	0.3828	0.2616	0.1673	0.0996	0.0547
3	0.9990	0.9872	0.9500	0.8791	0.7759	0.6496	0.5138	0.3823	0.2660	0.1719
4	0.9999	0.9984	0.9901	0.9672	0.9219	0.8497	0.7515	0.6331	0.5044	0.3770
5	1.0000	0.9999	0.9986	0.9936	0.9803	0.9527	0.9051	0.8338	0.7384	0.6230
6	1.0000	1.0000	0.9999	0.9991	0.9965	0.9894	0.9740	0.9452	0.8980	0.8281
7	1.0000	1.0000	1.0000	0.9999	0.9996	0.9984	0.9952	0.9877	0.9726	0.9453
8	1.0000	1.0000	1.0000	1.0000	1.0000	0.9999	0.9995	0.9983	0.9955	0.9893
9	1.0000	1.0000	1.0000	1.0000	1.0000	1.0000	1.0000	0.9999	0.9997	0.9990

$p =$	0.05	0.10	0.15	0.20	0.25	0.30	0.35	0.40	0.45	0.50
$n = 12, x = 0$	0.5404	0.2824	0.1422	0.0687	0.0317	0.0138	0.0057	0.0022	0.0008	0.0002
1	0.8816	0.6590	0.4435	0.2749	0.1584	0.0850	0.0424	0.0196	0.0083	0.0032
2	0.9804	0.8891	0.7358	0.5583	0.3907	0.2528	0.1513	0.0834	0.0421	0.0193
3	0.9978	0.9744	0.9078	0.7946	0.6488	0.4925	0.3467	0.2253	0.1345	0.0730
4	0.9998	0.9957	0.9761	0.9274	0.8424	0.7237	0.5833	0.4382	0.3044	0.1938
5	1.0000	0.9995	0.9954	0.9806	0.9456	0.8822	0.7873	0.6652	0.5269	0.3872
6	1.0000	0.9999	0.9993	0.9961	0.9857	0.9614	0.9154	0.8418	0.7393	0.6128
7	1.0000	1.0000	0.9999	0.9994	0.9972	0.9905	0.9745	0.9427	0.8883	0.8062
8	1.0000	1.0000	1.0000	0.9999	0.9996	0.9983	0.9944	0.9847	0.9644	0.9270
9	1.0000	1.0000	1.0000	1.0000	1.0000	0.9998	0.9992	0.9972	0.9921	0.9807
10	1.0000	1.0000	1.0000	1.0000	1.0000	1.0000	0.9999	0.9997	0.9989	0.9968
11	1.0000	1.0000	1.0000	1.0000	1.0000	1.0000	1.0000	1.0000	0.9999	0.9998
$n = 15, x = 0$	0.4633	0.2059	0.0874	0.0352	0.0134	0.0047	0.0016	0.0005	0.0001	0.0000
1	0.8290	0.5490	0.3186	0.1671	0.0802	0.0353	0.0142	0.0052	0.0017	0.0005
2	0.9638	0.8159	0.6042	0.3980	0.2361	0.1268	0.0617	0.0271	0.0107	0.0037
3	0.9945	0.9444	0.8227	0.6482	0.4613	0.2969	0.1727	0.0905	0.0424	0.0176
4	0.9994	0.9873	0.9383	0.8358	0.6865	0.5155	0.3519	0.2173	0.1204	0.0592
5	0.9999	0.9978	0.9832	0.9389	0.8516	0.7216	0.5643	0.4032	0.2608	0.1509
6	1.0000	0.9997	0.9964	0.9819	0.9434	0.8689	0.7548	0.6098	0.4522	0.3036
7	1.0000	1.0000	0.9994	0.9958	0.9827	0.9500	0.8868	0.7869	0.6535	0.5000
8	1.0000	1.0000	0.9999	0.9992	0.9958	0.9848	0.9578	0.9050	0.8182	0.6964
9	1.0000	1.0000	1.0000	0.9999	0.9992	0.9963	0.9876	0.9662	0.9231	0.8491
10	1.0000	1.0000	1.0000	1.0000	0.9999	0.9993	0.9972	0.9907	0.9745	0.9408
11	1.0000	1.0000	1.0000	1.0000	1.0000	0.9999	0.9995	0.9981	0.9937	0.9824
12	1.0000	1.0000	1.0000	1.0000	1.0000	1.0000	0.9999	0.9997	0.9989	0.9963
13	1.0000	1.0000	1.0000	1.0000	1.0000	1.0000	1.0000	1.0000	0.9999	0.9995
14	1.0000	1.0000	1.0000	1.0000	1.0000	1.0000	1.0000	1.0000	1.0000	1.0000
$n = 20, x = 0$	0.3585	0.1216	0.0388	0.0115	0.0032	0.0008	0.0002	0.0000	0.0000	0.0000
1	0.7358	0.3917	0.1756	0.0692	0.0243	0.0076	0.0021	0.0005	0.0001	0.0000
2	0.9245	0.6769	0.4049	0.2061	0.0913	0.0355	0.0121	0.0036	0.0009	0.0002
3	0.9841	0.8670	0.6477	0.4114	0.2252	0.1071	0.0444	0.0160	0.0049	0.0013
4	0.9974	0.9568	0.8298	0.6296	0.4148	0.2375	0.1182	0.0510	0.0189	0.0059
5	0.9997	0.9887	0.9327	0.8042	0.6172	0.4164	0.2454	0.1256	0.0553	0.0207
6	1.0000	0.9976	0.9781	0.9133	0.7858	0.6080	0.4166	0.2500	0.1299	0.0577
7	1.0000	0.9996	0.9941	0.9679	0.8982	0.7723	0.6010	0.4159	0.2520	0.1316
8	1.0000	0.9999	0.9987	0.9900	0.9591	0.8867	0.7624	0.5956	0.4143	0.2517
9	1.0000	1.0000	0.9998	0.9974	0.9861	0.9520	0.8782	0.7553	0.5914	0.4119
10	1.0000	1.0000	1.0000	0.9994	0.9961	0.9829	0.9468	0.8725	0.7507	0.5881
11	1.0000	1.0000	1.0000	0.9999	0.9991	0.9949	0.9804	0.9435	0.8692	0.7483
12	1.0000	1.0000	1.0000	1.0000	0.9998	0.9987	0.9940	0.9790	0.9420	0.8684
13	1.0000	1.0000	1.0000	1.0000	1.0000	0.9997	0.9985	0.9935	0.9786	0.9423
14	1.0000	1.0000	1.0000	1.0000	1.0000	1.0000	0.9997	0.9984	0.9936	0.9793
15	1.0000	1.0000	1.0000	1.0000	1.0000	1.0000	1.0000	0.9997	0.9985	0.9941
16	1.0000	1.0000	1.0000	1.0000	1.0000	1.0000	1.0000	1.0000	0.9997	0.9987
17	1.0000	1.0000	1.0000	1.0000	1.0000	1.0000	1.0000	1.0000	1.0000	0.9998
18	1.0000	1.0000	1.0000	1.0000	1.0000	1.0000	1.0000	1.0000	1.0000	1.0000

$p =$	0.05	0.10	0.15	0.20	0.25	0.30	0.35	0.40	0.45	0.50
$n = 25, x = 0$	0.2774	0.0718	0.0172	0.0038	0.0008	0.0001	0.0000	0.0000	0.0000	0.0000
1	0.6424	0.2712	0.0931	0.0274	0.0070	0.0016	0.0003	0.0001	0.0000	0.0000
2	0.8729	0.5371	0.2537	0.0982	0.0321	0.0090	0.0021	0.0004	0.0001	0.0000
3	0.9659	0.7636	0.4711	0.2340	0.0962	0.0332	0.0097	0.0024	0.0005	0.0001
4	0.9928	0.9020	0.6821	0.4207	0.2137	0.0905	0.0320	0.0095	0.0023	0.0005
5	0.9988	0.9666	0.8385	0.6167	0.3783	0.1935	0.0826	0.0294	0.0086	0.0020
6	0.9998	0.9905	0.9305	0.7800	0.5611	0.3407	0.1734	0.0736	0.0258	0.0073
7	1.0000	0.9977	0.9745	0.8909	0.7265	0.5118	0.3061	0.1536	0.0639	0.0216
8	1.0000	0.9995	0.9920	0.9532	0.8506	0.6769	0.4668	0.2735	0.1340	0.0539
9	1.0000	0.9999	0.9979	0.9827	0.9287	0.8106	0.6303	0.4246	0.2424	0.1148
10	1.0000	1.0000	0.9995	0.9944	0.9703	0.9022	0.7712	0.5858	0.3843	0.2122
11	1.0000	1.0000	0.9999	0.9985	0.9893	0.9558	0.8746	0.7323	0.5426	0.3450
12	1.0000	1.0000	1.0000	0.9996	0.9966	0.9825	0.9396	0.8462	0.6937	0.5000
13	1.0000	1.0000	1.0000	0.9999	0.9991	0.9940	0.9745	0.9222	0.8173	0.6550
14	1.0000	1.0000	1.0000	1.0000	0.9998	0.9982	0.9907	0.9656	0.9040	0.7878
15	1.0000	1.0000	1.0000	1.0000	1.0000	0.9995	0.9971	0.9868	0.9560	0.8852
16	1.0000	1.0000	1.0000	1.0000	1.0000	0.9999	0.9992	0.9957	0.9826	0.9461
17	1.0000	1.0000	1.0000	1.0000	1.0000	1.0000	0.9998	0.9988	0.9942	0.9784
18	1.0000	1.0000	1.0000	1.0000	1.0000	1.0000	1.0000	0.9997	0.9984	0.9927
19	1.0000	1.0000	1.0000	1.0000	1.0000	1.0000	1.0000	0.9999	0.9996	0.9980
20	1.0000	1.0000	1.0000	1.0000	1.0000	1.0000	1.0000	1.0000	0.9999	0.9995
21	1.0000	1.0000	1.0000	1.0000	1.0000	1.0000	1.0000	1.0000	1.0000	0.9999
22	1.0000	1.0000	1.0000	1.0000	1.0000	1.0000	1.0000	1.0000	1.0000	1.0000
$n = 30, x = 0$	0.2146	0.0424	0.0076	0.0012	0.0002	0.0000	0.0000	0.0000	0.0000	0.0000
1	0.5535	0.1837	0.0480	0.0105	0.0020	0.0003	0.0000	0.0000	0.0000	0.0000
2	0.8122	0.4114	0.1514	0.0442	0.0106	0.0021	0.0003	0.0000	0.0000	0.0000
3	0.9392	0.6474	0.3217	0.1227	0.0374	0.0093	0.0019	0.0003	0.0000	0.0000
4	0.9844	0.8245	0.5245	0.2552	0.0979	0.0302	0.0075	0.0015	0.0002	0.0000
5	0.9967	0.9268	0.7106	0.4275	0.2026	0.0766	0.0233	0.0057	0.0011	0.0002
6	0.9994	0.9742	0.8474	0.6070	0.3481	0.1595	0.0586	0.0172	0.0040	0.0007
7	0.9999	0.9922	0.9302	0.7608	0.5143	0.2814	0.1238	0.0435	0.0121	0.0026
8	1.0000	0.9980	0.9722	0.8713	0.6736	0.4315	0.2247	0.0940	0.0312	0.0081
9	1.0000	0.9995	0.9903	0.9389	0.8034	0.5888	0.3575	0.1763	0.0694	0.0214
10	1.0000	0.9999	0.9971	0.9744	0.8943	0.7304	0.5078	0.2915	0.1350	0.0494
11	1.0000	1.0000	0.9992	0.9905	0.9493	0.8407	0.6548	0.4311	0.2327	0.1002
12	1.0000	1.0000	0.9998	0.9969	0.9784	0.9155	0.7802	0.5785	0.3592	0.1808
13	1.0000	1.0000	1.0000	0.9991	0.9918	0.9599	0.8737	0.7145	0.5025	0.2923
14	1.0000	1.0000	1.0000	0.9998	0.9973	0.9831	0.9348	0.8246	0.6448	0.4278
15	1.0000	1.0000	1.0000	0.9999	0.9992	0.9936	0.9699	0.9029	0.7691	0.5722
16	1.0000	1.0000	1.0000	1.0000	0.9998	0.9979	0.9876	0.9519	0.8644	0.7077
17	1.0000	1.0000	1.0000	1.0000	0.9999	0.9994	0.9955	0.9788	0.9286	0.8192
18	1.0000	1.0000	1.0000	1.0000	1.0000	0.9998	0.9986	0.9917	0.9666	0.8998
19	1.0000	1.0000	1.0000	1.0000	1.0000	1.0000	0.9996	0.9971	0.9862	0.9506
20	1.0000	1.0000	1.0000	1.0000	1.0000	1.0000	0.9999	0.9991	0.9950	0.9786
21	1.0000	1.0000	1.0000	1.0000	1.0000	1.0000	1.0000	0.9998	0.9984	0.9919
22	1.0000	1.0000	1.0000	1.0000	1.0000	1.0000	1.0000	1.0000	0.9996	0.9974
23	1.0000	1.0000	1.0000	1.0000	1.0000	1.0000	1.0000	1.0000	0.9999	0.9993
24	1.0000	1.0000	1.0000	1.0000	1.0000	1.0000	1.0000	1.0000	1.0000	0.9998
25	1.0000	1.0000	1.0000	1.0000	1.0000	1.0000	1.0000	1.0000	1.0000	1.0000

$p =$	0.05	0.10	0.15	0.20	0.25	0.30	0.35	0.40	0.45	0.50
$n = 40, x = 0$	0.1285	0.0148	0.0015	0.0001	0.0000	0.0000	0.0000	0.0000	0.0000	0.0000
1	0.3991	0.0805	0.0121	0.0015	0.0001	0.0000	0.0000	0.0000	0.0000	0.0000
2	0.6767	0.2228	0.0486	0.0079	0.0010	0.0001	0.0000	0.0000	0.0000	0.0000
3	0.8619	0.4231	0.1302	0.0285	0.0047	0.0006	0.0001	0.0000	0.0000	0.0000
4	0.9520	0.6290	0.2633	0.0759	0.0160	0.0026	0.0003	0.0000	0.0000	0.0000
5	0.9861	0.7937	0.4325	0.1613	0.0433	0.0086	0.0013	0.0001	0.0000	0.0000
6	0.9966	0.9005	0.6067	0.2859	0.0962	0.0238	0.0044	0.0006	0.0001	0.0000
7	0.9993	0.9581	0.7559	0.4371	0.1820	0.0553	0.0124	0.0021	0.0002	0.0000
8	0.9999	0.9845	0.8646	0.5931	0.2998	0.1110	0.0303	0.0061	0.0009	0.0001
9	1.0000	0.9949	0.9328	0.7318	0.4395	0.1959	0.0644	0.0156	0.0027	0.0003
10	1.0000	0.9985	0.9701	0.8392	0.5839	0.3087	0.1215	0.0352	0.0074	0.0011
11	1.0000	0.9996	0.9880	0.9125	0.7151	0.4406	0.2053	0.0709	0.0179	0.0032
12	1.0000	0.9999	0.9957	0.9568	0.8209	0.5772	0.3143	0.1285	0.0386	0.0083
13	1.0000	1.0000	0.9986	0.9806	0.8968	0.7032	0.4408	0.2112	0.0751	0.0192
14	1.0000	1.0000	0.9996	0.9921	0.9456	0.8074	0.5721	0.3174	0.1326	0.0403
15	1.0000	1.0000	0.9999	0.9971	0.9738	0.8849	0.6946	0.4402	0.2142	0.0769
16	1.0000	1.0000	1.0000	0.9990	0.9884	0.9367	0.7978	0.5681	0.3185	0.1341
17	1.0000	1.0000	1.0000	0.9997	0.9953	0.9680	0.8761	0.6885	0.4391	0.2148
18	1.0000	1.0000	1.0000	0.9999	0.9983	0.9852	0.9301	0.7911	0.5651	0.3179
19	1.0000	1.0000	1.0000	1.0000	0.9994	0.9937	0.9637	0.8702	0.6844	0.4373
20	1.0000	1.0000	1.0000	1.0000	0.9998	0.9976	0.9827	0.9256	0.7870	0.5627
21	1.0000	1.0000	1.0000	1.0000	1.0000	0.9991	0.9925	0.9608	0.8669	0.6821
22	1.0000	1.0000	1.0000	1.0000	1.0000	0.9997	0.9970	0.9811	0.9233	0.7852
23	1.0000	1.0000	1.0000	1.0000	1.0000	0.9999	0.9989	0.9917	0.9595	0.8659
24	1.0000	1.0000	1.0000	1.0000	1.0000	1.0000	0.9996	0.9966	0.9804	0.9231
25	1.0000	1.0000	1.0000	1.0000	1.0000	1.0000	0.9999	0.9988	0.9914	0.9597
26	1.0000	1.0000	1.0000	1.0000	1.0000	1.0000	1.0000	0.9996	0.9966	0.9808
27	1.0000	1.0000	1.0000	1.0000	1.0000	1.0000	1.0000	0.9999	0.9988	0.9917
28	1.0000	1.0000	1.0000	1.0000	1.0000	1.0000	1.0000	1.0000	0.9996	0.9968
29	1.0000	1.0000	1.0000	1.0000	1.0000	1.0000	1.0000	1.0000	0.9999	0.9989
30	1.0000	1.0000	1.0000	1.0000	1.0000	1.0000	1.0000	1.0000	1.0000	0.9997
31	1.0000	1.0000	1.0000	1.0000	1.0000	1.0000	1.0000	1.0000	1.0000	0.9999
32	1.0000	1.0000	1.0000	1.0000	1.0000	1.0000	1.0000	1.0000	1.0000	1.0000

$p =$	0.05	0.10	0.15	0.20	0.25	0.30	0.35	0.40	0.45	0.50
$n = 50, x = 0$	0.0769	0.0052	0.0003	0.0000	0.0000	0.0000	0.0000	0.0000	0.0000	0.0000
1	0.2794	0.0338	0.0029	0.0002	0.0000	0.0000	0.0000	0.0000	0.0000	0.0000
2	0.5405	0.1117	0.0142	0.0013	0.0001	0.0000	0.0000	0.0000	0.0000	0.0000
3	0.7604	0.2503	0.0460	0.0057	0.0005	0.0000	0.0000	0.0000	0.0000	0.0000
4	0.8964	0.4312	0.1121	0.0185	0.0021	0.0002	0.0000	0.0000	0.0000	0.0000
5	0.9622	0.6161	0.2194	0.0480	0.0070	0.0007	0.0001	0.0000	0.0000	0.0000
6	0.9882	0.7702	0.3613	0.1034	0.0194	0.0025	0.0002	0.0000	0.0000	0.0000
7	0.9968	0.8779	0.5188	0.1904	0.0453	0.0073	0.0008	0.0001	0.0000	0.0000
8	0.9992	0.9421	0.6681	0.3073	0.0916	0.0183	0.0025	0.0002	0.0000	0.0000
9	0.9998	0.9755	0.7911	0.4437	0.1637	0.0402	0.0067	0.0008	0.0001	0.0000
10	1.0000	0.9906	0.8801	0.5836	0.2622	0.0789	0.0160	0.0022	0.0002	0.0000
11	1.0000	0.9968	0.9372	0.7107	0.3816	0.1390	0.0342	0.0057	0.0006	0.0000
12	1.0000	0.9990	0.9699	0.8139	0.5110	0.2229	0.0661	0.0133	0.0018	0.0002
13	1.0000	0.9997	0.9868	0.8894	0.6370	0.3279	0.1163	0.0280	0.0045	0.0005
14	1.0000	0.9999	0.9947	0.9393	0.7481	0.4468	0.1878	0.0540	0.0104	0.0013
15	1.0000	1.0000	0.9981	0.9692	0.8369	0.5692	0.2801	0.0955	0.0220	0.0033
16	1.0000	1.0000	0.9993	0.9856	0.9017	0.6839	0.3889	0.1561	0.0427	0.0077
17	1.0000	1.0000	0.9998	0.9937	0.9449	0.7822	0.5060	0.2369	0.0765	0.0164
18	1.0000	1.0000	0.9999	0.9975	0.9713	0.8594	0.6216	0.3356	0.1273	0.0325
19	1.0000	1.0000	1.0000	0.9991	0.9861	0.9152	0.7264	0.4465	0.1974	0.0595
20	1.0000	1.0000	1.0000	0.9997	0.9937	0.9522	0.8139	0.5610	0.2862	0.1013
21	1.0000	1.0000	1.0000	0.9999	0.9974	0.9749	0.8813	0.6701	0.3900	0.1611
22	1.0000	1.0000	1.0000	1.0000	0.9990	0.9877	0.9290	0.7660	0.5019	0.2399
23	1.0000	1.0000	1.0000	1.0000	0.9996	0.9944	0.9604	0.8438	0.6134	0.3359
24	1.0000	1.0000	1.0000	1.0000	0.9999	0.9976	0.9793	0.9022	0.7160	0.4439
25	1.0000	1.0000	1.0000	1.0000	1.0000	0.9991	0.9900	0.9427	0.8034	0.5561
26	1.0000	1.0000	1.0000	1.0000	1.0000	0.9997	0.9955	0.9686	0.8721	0.6641
27	1.0000	1.0000	1.0000	1.0000	1.0000	0.9999	0.9981	0.9840	0.9220	0.7601
28	1.0000	1.0000	1.0000	1.0000	1.0000	1.0000	0.9993	0.9924	0.9556	0.8389
29	1.0000	1.0000	1.0000	1.0000	1.0000	1.0000	0.9997	0.9966	0.9765	0.8987
30	1.0000	1.0000	1.0000	1.0000	1.0000	1.0000	0.9999	0.9986	0.9884	0.9405
31	1.0000	1.0000	1.0000	1.0000	1.0000	1.0000	1.0000	0.9995	0.9947	0.9675
32	1.0000	1.0000	1.0000	1.0000	1.0000	1.0000	1.0000	0.9998	0.9978	0.9836
33	1.0000	1.0000	1.0000	1.0000	1.0000	1.0000	1.0000	0.9999	0.9991	0.9923
34	1.0000	1.0000	1.0000	1.0000	1.0000	1.0000	1.0000	1.0000	0.9997	0.9967
35	1.0000	1.0000	1.0000	1.0000	1.0000	1.0000	1.0000	1.0000	0.9999	0.9987
36	1.0000	1.0000	1.0000	1.0000	1.0000	1.0000	1.0000	1.0000	1.0000	0.9995
37	1.0000	1.0000	1.0000	1.0000	1.0000	1.0000	1.0000	1.0000	1.0000	0.9998
38	1.0000	1.0000	1.0000	1.0000	1.0000	1.0000	1.0000	1.0000	1.0000	1.0000

PERCENTAGE POINTS OF THE NORMAL DISTRIBUTION

The values z in the table are those which a random variable $Z \sim N(0, 1)$ exceeds with probability p; that is, $P(Z > z) = 1 - \Phi(z) = p$.

p	z	p	z
0.5000	0.0000	0.0500	1.6449
0.4000	0.2533	0.0250	1.9600
0.3000	0.5244	0.0100	2.3263
0.2000	0.8416	0.0050	2.5758
0.1500	1.0364	0.0010	3.0902
0.1000	1.2816	0.0005	3.2905

POISSON CUMULATIVE DISTRIBUTION FUNCTION

The tabulated value is $P(X \leq x)$, where X has a Poisson distribution with parameter λ.

$\lambda =$	0.5	1.0	1.5	2.0	2.5	3.0	3.5	4.0	4.5	5.0
$x = 0$	0.6065	0.3679	0.2231	0.1353	0.0821	0.0498	0.0302	0.0183	0.0111	0.0067
1	0.9098	0.7358	0.5578	0.4060	0.2873	0.1991	0.1359	0.0916	0.0611	0.0404
2	0.9856	0.9197	0.8088	0.6767	0.5438	0.4232	0.3208	0.2381	0.1736	0.1247
3	0.9982	0.9810	0.9344	0.8571	0.7576	0.6472	0.5366	0.4335	0.3423	0.2650
4	0.9998	0.9963	0.9814	0.9473	0.8912	0.8153	0.7254	0.6288	0.5321	0.4405
5	1.0000	0.9994	0.9955	0.9834	0.9580	0.9161	0.8576	0.7851	0.7029	0.6160
6	1.0000	0.9999	0.9991	0.9955	0.9858	0.9665	0.9347	0.8893	0.8311	0.7622
7	1.0000	1.0000	0.9998	0.9989	0.9958	0.9881	0.9733	0.9489	0.9134	0.8666
8	1.0000	1.0000	1.0000	0.9998	0.9989	0.9962	0.9901	0.9786	0.9597	0.9319
9	1.0000	1.0000	1.0000	1.0000	0.9997	0.9989	0.9967	0.9919	0.9829	0.9682
10	1.0000	1.0000	1.0000	1.0000	0.9999	0.9997	0.9990	0.9972	0.9933	0.9863
11	1.0000	1.0000	1.0000	1.0000	1.0000	0.9999	0.9997	0.9991	0.9976	0.9945
12	1.0000	1.0000	1.0000	1.0000	1.0000	1.0000	0.9999	0.9997	0.9992	0.9980
13	1.0000	1.0000	1.0000	1.0000	1.0000	1.0000	1.0000	0.9999	0.9997	0.9993
14	1.0000	1.0000	1.0000	1.0000	1.0000	1.0000	1.0000	1.0000	0.9999	0.9998
15	1.0000	1.0000	1.0000	1.0000	1.0000	1.0000	1.0000	1.0000	1.0000	0.9999
16	1.0000	1.0000	1.0000	1.0000	1.0000	1.0000	1.0000	1.0000	1.0000	1.0000
17	1.0000	1.0000	1.0000	1.0000	1.0000	1.0000	1.0000	1.0000	1.0000	1.0000
18	1.0000	1.0000	1.0000	1.0000	1.0000	1.0000	1.0000	1.0000	1.0000	1.0000
19	1.0000	1.0000	1.0000	1.0000	1.0000	1.0000	1.0000	1.0000	1.0000	1.0000
$\lambda =$	5.5	6.0	6.5	7.0	7.5	8.0	8.5	9.0	9.5	10.0
$x = 0$	0.0041	0.0025	0.0015	0.0009	0.0006	0.0003	0.0002	0.0001	0.0001	0.0000
1	0.0266	0.0174	0.0113	0.0073	0.0047	0.0030	0.0019	0.0012	0.0008	0.0005
2	0.0884	0.0620	0.0430	0.0296	0.0203	0.0138	0.0093	0.0062	0.0042	0.0028
3	0.2017	0.1512	0.1118	0.0818	0.0591	0.0424	0.0301	0.0212	0.0149	0.0103
4	0.3575	0.2851	0.2237	0.1730	0.1321	0.0996	0.0744	0.0550	0.0403	0.0293
5	0.5289	0.4457	0.3690	0.3007	0.2414	0.1912	0.1496	0.1157	0.0885	0.0671
6	0.6860	0.6063	0.5265	0.4497	0.3782	0.3134	0.2562	0.2068	0.1649	0.1301
7	0.8095	0.7440	0.6728	0.5987	0.5246	0.4530	0.3856	0.3239	0.2687	0.2202
8	0.8944	0.8472	0.7916	0.7291	0.6620	0.5925	0.5231	0.4557	0.3918	0.3328
9	0.9462	0.9161	0.8774	0.8305	0.7764	0.7166	0.6530	0.5874	0.5218	0.4579
10	0.9747	0.9574	0.9332	0.9015	0.8622	0.8159	0.7634	0.7060	0.6453	0.5830
11	0.9890	0.9799	0.9661	0.9467	0.9208	0.8881	0.8487	0.8030	0.7520	0.6968
12	0.9955	0.9912	0.9840	0.9730	0.9573	0.9362	0.9091	0.8758	0.8364	0.7916
13	0.9983	0.9964	0.9929	0.9872	0.9784	0.9658	0.9486	0.9261	0.8981	0.8645
14	0.9994	0.9986	0.9970	0.9943	0.9897	0.9827	0.9726	0.9585	0.9400	0.9165
15	0.9998	0.9995	0.9988	0.9976	0.9954	0.9918	0.9862	0.9780	0.9665	0.9513
16	0.9999	0.9998	0.9996	0.9990	0.9980	0.9963	0.9934	0.9889	0.9823	0.9730
17	1.0000	0.9999	0.9998	0.9996	0.9992	0.9984	0.9970	0.9947	0.9911	0.9857
18	1.0000	1.0000	0.9999	0.9999	0.9997	0.9993	0.9987	0.9976	0.9957	0.9928
19	1.0000	1.0000	1.0000	1.0000	0.9999	0.9997	0.9995	0.9989	0.9980	0.9965
20	1.0000	1.0000	1.0000	1.0000	1.0000	0.9999	0.9998	0.9996	0.9991	0.9984
21	1.0000	1.0000	1.0000	1.0000	1.0000	1.0000	0.9999	0.9998	0.9996	0.9993
22	1.0000	1.0000	1.0000	1.0000	1.0000	1.0000	1.0000	0.9999	0.9999	0.9997

PERCENTAGE POINTS OF THE χ^2 DISTRIBUTION

The values in the table are those which a random variable with the χ^2 distribution on ν degrees of freedom exceeds with the probability shown.

ν	0.995	0.990	0.975	0.950	0.900	0.100	0.050	0.025	0.010	0.005
1	0.000	0.000	0.001	0.004	0.016	2.705	3.841	5.024	6.635	7.879
2	0.010	0.020	0.051	0.103	0.211	4.605	5.991	7.378	9.210	10.597
3	0.072	0.115	0.216	0.352	0.584	6.251	7.815	9.348	11.345	12.838
4	0.207	0.297	0.484	0.711	1.064	7.779	9.488	11.143	13.277	14.860
5	0.412	0.554	0.831	1.145	1.610	9.236	11.070	12.832	15.086	16.750
6	0.676	0.872	1.237	1.635	2.204	10.645	12.592	14.449	16.812	18.548
7	0.989	1.239	1.690	2.167	2.833	12.017	14.067	16.013	18.475	20.278
8	1.344	1.646	2.180	2.733	3.490	13.362	15.507	17.535	20.090	21.955
9	1.735	2.088	2.700	3.325	4.168	14.684	16.919	19.023	21.666	23.589
10	2.156	2.558	3.247	3.940	4.865	15.987	18.307	20.483	23.209	25.188
11	2.603	3.053	3.816	4.575	5.580	17.275	19.675	21.920	24.725	26.757
12	3.074	3.571	4.404	5.226	6.304	18.549	21.026	23.337	26.217	28.300
13	3.565	4.107	5.009	5.892	7.042	19.812	22.362	24.736	27.688	29.819
14	4.075	4.660	5.629	6.571	7.790	21.064	23.685	26.119	29.141	31.319
15	4.601	5.229	6.262	7.261	8.547	22.307	24.996	27.488	30.578	32.801
16	5.142	5.812	6.908	7.962	9.312	23.542	26.296	28.845	32.000	34.267
17	5.697	6.408	7.564	8.672	10.085	24.769	27.587	30.191	33.409	35.718
18	6.265	7.015	8.231	9.390	10.865	25.989	28.869	31.526	34.805	37.156
19	6.844	7.633	8.907	10.117	11.651	27.204	30.144	32.852	36.191	38.582
20	7.434	8.260	9.591	10.851	12.443	28.412	31.410	34.170	37.566	39.997
21	8.034	8.897	10.283	11.591	13.240	29.615	32.671	35.479	38.932	41.401
22	8.643	9.542	10.982	12.338	14.042	30.813	33.924	36.781	40.289	42.796
23	9.260	10.196	11.689	13.091	14.848	32.007	35.172	38.076	41.638	44.181
24	9.886	10.856	12.401	13.848	15.659	33.196	36.415	39.364	42.980	45.558
25	10.520	11.524	13.120	14.611	16.473	34.382	37.652	40.646	44.314	46.928
26	11.160	12.198	13.844	15.379	17.292	35.563	38.885	41.923	45.642	48.290
27	11.808	12.879	14.573	16.151	18.114	36.741	40.113	43.194	46.963	49.645
28	12.461	13.565	15.308	16.928	18.939	37.916	41.337	44.461	48.278	50.993
29	13.121	14.256	16.047	17.708	19.768	39.088	42.557	45.722	49.588	52.336
30	13.787	14.953	16.791	18.493	20.599	40.256	43.773	46.979	50.892	53.672

Answers

Prior knowledge 1

1 a 0.296 b 0.677 c 1.34×10^{-5}
2 a $k = \frac{1}{146}$ b $\frac{29}{146}$
3 $x = 3, y = -2, z = 0$

Exercise 1A

1 a $E(X) = 4.6, E(X^2) = 26$
b $E(X) = 0.3, E(X^2) = 2.5$
2 $E(X) = 4, E(X^2) = 18.2$
3 a

x	2	3	6
$P(X = x)$	$\frac{1}{2}$	$\frac{1}{3}$	$\frac{1}{6}$

b $E(X) = 3, E(X^2) = 11$
c No
4 a

X	1	2	3	4	5
$P(X = x)$	$\frac{1}{2}$	$\frac{1}{4}$	$\frac{1}{8}$	$\frac{1}{16}$	$\frac{1}{16}$

b $E(X) = 1.9375, E(X^2) = 5.1875$
c No
5 $a = 0.3, b = 0.3$
6 $a = 0.3, b = 0.4, c = 0.2$
7 $a = 0.1, b = 0.4$
8 a

X	1	2	3	4	5	6
$P(X = x)$	$\frac{1}{8}$	$\frac{1}{8}$	$\frac{1}{8}$	$\frac{1}{8}$	$\frac{3}{20}$	$\frac{7}{20}$

b X = number of 6s in 10 rolls, then $X \sim B(10, \frac{7}{20})$
$P(X \geqslant 3) = 0.738$
9 $P = 0.5$

Challenge
$E(X) = \frac{119}{24}$

Exercise 1B

1 a 1 b 2
2 a $E(X) = \frac{11}{6} = 1.83, Var(X) = \frac{17}{36} = 0.472$
b $E(X) = 0, Var(X) = 0.5$
c $E(X) = -0.5, Var(X) = 2.25$
3 $E(Y) = 4.5, Var(X) = 5.25$
4 a

s	$P(S = s)$
2	$\frac{1}{36}$
3	$\frac{2}{36}$
4	$\frac{3}{36}$
5	$\frac{4}{36}$
6	$\frac{5}{36}$
7	$\frac{6}{36}$
8	$\frac{5}{36}$
9	$\frac{4}{36}$
10	$\frac{3}{36}$
11	$\frac{2}{36}$
12	$\frac{1}{36}$

b 7
c 5.833
d 2.415

5 a

d	0	1	2	3
$P(D = d)$	$\frac{1}{4}$	$\frac{3}{8}$	$\frac{1}{4}$	$\frac{1}{8}$

b 1.25
c $\frac{15}{16} = 0.9375$
6 a $P(T = 1) = P(\text{head}) = 0.5,$
$P(T = 2) = P(\text{tail, head}) = 0.5 \times 0.5 = 0.25,$
$P(T = 3) = 1 - P(T = 1) - P(T = 2) = 0.25$
b $E(T) = 1.75, Var(T) = \frac{11}{16} = 0.688.$
7 a $E(X) = 4a + 2b$
b $a = 0.375, b = 0.25.$

Exercise 1C

1 a

y	−1	1	3	5
$P(Y = y)$	0.1	0.3	0.2	0.4

b $E(Y) = 2.8$
c $E(X) = 2.9$ and $2E(X) - 3 = 5.8 - 3 = 2.8 = E(Y)$.
2 a

y	−8	−1	0	1	8
$P(Y = y)$	0.1	0.1	0.2	0.4	0.2

b $E(Y) = 1.1$
3 a 8 b 4 c 2 d 18
e 8 f 3
4 a 6 b −9 c −2 d 1
e 9
5 a 4μ b $2\mu + 2$ c $2\mu - 2$ d $4\sigma^2$
e $4\sigma^2$
6 a 3.5
b $Y = 200 + 100X$
c $E(Y) = £550$
7 726.5 cm^3
8 a $E(X) = 1.25, Var(X) = 0.9375$
b $E(Y) = \frac{1}{4} \times 1 + \frac{3}{8} \times 2 + \frac{1}{4} \times 4 + \frac{1}{8} \times 8 = 3$
$E(Z) = 2E(X) + \frac{1}{2} = 3$
c $Var(Z) = 4Var(X) = 3.75$

Challenge
$E((X - E(X))^2) = E(X^2 - 2E(X)X + (E(X))^2)$
$= E(X^2) - 2E(X)E(X) + (E(X))^2 = E(X^2) - E(X)^2$

Exercise 1D

1 a $E(X) = 2$ b $Var(X) = 2$ c 1.414
2 a $E(X) = 2$ b $Var(X) = 4$ c $E(X^2) = 8$
3 $a = 0.1, b = 0.4$
4 a $-0.3 \leqslant E(Y) \leqslant 0.4$
b $a = 0.5, b = 0.2$
5 a $1 = \sum P(X = x) = 2a + 2b + c$
$2.4 = E(Y) = a + c + 4b + 9a = 10a + 4b + c$
$0.4 = P(Y > 2) = a + b$
b $a = 0.1, b = 0.3, c = 0.2$
c $P(2X + 3 \leqslant Y) = P(2 \leqslant X^2) = 2a = 0.2$
6 a $E(X) = 3.3$
b $1 = \sum P(X = x) = 3a + 2b + c$
$3.3 = E(X) = 6a + 9b + 6c$
$0.6 = P(Y \leqslant -5) = P(X \geqslant 3) = a + 2b + c$
c $a = 0.2, b = 0.1, c = 0.2$
d $P(X > 5 + Y) = P(X > 2) = 0.6$

Mixed exercise 1

1 a

x	$P(X = x)$
1	$\frac{1}{21}$
2	$\frac{2}{21}$
3	$\frac{3}{21}$
4	$\frac{4}{21}$
5	$\frac{5}{21}$
6	$\frac{6}{21}$

b $\frac{12}{21}$ or $\frac{4}{7}$ c $\frac{91}{21}$ or $\frac{13}{3}$ d $\frac{20}{9} = 2.22$
e $\frac{80}{9} = 8.89$ f $\frac{325}{3} = 108.3$

2 a 0.2 b 0.7 c 3.6 d 8.04

3 a 0.3 b $E(X) = 0 \times 0.2 + 1 \times 0.3 + 2 \times 0.5 = 1.3$
c 0.61 d 0.5

4 a $k + 0 + k + 2k = 1$,
so $4k = 1$, so $k = 0.25$
b $E(X) = 2$
$E(X^2) = 0^2 \times 0.25 + 1^2 \times 0 + 2^2 \times 0.25 + 3^2 \times 0.5$
$= 1 + 4.5 = 5.5$
c 6

5 a $\frac{1}{8}$ b $\frac{9}{8}$ c $\frac{19}{8}$ d $\frac{55}{64}$
e 0.2854

6 a 0.3 b 2.3 c 1.61 d 0.35
e 1.46 f 0.281

7 a Discrete uniform distribution
b Any discrete distribution where all the probabilities are the same.
c 2
d 2

8 a $p + q = 0.5$, $2p + 3q = 1.3$
b $p = 0.2$, $q = 0.3$.
c 1.29
d 5.16

9 a $\frac{1}{9}$ b $\frac{31}{9}$
c $\text{Var}(X) = E(X^2) - E(X)^2$
$= \frac{125}{9} - \left(\frac{31}{9}\right)^2 = \frac{164}{81} = 2.02$ (3 s.f.)
d 8.1 (1 d.p.)

10 a $E(X) = 3.5$
$\text{Var}(X) = E(X^2) - E(X)^2$
$= \frac{91}{6} - \frac{49}{4} = \frac{35}{12}$
b 6
c $\frac{35}{3}$ or 11.67
d 21

11 a

x	1	2	3	4
$P(X = x)$	$\frac{2}{26}$	$\frac{5}{26}$	$\frac{8}{26}$	$\frac{11}{26}$

b $\frac{19}{26}$ or 0.731 c $\frac{40}{13}$ or 3.077
d $\text{Var}(X) = E(X^2) - E(X)^2$
$\frac{270}{26} - \left(\frac{40}{13}\right)^2 \approx 0.92$ (2 s.f.)
e 8.3 (2 s.f.)

12 a 7 b −4 c 81 d 81
e 13 f 12

13 $E(S) = 64$, $\text{Var}(S) = 225$

14 a

x	1	2	3
$P(X = x)$	0.25	0.375	0.375

b 2.125 c 0.609 d 5.25 e 5.48

15 a 0.2 b 0.76 c 1.07 d 0.0844

16 a $a = 0.4$, $b = 0.2$
b $E(X^2) = 1.3$, $\text{Var}(X) = 0.81$
c $\text{Var}(Y) = 7.29$
d $P(Y + 2 > X) = P(3X - 1 + 2 > X) = P(X > -0.5) = 0.9$

17 a $2a + 2b + c = 1$, $3b + 4c = 2.3$, $2a + b = 0.4$
b $a = 0.15$, $b = 0.1$, $c = 0.5$
c $P(-2X > 10Y) = P(X > 1) = 0.75$

Challenge

$$E(X) = \sum_{i=1}^{n} \frac{i}{n} = \frac{n(n+1)}{2n} = \frac{n+1}{2}$$

$$E(X^2) = \sum_{i=1}^{n} \frac{i^2}{n} = \frac{n(n+1)(2n+1)}{6n} = \frac{(n+1)(2n+1)}{6}$$

$$\text{Var}(X) = E(X^2) - (E(X))^2 = \frac{(n+1)(2n+1)}{6} - \frac{(n+1)^2}{4}$$

$$= \frac{4n^2 + 6n + 2 - 3n^2 - 6n - 3}{12} = \frac{(n+1)(n-1)}{12}$$

Prior knowledge 2

1 a 0.0168 b 0.001 c 0.3972

2 a $E(X) = 3.8$ b $E(X^2) = 18$ c $\text{Var}(X) = 3.56$

Exercise 2A

1 a 0.2138 b 0.7127 c 0.4703

2 a 0.1733 b 0.8153 c 0.7531

3 a 0.1323 b 0.3954 c 0.5429

4 a 0.3626 b 0.5683 c 0.1950

5 $\lambda = 3$

6 $\lambda = 6$

Exercise 2B

1 a 0.2017 b 0.4711 c 0.7211

2 a 0.7798 b 0.6615 c 0.3035

3 a 0.8641 b 0.6139 c 0.5368

4 a 0.4679 b 0.3606 c 0.8200

5 a 6 b 9 c 5 d 5

6 a 5 b 2 c All values of $c > 6$
d All values of $d > 8$

Exercise 2C

1 a i 0.1680 ii 0.0839
b i 0.1606 ii 0.2851

2 a (1) Weeds grow independently
(2) Weeds grow at a constant rate/unit of area
b 0.1088 c 0.2084

3 a $X \sim \text{Po}(2.5)$
b (1) Faults occur independently
(2) Faults occur at a constant rate
c 0.2565 d 0.7586 e 0.8699

4 a i 0.1755 ii 0.0681
b i 0.7586 ii 0.8622

5 a 0.0839 b 0.1512

6 a 0.1247 b 0.0137

7 a i 0.1653 ii 0.1607 iii 0.2694
b 0.4202

8 a i 0.1336 ii 0.4562
b 0.7135

9 a i 0.5276 ii 0.1329
b 0.5276 – breakdowns occur independently of each other

10 a 0.1255 b 0.1512 c 0.1670

11 a 0.1247 b 0.0260

12 a 0.2650 b 11 minibuses

13 a 0.2971
b $Y \sim \text{Po}(3)$. $P(X > 8) = 1 - P(X \leqslant 8) = 1 - 0.9962 = 0.0038 = 0.38\%$
c 10
14 a 0.8088 **b** 0.1847 **c** 14

Exercise 2D

1 a 0.1606 **b** 0.7440 **c** 0.7149
2 a 0.1465 **b** 0.2414 **c** 0.2236
3 a 0.0474 **b** 0.6159 **c** 0.3099 **d** 0.2851
4 a 0.5049 **b** 0.3134
5 a i 0.1213 **ii** 0.7166
b Events occur at a constant average rate – the mean number of an interval is proportional to the length of the interval.
6 a 0.2238 **b** 0.2707 **c** 0.4579
7 a 0.6988 **b** 0.3153 **c** 0.1607
8 a 0.2090 **b** 0.3374 **c** 0.4457
9 a 0.0158 **b** 0.7534 **c** 0.2417
10 a 0.2639 **b** 0.7657 **c** 0.0754

Challenge

a $Q \sim \text{Po}(\lambda + \mu)$

$$P(Q = 0) = \frac{(e^{-(\lambda+\mu)} \times (\lambda + \mu)^0)}{0!}$$

$(\lambda + \mu)^0 = 1$ and $(0!) = 1$
therefore $P(Q = 0) = e^{(-\lambda + \mu)}$
$Q \sim \text{Po}(\lambda + \mu)$

b $$P(Q = 1) = \frac{(e^{-(\lambda+\mu)} \times (\lambda + \mu)^1)}{1!}$$

$(\lambda + \mu)^1 = (\lambda + \mu)$ and $1! = 1$
therefore $P(Q = 1) = e^{-(\lambda+\mu)} \times (\lambda + \mu)$

Exercise 2E

1 a Mean = 1.43, Variance = 1.4251
b Mean ≈ variance
c Using $\lambda = 1.4$, P = 0.1128
2 a Mean = 3.64, Variance = 3.5604
b Mean ≈ variance
c Using $\lambda = 3.6$, $P(X \leqslant 2) = 0.3027$
d From the table relative frequency of obtaining no more than 2 cars per period is 0.29. Answer from **c** is very close to recorded value.
3 a Mean = 2.867, Variance = 2.897
b Mean ≈ variance
c Because the observed frequency for 8 or more flaws was 0.
d 99 (using $\lambda = 2.9$)

Challenge

Proof outline:

$$E(X) = \sum_{i=0}^{\infty} i \times P(X = i) = \sum_{i=0}^{\infty} \frac{i \times e^{-\lambda}\lambda^i}{i!} = e^{-\lambda}\sum_{i=1}^{\infty} \frac{i \times \lambda^i}{(i-1)!}$$

$$= e^{-\lambda}\sum_{i=0}^{\infty} \frac{\lambda^{i+1}}{i!} = e^{-\lambda} \times \lambda\sum_{i=0}^{\infty} \frac{\lambda^i}{i!} = e^{-\lambda} \times \lambda \times e^{\lambda} = \lambda$$

$\text{Var}(X) = E(X^2) - E^2(X)$
Using a similar approach to above to gain: $E(X^2) = \lambda^2 + \lambda$
$\text{Var}(X) = (\lambda^2 + \lambda) - \lambda^2 = \lambda$

Exercise 2F

1 a 8.4 **b** 2.52
2 a 8 **b** 0.1239 **c** 0.3154
3 0.4 or 0.6
4 0.2 or 0.8
5 $n = 12$, $p = 0.4$
6 a 0.3 **b** 0.1643
7 a i 0.1536 **ii** 0.9051
b i 120 **ii** 27.3
8 a 0.5772 **b** Mean = 69.26 Variance = 29.28
9 a 0.3 **b** $E(X) = 1.5$, $\text{Var}(X) = 1.05$
10 a Mean = 1, Variance = 0.8
b 0.2
c 164, 205, 102, 26, 3, 0
The values support the student's suggestion that the data can be modelled by a binomial distribution
d Variance = 5 × 0.2 × 0.8 = 0.8
The calculated variance matches the observed variance of the data and supports the use of a binomial distribution

Challenge

a $P(X = 0) = \binom{3}{0} \times p^0 \times (1 - p)^3 = (1 - p)^3$

$P(X = 1) = \binom{3}{1} \times p^1 \times (1 - p)^2 = 3p(1 - p)^2$

$P(X = 2) = \binom{3}{2} \times p^2 \times (1 - p)^1 = 3p^2(1 - p)$

$P(X = 3) = \binom{3}{3} \times p^3 \times (1 - p) = p^3$

$E(X) = \sum XP(X = x)$
$= (0 \times (1 - p)^3) + (1 \times 3p(1 - p)^2) + (2 \times 3p^2(1 - p)) + (3 \times p^3)$
$= 3p - 6p^2 + 3p^3 + 6p^2 - 6p^3 + 3p^3 = 3p$

b $E(X^2) = (0^2 \times (1 - p)^3) + (1^2 \times 3p(1 - p)^2) + (2^2 \times 3p^2(1 - p)) + (3^2 \times p^3)$
$= 3p - 6p^2 + 3p^3 + 12p^2 - 12p^3 + 9p^3 = 3p + 6p^2$
$\text{Var}(X) = E(X^2) - E^2(X) = 3p + 6p^2 - (3p)^2 = 3p(1 - p)$

Exercise 2G

1 a i 0.1781 **ii** 0.1183
b i 0.1755 **ii** 0.1247
2 a i 0.1628 **ii** 0.1458
b i 0.1606 **ii** 0.1512
3 a i 0.1963 **ii** 0.2351
b i 0.1954 **ii** 0.2381
4 a 0.1075 **b** 0.1074
c The two values are similar, so a Poisson distribution is a good approximation in this case.
5 a 0.6472 **b** 0.2240
6 a 0.0984 **b** 0.8743
7 a 0.1422 **b** 0.3782
8 a 0.1991 **b** 0.6472
9 a $X \sim B(10, 0.05)$ **b** 0.0105 **c** 0.5298
10 0.3498
11 a $X \sim B(1200, 0.005)$
b Mean = 6, Variance = 5.97
c 0.2851
12 a 0.2378 **b** 0.0315
13 a 0.7350 **b** 0.2788
14 a 0.2068 **b** 0.1242
15 a 0.0186 **b** 0.0863

Mixed exercise 2

1 a 0.4966 **b** 0.2700 **c** 0.2376
2 a (1) Misprints occur independently
(2) Misprints occur at a constant rate
b 0.3425 **c** 0.1689
3 $\lambda = 5$
4 a (1) Emails arrive independently
(2) Emails arrive at a constant rate

b **i** 0.1377
ii 0.2560
5 **a** n is large, p is small
b 0.4253 **c** 0.4335 **d** 1.93%
6 $\lambda = 7$
7 **a** **i** 0.2627 **ii** 0.0582
b 0.2560
8 0.2022
9 **a** **i** 0.1336 **ii** 0.4562
b 0.2084 **c** 0.2992
10 **a** $X \sim \text{Po}(6)$, properties are sold independently and at a constant rate
b 0.1606 **c** 0.1090 or 0.1091 (using unrounded answer for part **a**)
11 **a** 0.3848 **b** 0.1804 **c** 0.0440
12 **a** 0.3285 **b** 0.1042 **c** 0.3134
13 **a** $X \sim \text{B}(150, 0.04)$
b 0.0174 **c** 0.9380
14 **a** 0.0162 **b** Mean = 10, Variance = 9.8
c 0.2202
15 **a** 0.0230 **b** Mean = 3, Variance = 2.925
c 0.0335
16 **a** 0.1321 **b** 0.7135 **c** 0.3191
17 **a** 0.1141 **b** 0.0103
18 **a** $X \sim \text{Po}(4)$ Website visits occur independently of each other and at a constant average rate.
b 0.0298 **c** 0.2834
19 **a** Mean = 2.86, Variance = 2.867
b Mean $\approx$ variance
c Using $\lambda = 2.86$, P = 0.4553

Challenge

a $\frac{63}{256}$ or 0.2461 (4 d.p.)
b $\frac{7}{128}$ or 0.0547 (4 d.p.)

Prior knowledge 3

1 **a** **i** 0.1171 **ii** 0.4159 **iii** 0.0565
b **i** 8 **ii** 4.8
2 $\frac{25}{216}$

Exercise 3A

1 **a** 0.0347 **b** 0.6794 **c** 0.5803
2 **a** 0.0623 **b** 0.4565 **c** 0.4324
3 **a** 0.0965 **b** 0.4213 **c** 0.4823 **d** 0.4984
4 **a** **i** 0.147 **ii** 0.3430
b Attempts are independent; probability stays constant
5 **a** **i** $\frac{1}{4}$ **ii** 0.0791 **iii** 0.6836
b The probability is the same on each attempt
6 **a** 3 **b** 0.9085
7 **a** 15 **b** 3 **c** 94
8 **a** 0.0531 **b** 0.5905
9 **a** 0.0315 **b** 0.4877
10 **a** 0.0469 **b** 0.0156
11 **a** 0.3712 **b** 0.0580 **c** 0.0244

Exercise 3B

1 **a** 5 **b** 20
2 **a** 3 **b** 6
3 **a** 0.0796 **b** 0.0429
c **i** $\frac{20}{13}$ or 1.538 **ii** $\frac{140}{169}$ or 0.8284
4 **a** $\frac{1}{4}$ **b** 12
5 **a** $\frac{1}{5}$ **b** 5
6 **a** $\frac{1}{20}$ **b** 20
7 **a** Geometric **b** Constant p; independent trials
c 0.2 **d** 5 **e** 20
8 **a** **i** $\frac{20}{3}$ **ii** $\frac{340}{9}$ **b** 0.0921 **c** 0.3206
d 0.6229
9 **a** Constant p; independent trials
b **i** 0.1056 **ii** 0.7744
c $\text{E}(X) = \frac{25}{3}$, $\text{Var}(X) = \frac{550}{9}$
d 0.0086 **e** 0.0098
10 **a** Poisson; faults occur independently and at random, long term average is constant.
b 0.0474 **c** 0.0354
d $\text{E}(X) = 21$, $\text{Var}(X) = 424$ (both nearest whole number)
e 0.1196

Exercise 3C

1 0.0659
2 0.1668
3 0.0552
4 **a** 0.1406 **b** 0.0330 **c** 0.2816 **d** 0.4744
5 **a** 0.1029 **b** 0.6496 **c** 0.0953 **d** 0.2763
6 **a** 0.1853
b Games are independent and probability of success is same in each game.
c 0.1838 **d** 0.2660
7 **a** 0.1409
b Trials are independent and probability of success is same in each trial.
c 0.9806 **d** 0.1958
8 **a** 0.0285 **b** 0.6474 **c** 0.0272 **d** 0.4629
9 **a** 0.0515 **b** 0.4202 **c** 0.9993 **d** 0.0095
10 **a** $D \sim$ Negative binomial $(3, p)$
b **i** 0.1148 **ii** 0.5323 **iii** 0.0647
c The probability of success might change as she gets more practice.

Challenge

a $\text{P}(Y \leqslant 8) = 1 - \text{P}(Y > 8)$
$\text{P}(Y > 8)$ is the probability of 2 or fewer successes in first 8 trials
So $\text{P}(Y \leqslant 8) = 1 - \text{P}(X \leqslant 2)$ where $X \sim \text{B}(8, 0.4)$
So $\text{P}(Y \leqslant 8) = 1 - \text{F}_{8,\,0.4}(2)$
b $\text{P}(Y \leqslant y) = 1 - \text{P}(Y > y)$
$\text{P}(Y > y)$ is the probability of $r - 1$ or fewer successes in first y trials
So $\text{P}(Y \leqslant y) = 1 - \text{P}(X \leqslant r - 1)$ where $X \sim \text{B}(y, p)$
So $\text{P}(Y \leqslant 8) = 1 - \text{F}_{y,\,p}(r - 1)$

Exercise 3D

1 **a** $\frac{15}{2}$ **b** $\frac{45}{4}$
2 **a** $\frac{40}{3}$ **b** $\frac{40}{9}$
3 **a** $\frac{1}{4}$ **b** 0.1055 **c** 24
4 **a** $\frac{2}{5}$
b **i** 0.0280 **ii** 0.0250
5 **a** 6
b **i** 0.0669 **ii** 0.0498
6 **a** $\frac{2}{3}$, 4 **b** $\frac{16}{81}$ or 0.1975
7 **a** Attempts are independent and probability of success is same in each attempt
b $\frac{50}{7}$, 1.75 (3 s.f.)
8 **a** $\frac{1}{8}$ **b** 0.0131 **c** 96
9 **a** $\frac{4}{25}$ **b** $\frac{1575}{16}$ **c** $\frac{125}{6}$ **d** 0.4561
10 **a** Negative binomial

b i 25 ii 3
c Probability of success is not constant
d $\frac{50}{539}$

Mixed exercise 3

1 a $\frac{81}{1024}$ or 0.0791 b $\frac{9}{16}$
2 a Geo(0.1) b 10, 90 c 0.3138
3 a Geometric b 0.0804 c 30
d Throws are independent, probability is the same on each throw
4 6
5 a 0.001234 b 0.0012
6 a 0.0655 b 0.0604 c 25, 10 d $\frac{3}{7}$
7 a $\frac{2}{5}$ b 0.1003 c 0.5941 d 0.1045
8 a $0.65 \times 0.35 = 0.2275$
b i 0.2786 ii 0.1741 iii 0.1717 iv 0.1811
c Mean $= \frac{100}{13}$, Standard deviation $= \sqrt{\frac{700}{169}}$
d 0.39 e 0.084
f 0.009261

Challenge

1 a Negative binomial$(2, p)$
b Both are geometric distributions
c $X = Y_1 + Y_2$
d $\text{E}(X) = \text{E}(Y_1) + \text{E}(Y_2) = \frac{1}{p} + \frac{1}{p} = \frac{2}{p}$
2 $X \sim$ Negative binomial(r, p)
Allocate random variables $Y_1,\ldots,Y_r$ with $Y_1,\ldots, Y_r \sim \text{Geo}(p)$ such that $X = Y_1 + \ldots + Y_r$
Then, $\text{E}(X) = \text{E}(Y_1) + \ldots + \text{E}(Y) = \frac{1}{p} + \ldots + \frac{1}{p} = r \times \frac{1}{p}$
Also, $\text{Var}(X) = \text{Var}(Y_1) + \ldots + \text{Var}(Y_r)$
$= \frac{1(1-p)}{p^2} + \ldots + \frac{1(1-p)}{p^2} = r \times \frac{1(1-p)}{p^2}$

Prior knowledge 4

1 a 0.1563 b 0.4335 c 0.1079 d 4
2 a 0.1052 b 0.8308
c 270 (nearest whole number)
3 $x \geqslant 8$

Exercise 4A

1 Reject H_0
2 Reject H_0
3 Fail to reject H_0
4 Fail to reject H_0
5 Reject H_0: Evidence suggests the mean number of misprints has increased.
6 Reject H_0: Evidence suggests there is an increase in rate of accidents.
7 Fail to reject H_0: There is no evidence of increase in the rate at which the coffee machine seizes up.
8 Fail to reject H_0: There is no evidence to suggest the rate of sales has changed.
9 Fail to reject H_0: There is no evidence to suggest a reduction in the rate of accidents occurring at the crossroads.
10 Fail to reject H_0: There is no evidence to suggest the average number of flaws has changed.
11 a 0.1606 b 0.8472
c Fail to reject H_0: There is no evidence to suggest the mean number of breakdowns has decreased.
12 Fail to reject H_0: There is no evidence to suggest reductions of times the doctor sees patients with the condition.
13 Fail to reject H_0: There is no evidence to suggest manager's suspicion is correct.
14 a i 0.1251 ii 0.2202
b n large, p small
c Fail to reject H_0: There is no evidence to suggest the servicing has reduced the number of defective components.

Exercise 4B

1 a Critical region: $X \leqslant 1$; Significance level = 0.0266
b Critical region: $X \geqslant 16$; Significance level = 0.0082
c Critical region: $X \geqslant 9$; Significance level = 0.0214
2 Critical region: $X \geqslant 16$
3 Critical region: $X \geqslant 14$
4 Critical region: $X \leqslant 2$
5 Critical region: $X \leqslant 1$
6 Critical region: $X \geqslant 13$
7 a Critical region: $X = 0$ or $X \geqslant 9$; Significance level = 0.0397
b Critical region: $X \leqslant 2$ or $X \geqslant 15$; Significance level = 0.0311
c Critical region: $X \leqslant 3$ or $X \geqslant 17$ Significance level = 0.0326
8 a Critical region: $X \leqslant 2$ or $X \geqslant 14$
b 0.0419
c $X = 11$ not in critical region, hence no change in the rate.
9 a Critical region: $X \leqslant 1$ or $X \geqslant 10$
b 0.0722
10 a e-mails arrive at random and at a constant average rate.
b Critical region: $X \leqslant 3$ or $X \geqslant 16$
c 0.0432
d $X = 13$ not in critical region hence no evidence to suggest the mean rate is different to 9.
11 a $c = 8$ b 0.0311

Exercise 4C

1 Not significant. Fail to reject H_0.
2 Significant. Reject H_0.
3 Significant. Reject H_0.
4 Not significant. Fail to reject H_0.
5 Significant. Reject H_0.
6 Reject H_0. There is evidence that the probability of getting a 6 is less than $\frac{1}{6}$.
7 Reject H_0. There is evidence that the probability of getting an A is less than $\frac{1}{5}$.
8 a $X \sim \text{Geo}(\frac{1}{4})$
b 0.0791
c Fail to reject H_0. There is no evidence to suggest the probability of Lucy scoring a goal from a free kick is now less than $\frac{1}{4}$.
9 Reject H_0. There is evidence the student's suspicion is correct.
10 Reject H_0. There is evidence that *Wisetalk* are over-stating their percentage.
11 H_0: $p = 0.3$, H_1: $p < 0.3$.
Reject H_0. There is evidence to suggest rival's claim is correct.
12 a $X \sim \text{Geo}(\frac{1}{6})$.
Fixed probability of seeing a robin. The probability of seeing a robin on one day is independent of the probability of seeing a robin on another day.
b i 0.1157 ii 0.4823
c Reject H_0. There is evidence to suggest Imelda is over-stating the probability.

Exercise 4D

1 a Critical region: $X \geqslant 10$ b 0.0404
2 a Critical region: $X \geqslant 8$ b 0.0490
3 a Critical region: $X \leqslant 2$ b 0.0975
4 a Critical region: $X \geqslant 13$ b 0.0434
c As $X = 11$ is not in the critical region, we do not reject H_0.
5 a Critical region: $X \geqslant 9$ b 0.0390
6 Critical region: $X \geqslant 5$

Challenge
a Critical region: $X \leqslant 3$ or $X \geqslant 409$ b 0.0518
c $X = 5$ is not in the critical region, hence do not reject H_0.

Mixed exercise 4

1 a 0.1575 b 0.3272
c Reject H_0. There is evidence to suggest that the number of vehicles has reduced.
2 Fail to reject H_0. There is no evidence to suggest a decrease in the number of deformed red blood cells.
3 Fail to reject H_0. There is no evidence to suggest that the crosswords are more difficult.
4 a 0.3025
b Reject H_0. There is evidence to suggest that the meteorologist is correct.
5 a 0.1708 b 0.3423
c Fail to reject H_0. There is no evidence to suggest that Waldo has decreased the rate.
6 a 0.1552 b 0.1424
c 0.1031 or 0.1032 (using unrounded answer from part **a**)
d $X \geqslant 2$ e 0.0296
7 Fail to reject H_0. There is no evidence to suggest an increase in sales
8 a Fail to reject H_0. There is no evidence to suggest the rate of visits is greater on a Saturday.
b $X = 15$
9 Fail to reject H_0. There is no evidence to suggest the percentage is higher than the manager thinks.
10 a 0.0783 b 0.7225 c $X \geqslant 20$ d 0.0456
11 a H_0: $\lambda = 4$, H_1: $\lambda > 4$ b $X = 9$
c As $X = 8$ is not in the critical region, the scientist's suggestion is rejected.
12 a $X \leqslant 8$ b $Y \geqslant 7$
c The probability that Alison has incorrectly rejected H_0 is 0.0081 and Paul is 0.0156.

Challenge
a Negative binomial, successive trials each with the same probability of success where p is the number of trials needed for r successes.
b Critical region: $X \leqslant 5$ c 0.0426

Prior knowledge 5

1 a $P(X > 115) = 1 - 0.2660 = 0.7340$
b $P(120 < X < 130) = 0.3944$
c $a = 114.60$
2 a -9.5 b $\frac{105}{4}$ c $\frac{2}{3}$
3 0.2744

Exercise 5A

1 a 0.0072
b Sample taken from a population that was normally distributed, so answer is not an approximation.
2 a 0.2525 b 0.0098 ~ 0.0096
3 a 0.0668 or 0.066807 b $n = 241$
4 0.1855
5 a 0.0416 b 0.0130
6 0.1103
7 a $k = 0.15$ b 0.1727
c Answer is an approximation, n is large, so fairly accurate.
8 Need n at least 1936
9 a Salaries are unlikely to be symetrically distributed so normal distribution would not be a good model.
b i 0.0231 ii 0.7804
c Estimate likely to be inaccurate, small sample size and unknown if original distribution was normal.
10 96

Exercise 5B

1 a 0.2084
b 0.1807, this estimate is inaccurate, sample size not big enough
2 a $E(X) = 4$, $Var(X) = 12$
b 0.1587
3 0.9214
4 a 10 b 0.0786
5 a 0.1680 b 0.1587
6 a 5 b 0.3085
7 a 0.0352 b 0.9981
8 a 0.0019 b 0.0416

Mixed exercise 5

1 0.0228
2 0.1855
3 $n \geqslant 3$
4 0.1030
5 a 0.1804 b 0.4191
6 a 0.25 b 0.1855
7 a 0.9171
b Sample taken from a population that was normally distributed, so answer is not an estimate.
c 0.7009
8 0.8944
9 $n = 7$
10 0.0014
11 a

x	0	1
$P(X = x)$	0.4	0.6

$E(X) = 0.6$
$Var(X) = 0.24$

b 0.1709
c $n > 1025$

Challenge
$X_1 + \dots + X_n \sim N(n\mu, n\sigma^2)$ and so
$$\bar{X} = \frac{1}{n}(X_1 + \dots + X_n) \sim N\left(\frac{n\mu}{n}, \frac{n\sigma^2}{n^2}\right) = N\left(\mu, \frac{\sigma^2}{n}\right)$$

Review exercise 1

1 a

x	$P(X = x)$
1	$\frac{1}{36}$
2	$\frac{3}{36}$
3	$\frac{5}{36}$
4	$\frac{7}{36}$
5	$\frac{9}{36}$
6	$\frac{11}{36}$

Online Full worked solutions are available in SolutionBank.

b $\frac{7}{12}$ or 0.583 **c** $\frac{161}{36}$ or 4.47

d $\text{Var}(X) = \text{E}(X^2) - (\text{E}(X))^2 = \frac{791}{36} - \frac{25921}{1296} = 1.97$ (3 s.f.)

e 17.7

2 a $\frac{1}{17}$ or 0.0588 **b** $\frac{64}{17}$ or 3.76

c $\text{Var}(X) = \text{E}(X^2) - (\text{E}(X))^2 = \frac{266}{17} - \frac{4096}{289} = 1.47$ (3 s.f.)

d 13.3

3 a $p + q = 0.4,\ 2p + 4q = 1.3$

b $p = 0.15,\ q = 0.25$

c 1.75

d 7.00

4 a $p + q = 0.45,\ 3p + 7q = 1.95$

b $p = 0.3,\ q = 0.15$

c 0.35 **d** 7.15 **e** 1 **f** 114.4

5 a 0.1, 0.4 **b** 1.5, 1.41 **c** 12.69 **d** 0.4

6 a

x	−3	−2	0	1	3
$\text{P}(X = x)$	0.1	0.2	0.2	0.1	0.4

b 0.9

7 a $X \sim \text{Po}(1.5)$ **b** 0.251 (3 s.f.)

c 0.469 (3 s.f.) **d** 0.185 (3 s.f.)

8 a Events occur at a constant rate.
Events occur independently or randomly.
Events occur singly.

b i 0.134 (3 s.f.) **ii** 0.715 (3 s.f.)

c 0.149 (3 s.f.)

9 a 0.0816 **b** 0.1931 **c** 0.5673

10 a 1.45; 1.4075

b Mean ≈ Variance

c $\text{P}(X > 2) = 0.4253$

11 a If $X \sim \text{B}(n, p)$ and
n is large
p is small
then X can be approximated by $\text{Po}(np)$.

b 0.0001

c 0.00098

d mean $= np = 10$
variance $= np(1 - p) = 9.9$

e 0.870 (3 s.f.)

12 a 0.226 (3 s.f.)

b $\lambda = 3$, if $X \sim \text{B}(n, p)$ and n is large p is small then X can be approximated by $\text{Po}(np)$.

13 a $X \sim \text{B}(200, 0.015)$

b 0.1693

c $X \sim \text{B}(n, p)$ and n is large p is small then X can be approximated by $\text{Po}(np)$.
$\lambda = 3$

d $\text{P}(X = 4) = 0.1680$
% error = 0.77

14 a Geo(0.05) **b** 20, 380 **c** 0.4877

15 a Geometric **b** 0.0531 **c** 90

d Attempts are independent and probability of success is constant.

16 a 0.5811 **b** 16 spins

17 a 0.0529

b Games are independent and the probability of winning is constant.

c 22.2; 10.1 (both 3 s.f.)

d 0.15

18 a 0.3 **b** 0.0889 **c** 0.0467

19 a i An hypothesis test is a mathematical procedure to examine a value of a population parameter proposed by the null hypothesis H_0, compared to the alternative hypothesis H_1.

ii The critical region is the range of values of a test statistic that would lead you to reject H_0.

b $\lambda < 0.45$ critical region is 3 or fewer ($X \leqslant 3$) calls in the 20-minute period.
$\lambda > 0.45$ critical region is 16 or more ($X \geqslant 16$) calls in the 20-minute period.

c 4.33%

d $\text{P}(X \leqslant 1) = 0.0611 > 0.05$ so result is not significant, do not reject H_0.

20 a $\text{P}(X \geqslant 11) = 0.138 > 0.05$ so result is not significant, do not reject H_0.

b Would reject H_0 at the 15% significance level.

21 a 16 or fewer, 33 or more

b 10.3%

c 18 does not lie in critical region so there is no evidence at the given level of significance that the mean rate of orders is different from that claimed.

22 a $X \sim \text{Geo}(0.2)$

b $0.0440 < 0.05$ so there is evidence that Mr Taylor's suspicion valid.

23 H_0: $p = 0.5$ H_1: $p < 0.5$
Assume H_0, so that $X \sim \text{Geo}(0.5)$
Significance level 10%
$\text{P}(X \geqslant 5) = (1 - 0.5)^4 = (0.5)^4 = 0.0625$
$0.0625 < 0.1$
There is sufficient evidence to reject H_0, and conclude that Xander is overestimating his shooting accuracy.

24 a $X \leqslant 102$

b 115 is not in the critical region so there is no evidence to doubt Brian's claim.

25 a N(90, 0.25)
Application of central limit theorem (as sample large)

b 0.0228

26 0.4875

27 a 0.1

b 0.5539

c Accurate since n is large.

28 a 0.0699 **b** 0.0810

c 0.0787; values are close so the central limit theorem does provides a reasonable estimate.

29 0.2593

30 a 0.0443 **b** 20 **c** 0.3325

Challenge

1 a

x	0	1	2	3
$\text{P}(X = x)$	$\frac{1}{16}$	$\frac{9}{32}$	$\frac{12}{32}$	$\frac{9}{32}$

b $\text{E}(X) = \Sigma x\text{P}(X = x) = 0 \times \frac{1}{16} + 1 \times \frac{9}{32} + 2 \times \frac{12}{32} + 3 \times \frac{9}{32} = \frac{15}{8}$

2 a Each term is $\binom{x}{r} p^r q^{x-r}$

b $\text{E}(X) = \Sigma x\text{P}(X = x) = 0q^4 + 4q^3p + 12q^2p^2 + 12qp^3 + 4p^4,\ q = 1 - p$
$\Rightarrow 4p - 12p^2 + 12p^3 - 4p^4 + 12p^2 - 24p^3 + 12p^4 + 12p^3 - 12p^4 + 4p^4 = 4p$
$\text{Var}(X) = \text{E}(X^2) - (\text{E}(X))^2 = 0q^4 + 4q^3p + 24q^2p^2 + 36qp^3 + 16p^4 - 16p^2,\ q = 1 - p$
$\Rightarrow 4p - 12p^2 + 12p^3 - 4p^4 + 24p^2 - 48p^3 + 24p^4 + 36p^3 - 36p^4 + 16p^4 - 16p^2$
$= 4p - 4p^2 = 4p(1 - p)$

3 a $X \leqslant 10$

b 2.39%

Prior knowledge 6

1 0.4043
2 0.132
3 H_0: $p = 0.6$, H_1: $p \neq 0.6$
Here n is large and p is near 0.5 so a normal approximation can be used.
$B(100, 0.6) \approx N(60, 24)$
$P(X \geqslant 70) = P(W > 69.5) = P\left(Z > \frac{69.5 - 60}{\sqrt{24}}\right)$
$= P(Z > 1.94) = 0.026$
$0.025 > 0.025$, therefore we fail to reject H_0. There is no evidence that David is wrong.

Exercise 6A

1 H_0: There is no difference between the observed and expected distributions.
H_1: There is a difference between the observed and expected distributions.
2 a H_0: The observed distribution is the same as the discrete uniform distribution.
H_1: The observed distribution is not the discrete uniform distribution.
b $X^2 = 1.6$
3 a H_0: There is no difference between the observed distribution and the discrete uniform distribution.
H_1: There is a difference between the observed distribution and the discrete uniform distribution.
b 150 c $X^2 = 14.33$
4 a

Mutation present	Yes	No
Expected frequency	120	40

b H_0: There is no difference between the observed and expected distributions.
H_1: There is a difference between the observed and expected distributions.
c $X^2 = 0.3$
5 a

Result	H	T
Expected frequency for fair coin	25	25
Expected frequency for biased coin	30	20

b $X^2_{\text{fair}} = 0.72$, $X^2_{\text{bias}} = 0.33$
c Since a lower goodness of fit score is better, it is more likely John was flipping the biased coin.
6 Goodness of fit for Welsh men: 2.074, goodness of fit for Welsh women: 4.076. Therefore the distribution for English adults is a closer match for Welsh men than women.

Exercise 6B

1 6 degrees of freedom (7 observations, 1 constraint)
2 11.070
3 a 11.070 b 20.090 c 15.987
4 18.307
5 13.362
6 1.646
7 1.145
8 a 5.226 b 21.026
9 a Combine $x = 4$ and $x = 5$ cells into a $x \geqslant 4$ cell, so that expected value is $6.25 > 5$.
b With 4 cells there are 3 degrees of freedom (4 observations, 1 constraint), and so χ^2_3 is a suitable distribution to model the goodness of fit. We have $\chi^2_3\,(1\%) = 11.345$.

Exercise 6C

1 We calculate $X^2 = 4.33$.There are 5 degrees of freedom. $\chi^2_5\,(5\%) = 11.070$, so there is insufficient evidence to reject the null hypothesis.
2 We expect 24 winning tickets and 96 losing. $X^2 = 4.21875$, whereas $\chi^2_1\,(5\%) = 3.841$, so we reject the null hypothesis, the tombola is unfair.
3 $X^2 = 8.05$, whereas $\chi^2_2\,(2.5\%) = 7.378$, so we reject the null hypothesis, the expected distribution does not fit the data.
4 a Group together observations for '4 dogs', '5 dogs' and '>5 dogs', so that the expected frequency exceeds 5. There are then 5 observations, and 1 constraint, so 4 degrees of freedom.
b $X^2 = 12.236$, whereas $\chi^2_4\,(5\%) = 9.488$. Therefore we reject the null hypothesis, expected distribution doesn't fit the data.
5 $X^2 = 737.6$, whereas $\chi^2_5\,(5\%) = 11.071$, so we reject the null hypothesis that the distribution from 2000 is a good model for the data from 2015.

Exercise 6D

1 a H_0: the data may be modelled by Po(2).
$\chi^2_5(5\%) = 11\,070$,
$X^2 = 4.10$
No reason to reject H_0.
b Reduction by 1
2 Expected values 17, H_0: deliveries are uniformly distributed.
$\chi^2_5\,(5\%) = 11.070$, $X^2 = 5.765$
No reason to reject H_0,
3 a 1.4
b $\chi^2_2\,(10\%) = 4.605$, $X^2 = 5.04$
Reject H_0.
These data do not come from a Poisson distribution with $\lambda = 1.4$.
4 a 0.4
b $\chi^2_4\,(5\%) = 5.991$, $X^2 = 3.19$
No reason to reject H_0
5 Expected values: 21.6, 16.2, 27, 5.4, 10.8
$\chi^2_4\,(5\%) = 9.488$, $X^2 = 1.84$
No reason to reject H_0.
The number of accidents might well be constant at each factory.
6 $\lambda = 3.45$, $\chi^2_4(5\%) = 9.488$, $X^2 = 0.990$
No reason to reject H_0.
There is not sufficient evidence to suggest the data are not modelled by Po(3.45).
7 a Breakdowns are independent of each other, occur singly at random and at a constant rate.
b $\lambda = 0.95$, H_0: the data can be modelled by Po(0.95)
Expected values; 38.67, 36.74, 17.45, 7.14
$\chi^2_2\,(5\%) = 5.991$, $X^2 = 16.04$.
Reject H_0. The breakdowns are not modelled by Po(0.95).
8 H_0: prizes are uniformly distributed
H_1: prizes are not uniformly distributed
$\chi^2_9\,(5\%) = 16.919$, $X^2 = 10.74$
Do not reject H_0. There is no reason to believe the distribution of prizes is not uniform.
9 a $R = 43.75$ $S = 54.69$ $T = 43.75$
b H_0: A binomial model is a suitable model
H_1: A binomial model is not a suitable model
$\chi^2 <$ c.v. so accept H_0.

Online Full worked solutions are available in SolutionBank.

Conclude no reason to doubt data are from B(8, 0.5).

c Mean would have to be calculated, an extra restriction.
c.v. would be χ_5^2 (5%) = 11.070.
$\chi^2 <$ c.v. so no change in conclusion.

10 a Unbiased estimator of variance = 2.4
b Mean is close to variance.
c $s = 27.2$ $t = 78.4$
d H_0: the data are from Po(2.4)
H_1: the data aren't from Po(2.4)
e 3.5
f This expected frequency of 3.5 < 5 so must be combined with $E(X = 6)$ to give class '6 or more goals' which now has expected frequency 7.2 + 3.5 = 10.7
We now have 7 classes after pooling and 2 restrictions so degrees of freedom = 7 − 2 = 5
g $\chi^2 = 15.7$ c.v. = 11.070
$\chi^2 >$ c.v. so reject H_0.
Conclude there is evidence that the data can not be modelled by Po(2.4).

11 a $\frac{383}{148} = 2.59$ (2 d.p.)
b It is assumed that plants occur at a constant average rate and occur independently and at random in the meadow.
c $s = 37.24$ (2 d.p.) $t = 2.50$ (2 d.p.)
d $x^2 <$ c.v. so accept H_0.
Conclude there is no reason to doubt the data can be modelled by Po(2.59).

Exercise 6E

1 $\nu = 2$, χ_2^2 (5%) = 5.991

2 H_0: Ownership is not related to locality
H_1: Ownership is related to locality
χ_2^2 (5%) = 5.991, $X^2 = 13.1$
Reject H_0.

3 a (3 − 1)(3 − 1) = 4
b χ_4^2 (5%) = 9.488.
Reject H_0. There is an association between groups and grades.

4 H_0: There is no relationship between results
χ_4^2 (5%) = 9.488, $X^2 = 8.56$
Do not reject H_0. There is no reason to believe there is a relationship between results

5 χ_2^2 (5%) = 5.991, $X^2 = 1.757$
Do not reject H_0. There is no evidence to suggest association between station and lateness.

6 χ_4^2 (1%) = 13.277
Reject H_0. Gender and grade appear to be associated.

7 a Observed

	A	*B*	**Total**
OK	47	28	**75**
Defective	13	12	**25**
Total	**60**	**40**	**100**

Expected

A	*B*
45	30
15	10
–	–

b H_0: Factory and quality are not associated.
H_1: Factory and quality are associated.
χ_1^2 (0.05) = 3.841,

$$\sum\frac{(O_i - E_i)^2}{E_i} = \frac{2^2}{45} + \frac{2^2}{30} + \frac{2^2}{15} + \frac{2^2}{10} = 0.8888$$

0.8888 < 3.841
Do not reject H_0. There is no evidence between factory involved and quality.

8 H_0: Gender and susceptibility to flu are not associated.
H_1: Gender and susceptibility to flu are associated.
Observed

	Boys	**Girls**	**Total**
Flu	15	8	**23**
No flu	7	20	**27**
Total	**22**	**28**	**50**

Expected

Boys	**Girls**
10.12	12.88
11.88	15.12
–	–

χ_1^2 (5%) = 3.841, $X^2 = 7.78$
Reject H_0. There is evidence for an association between gender and susceptibility to influenza.

9 χ_2^2 (5%) = 5.991, $X^2 = 27.27$
Reject H_0. There is evidence of an association between the gender of an organism and the beach on which it is found.

10 H_0: There is no association between age and number of credit cards.
H_1: There is an association between age and number of credit cards.
χ_1^2 (5%) = 3.841, $X^2 = 8.31$
Reject H_0. There is an association between age and the number of credit cards possessed.

11 a H_0: There is no association between gym and whether or not a member got injured.
H_1: There is an association between gym and whether or not a member gets injured.
b $\frac{34}{865} \times 175 = 6.88$ (2 d.p.)
c Expected frequencies are

Gym	*A*	*B*	*C*	*D*
Expected injured	9.32	10.14	6.88	7.66
Expected uninjured	227.68	247.86	168.12	187.34

From which we calculate $X^2 = 7.732$
There are 3 degrees of freedom and χ_3^2, so we do not reject the null hypothesis, there is not enough evidence at the 5% significance level to think that any gym is more dangerous than the others.

12 a H_0: There is no association between science studied and pay.
H_1: There is an association between science studied and pay.
b The expected frequencies are

Science studied	**Salary**		
	£0–£40k	**£40k–£60k**	**>£60k**
Biology	70.19	26.40	7.41
Chemistry	72.89	27.42	7.69
Physics	74.92	28.18	7.90

From which we calculate $X^2 = 2.031$.
The number of degrees of freedom is (3 − 1) × (3 − 1) = 4 and χ_4^2 (5%) = 9.49, therefore we do not reject the null hypothesis, there is not enough evidence to at the 5% significance level to think that the subject studied has an effect on pay.

Exercise 6F

1 The expected frequencies are (grouping to ensure each expected frequency is at least 5)

k	1	2	3	4	$\geqslant 5$
Expected frequency	180	72	28.8	11.52	7.68

From which we calculate $X^2 = 14.705$.
$\chi^2_4(1\%) = 13.277$ therefore we reject the null hypothesis.

2 The expected frequencies are

k	1	2	3	4	5	$\geqslant 6$
Expected frequency	40	24	14.4	8.64	5.184	7.776

From this we calculate $X^2 = 7.966$.
$\chi^2_5(5\%) = 11.071$, so we do not reject the null hypothesis. Geo(0.4) is a good model for the data.

3 **a** We have $\frac{1}{p} \approx \frac{1}{100}\sum k \times O_k = 1.62$, so $p \approx 0.617$.

b The expected frequencies are

k	1	2	3	$\geqslant 4$
Expected frequency	61.7	23.63	9.05	5.62

From these we calculate $X^2 = 0.901$. There are 2 degrees of freedom (since we estimated p from the data) and $\chi^2_2(5\%) = 5.991$, so we do not reject the null hypothesis. Geo(0.617) is a good model for the data.

4 **a** We have $\frac{1}{p} \approx \frac{1}{100}\sum k \times O_k = 1.35$, so $p \approx 0.741$.

b The expected frequencies are

k	1	2	$\geqslant 3$
Expected frequency	74.1	19.19	6.71

Where we have grouped the observations for 3 and above to ensure the expected values are all at least 5. From this we can calculate $X^2 = 0.312$. There is only one degree of freedom, and $\chi^2_1(2.5\%) = 5.024$. Therefore we do not reject the null hypothesis, Geo(0.741) is a good model for the data.

5 **a** Assuming consecutive letters are independent and equally likely, we could model the number of characters until the next vowel using a Geo($\frac{5}{26}$) distribution.

b The expected frequencies are

k	1	2	3	4	5	$\geqslant 6$
Expected frequency	14.42	11.65	9.41	7.60	6.14	25.78

From this we calculate $X^2 = 2.418$. Here are 5 degrees of freedom, and $\chi^2_5(5\%) = 11.071$, so we do not reject the null hypothesis, the data could be modelled by a Geo($\frac{5}{26}$) distribution.

c e.g. Experiment does not tell us anything about the distribution within either the consonants or the vowels

Challenge

The number of people is the number of trials until 10 successes with fixed probability of success, so a negative binomial distribution is a natural choice of model.
H_0: The data can be modelled by a negative binomial distribution
H_1: The data can not be modelled by a negative binomial distribution
We estimate the parameter p from the data
$\frac{1}{p} \approx \frac{\bar{X}}{r} = \frac{1}{10} \times \frac{1}{104}\sum k \times O_k = 1.229$
so $p \approx 0.814$.
Using this, we calculate the expected frequencies

Number of people	10	11	12	13	14	$\geqslant 15$
Expected frequency	13.28	24.71	25.27	18.80	11.37	10.57

From which we can calculate $X^2 = 3.32$. Since $\chi^2_4(5\%) = 9.49$ (4 degrees of freedom, 6 observations and 2 constraints, i.e. 1 parameter estimation and 1 total) there is not enough evidence to reject the null hypothesis. We can model the data with a negative binomial distribution.

Mixed exercise 6

1 23.209
2 15.507
3 $\nu = 8$, critical region $\chi^2 > 15.507$
4 $\nu = 6$, 12.592
5 H_0: Taking drug and catching a cold are independent (not associated)
H_1: Taking drug and catching a cold are not independent (associated)
$\sum\frac{(O-E)^2}{E} = 2.53$
$\nu = 1 \quad \chi^2_1(5\%) = 3.841 > 2.53$
No reason to believe that the chance of catching a cold is affected by taking the new drug
6 H_0: Poisson distribution is a suitable model
H_1: Poisson distribution is not a suitable model
From these data $\lambda = \frac{52}{80} = 0.65$
Expected frequencies 41.76, 27.15, $\underbrace{8.82, 2.27}_{11.09}$
$\alpha = 0.05, \nu = 3 - 1 - 1 = 1$; critical value = 3.841
$\sum\frac{(O-E)^2}{E} = 1.312$
Since 1.312 is not the critical region there is insufficient evidence to reject H_0 and we can conclude that the Poisson model is a suitable one.
7 27.5, 22.5; 27.5, 22.5
$\sum\frac{(O-E)^2}{E} = \frac{(23-27.5)^2}{27.5} + \ldots \frac{(18-22.5)^2}{22.5} = 3.27$
$\alpha = 0.10 \Rightarrow \chi^2 > 2.705$
$3.27 > 2.705$
Since 3.27 is in the critical region there is evidence of association between gender and test result.
8 **a** Each box has an equal chance of being opened – we would expect each box to be opened 20 times.
b $\chi^2_4(5\%) = 9.488, \chi^2 = 2.3$
No reason to reject H_0, A discrete uniform distribution could be a good model.
9 **a** 0.72
b $\chi^2_2(5\%) = 5.991, X^2 = 2.62$
No reason to reject H_0. The B(5, 0.72) could be a good model.
10 $\lambda = 0.654, \nu = 2, X^2 = 21.506$,
$\chi^2_2(5\%) = 5.991$. Reject H_0.
Po(0.654) distribution is not a suitable model.
11 $\chi^2_2(5\%) = 5.991, X^2 = 4.74\ldots$
12 **a** 4.28
b $\chi^2_4(5\%) = 9.488, X^2 = 1.18$

Online Full worked solutions are available in SolutionBank.

No reason to reject H_0. Po(4.28) could be a good model.

13 $\chi_1^2(5\%) = 3.841$. $X^2 = 10.42$. Reject H_0.
There is evidence to suggest association between left-handedness and gender in this population.

14 a H_0: There is no association between gender and preferred subject.
H_1: There is an association between gender and preferred subject.

b $\frac{(28 + 40) \times (45 + 40 + 45)}{300} \approx 29.47$

c The expected frequencies are

		Subject		
		Physics	Biology	Chemistry
Gender	Male	67.43	38.53	64.03
	Female	51.57	29.47	48.97

From which we calculate $X^2 = 8.685$.

d There are $(3 - 1) \times (2 - 1) = 2$ degrees of freedom, and $\chi_2^2(1\%) = 9.21$. Therefore we do not reject the hypothesis.

e Since $\chi_2^2(5\%) = 5.991$, we would reject the null hypothesis at the 5% significance level.

15 a i $P(X = 1) \approx 0.2504$
ii $P(X > 2) \approx 0.3639$

b $\sum k \times O_k = \frac{129}{60} = 2.15$

c $a = 16.15$, $b = 1.36$

d H_0: The data can be modelled by a Poisson distribution with mean 2.15.
H_1: The data cannot be modelled by a Poisson distribution with mean 2.15.

e The chi-squared test is not very effective if the expected values are below 5, so we combine expected values in order to make sure each of the expected values are at least 5.

f The test statistic is $X^2 = 2.507$. There are 3 degrees of freedom (5 observations, and 2 constraints since we estimated the mean from the data). From the tables we see $\chi_3^2(5\%) = 7.815$. Therefore we do not reject the null hypothesis.

16 a A geometric distribution would be a good choice.

b We have $\frac{1}{p} \approx \frac{1}{255}\sum k \times O_k = \frac{524}{255}$, so $p \approx \frac{255}{524}$.

c The expected values are

Number of calls	1	2	3	4	5	≥6
Expected frequency	124.09	63.70	32.70	16.79	8.62	9.09

From which we calculate $X^2 = 14.84$. There are 4 degrees of freedom and $\chi_4^2(5\%) = 9.488$, so we reject the null hypothesis, the data is not well modelled by a geometric distribution.

17 a e.g. It is unlikely that each guess will be independent of the others.

b If Wilfred is equally likely to select the digits 2, 5 or 7, then he has a $\frac{1}{3}$ chance of getting the right number each time he calls his parents. So the number of calls can be modelled by a Geo($\frac{1}{3}$) distribution. The expected values are

Attempts	1	2	3	4	≥5
Expected frequency	17.33	11.56	7.7	5.14	10.27

From which we calculate $X^2 = 14.914$. There are four degrees of freedom and $\chi_4^2(5\%) = 9.488$, so we reject the null hypothesis. The evidence suggests Wilfred may not be equally likely to pick any of the numbers 2, 5 or 7.

Challenge

a Using the midpoints of each range, we calculate the mean and variance. Mean = 14.14, Var = 17.11

b We calculate expected frequencies as follows

Length of call (l)	$l < 5$	$5 \leqslant l < 10$	$10 \leqslant l < 15$	$15 \leqslant l < 20$	$20 \leqslant l$
Expected frequency	6.78	72.44	211.95	169.68	39.14

The null and alternative hypotheses are:
H_0: The call length can be modelled by a normal distribution
H_1: The call length can't be modelled by a normal distribution
From this we can calculate $X^2 = 3.242$. There are 2 degrees of freedom, since there are 5 observations, we estimated 2 parameters, and we have the constraint of the total summing to 500. Thus the critical value is $\chi_2^2(5\%) = 5.991$. Hence we do not reject the null hypothesis, the length of call can be modelled by a normal distribution.

Prior knowledge check 7

1 a 0.1708 **b** 0.3423 **c** 4.5 **d** 4.5

2 a 0.0720 **b** 0.49 **c** $\frac{10}{3}$ **d** $\frac{70}{9}$

3 a 0.1274 **b** 0.4167 **c** $\frac{100}{7}$ **d** $\frac{1300}{49}$

4 a $4e^x(3 + e^x)^3$ **b** $2e^{2x}(1 - 2x^2 - 2x)$

Exercise 7A

1 a 0, 1, 2
b i 0.3 **ii** 1

2 a 0, 1, 2, 3
b i $\frac{3}{8}$ **ii** $\frac{7}{8}$

3 a 0 **b** 0.8 **c** 0.2

4 $\frac{1}{6}(t + t^2 + t^3 + t^4 + t^5 + t^6)$

5 $\frac{2}{5}t + \frac{1}{5}t^2 + \frac{1}{5}t^3 + \frac{1}{5}t^4$

6 a $\frac{1}{10}(t + 2t^2 + 3t^3 + 4t^4)$ **b** $\frac{1}{14}(t + 4t^2 + 9t^3)$

7 a $\frac{1}{16}$ **b** $\frac{1}{4}$

8 a $\frac{1}{25}$

b

x	0	1	2	3	4
$P(X = x)$	$\frac{1}{25}$	$\frac{4}{25}$	$\frac{8}{25}$	$\frac{8}{25}$	$\frac{4}{25}$

9 a

x	2	3	4	5	6	7	8
$P(X = x)$	$\frac{1}{16}$	$\frac{2}{16}$	$\frac{3}{16}$	$\frac{4}{16}$	$\frac{3}{16}$	$\frac{2}{16}$	$\frac{1}{16}$

b $\frac{1}{16}(t^2 + 2t^3 + 3t^4 + 4t^5 + 3t^6 + 2t^7 + t^8)$

10 $G_X(1) \neq 1$

11 a $k = \frac{1}{1024}$ **b** 10; $\frac{1}{1024}$ **c** $\frac{63}{256}$ **d** $Y \sim B(10, 0.5)$

Exercise 7B

1 a $(0.5 + 0.5t)^4$ **b** $(0.8 + 0.2t)^6$ **c** $(0.1 + 0.9t)^5$
d $e^{3(t-1)}$ **e** $e^{1.7(t-1)}$ **f** $e^{0.2(t-1)}$

2 **a** $\frac{0.3t}{1-0.7t}$ **b** $\frac{0.8t}{1-0.2t}$ **c** $\left(\frac{0.4t}{1-0.6t}\right)^3$
d $\left(\frac{0.9t}{1-0.1t}\right)^5$

3 **a** $(0.8+0.2t)^5$ **b** $\frac{0.2t}{1-0.8t}$ **c** $\left(\frac{0.2t}{1-0.8t}\right)^2$

4 **a** $X \sim \text{Po}(0.3)$ **b** 0.2222 **c** $e^{0.3(t-1)}$

5 **a** Geo(0.35) **b** 0.0406 **c** $\frac{0.35t}{1-0.65t}$

6 $X \sim \text{B}(4, 0.8)$
The probability distribution of X:
$0.2^4, 4 \times 0.8 \times 0.2^3, 6 \times 0.8^2 \times 0.2^2, 4 \times 0.8^3 \times 0.2, 0.8^4$
$G_x(t) = \sum t^x P(X = x)$
$G_x(t) = 0.2^4 + 4 \times 0.8 \times 0.2^3 t + 6 \times 0.8^2 \times 0.2^2 t^2 + 4 \times 0.8^3 \times 0.2t^3 + 0.8^4 t^4$
$G_x(t) = (0.2 + 0.8t)^4$

7 $X \sim \text{Po}(3.5)$
$P(X = x) = \frac{e^{-3.5} \times 3.5^x}{x!}$
$G_x(t) = \sum t^x P(X = x) = \sum t^x\left(\frac{e^{-3.5} \times 3.5^x}{x!}\right)$
$= e^{-3.5}\sum\frac{(3.5t)^x}{x!} = e^{-3.5}\left(1 + 3.5t + \frac{(3.5t)^2}{2!} + \frac{(3.5t)^3}{3!} + \ldots\right)$
The Maclaurin expansion of e^x with $x = 3.5t$
$= e^{-3.5} \times e^{3.5t} = e^{3.5(t-1)}$

8 $Y \sim \text{Geo}(0.7)$
$P(Y = y) = 0.3^{y-1} \times 0.7$
$G_Y(t) = \sum t^y P(Y = y) = \sum t^y \times 0.3^{y-1} \times 0.7$
$= 0.7t + 0.3 \times 0.7t^2 + 0.3^2 \times 0.7t^3 + 0.3^3 \times 0.7t^4 + \ldots$
$= 0.7t(1 + 0.3t + (0.3t)^2 + (0.3t)^3 + \ldots)$
The infinite geometric sum with first term 1 and common ratio $0.3t$
$= 0.7t\left(\frac{1}{1-0.3t}\right) = \frac{0.7t}{1-0.3t}$

9 $X \sim \text{B}(n, p)$
$P(X = x) = \binom{n}{x} \times p^x \times (1-p)^{n-x}$
$G_X(t) = \sum t^x P(X = x) = \sum t^x \times \binom{n}{x} \times p^x \times (1-p)^{n-x}$
$= \sum\binom{n}{x} \times (pt)^x \times (1-p)^{n-x}$
$= \left(\binom{n}{0} \times (pt)^0 \times (1-p)^n\right) + \left(\binom{n}{1} \times (pt)^1 \times (1-p)^{n-1}\right)$
$+ \left(\binom{n}{2} \times (pt)^2 \times (1-p)^{n-2}\right) + \ldots$
Binomial expansion of $(a + b)^n$ with $a = 1 - p$ and $b = pt$
Hence $G_X(t) = (1 - p + pt)^n$

10 $X \sim \text{Po}(\lambda)$
$P(X = x) = \frac{e^{-\lambda} \times \lambda^x}{x!}$
$G_x(t) = \sum t^x P(X = x) = \sum t^x\left(\frac{e^{-\lambda} \times \lambda^x}{x!}\right) = e^{-\lambda}\sum\frac{(\lambda t)^x}{x!}$
$= e^{-\lambda}\left(1 + \lambda t + \frac{(\lambda t)^2}{2!} + \frac{(\lambda t)^3}{3!} + \ldots\right)$
The Maclaurin expansion of e^x with $x = \lambda t$
Hence $G_X(t) = e^{-\lambda} \times e^{\lambda t}$
$= e^{\lambda(t-1)}$

11 $Y \sim \text{Geo}(p)$
$P(Y = y) = (1-p)^{y-1} \times p$
$G_Y(t) = \sum t^y P(Y = y) = \sum t^y \times (1-p)^{y-1} \times p$
$= pt + (1-p) \times pt^2 + (1-p)^2 \times pt^3 + (1-p) \times pt^4 + \ldots$
$= pt(1 + (1-p)t + ((1-p)t)^2 + ((1-p)t)^3 + \ldots)$
The infinite geometric sum with first term 1 and common ratio $(1-p)t$
Hence $G_Y(t) = pt\left(\frac{1}{1-(1-p)t}\right) = \frac{pt}{1-(1-p)t}$

Exercise 7C

1 $\frac{3}{4}; \frac{11}{16}$

2 $\frac{11}{6}; \frac{\sqrt{41}}{6} (= 1.067)$ (3 d.p.)

3 **a** $(0.5 + 0.5t)^3$ **b** $\frac{3}{2}; \frac{3}{4}$

4 **a** $(0.4 + 0.6t)^4$
b **i** 2.4 **ii** $\frac{2\sqrt{6}}{5} (= 0.980)$ (3 s.f.)

5 $\frac{10}{3}; \frac{8}{9}$

6 **a** $\frac{2}{3}; \frac{2\sqrt{2}}{3} (= 0.943)$ (3 s.f.)
b **i** $\frac{9}{16}$
ii $\frac{9}{32}$

7 **a** 2; 4
b **i** $\frac{1}{e}$ **ii** $\frac{1}{e}$ **iii** 0
iv $\frac{1}{2e}$

8 **a** $X \sim \text{Geo}(\frac{1}{6})$ **b** $\frac{\frac{1}{6}t}{1 - \frac{5}{6}t}$
c **i** 6 **ii** 30

9 $X \sim \text{B}(n, p)$
Hence has a pgf: $G_X(t) = (1 - p + pt)^n$
$G'_X(t) = np(1 - p - pt)^{n-1}$
$E(X) = G'_X(1) = np(1 - p + p)^{n-1} = np$
$Var(X) = G''_X(1) + G'_X(1) - (G'_X(1))^2$
$G''_X(t) = p^2n(n-1)(1 - p + pt)^{n-2}$
$G''_X(1) = p^2n(n-1)$
Substitute into equation for Var(X)
$Var(X) = p^2n(n-1) + np - (np)^2 = p^2n^2 - p^2n + np - n^2p^2$
$= np - np^2 = np(1-p)$

10 $X \sim \text{Po}(\lambda)$
Hence the pgf is $G_X(t) = e^{\lambda(t-1)}$
$E(X) = G'_X(1)$
$G'_X(t) = \lambda e^{\lambda(t-1)}$
$G'_X(1) = \lambda e^0 = \lambda$
$Var(X) = G''_X(1) + G'_X(1) - (G'_X(1))^2$
$G''_X(t) = \lambda^2 e^{\lambda(t-1)}$
$G''_X(1) = \lambda^2 e^0 = \lambda^2$
Substitute into equation for Var(X):
$Var(X) = \lambda^2 + \lambda - (\lambda)^2 = \lambda$

11 **a** $X \sim \text{P}(4)$ **b** $e^{4(t-1)}$
c **i** $E(X) = G'_X(1)$
$G'_X(t) = 4e^{4(t-1)}$
By substituting $t = 1$ into $G'_X(t)$: $E(X) = 4e^0 = 4$
ii $Var(X) = G''_X(1) + G'_X(1) - (G'_X(1))^2$
$G''_X(t) = 16e^{4(t-1)}$
By substituting $t = 1$ into $G''_X(t)$: $G''_X(1) = 16e^0 = 16$
Hence, $Var(X) = 16 + 4 - (4)^2 = 4$
standard deviation $= \sqrt{Var(X)} = \sqrt{4} = 2$

12 $\frac{1}{8}, \frac{1}{2}, \frac{3}{8}$

13 $a = 4, b = 5$

14 **a** $\frac{1}{16}$ **b** $X \sim \text{B}(4, 0.5)$
c $G_X(t) = \frac{1}{16}(1 + t)^4$
$G'_X(t) = \frac{1}{4}(1 + t)^3$
$E(X) = G'_X(1) = \frac{1}{4}(1 + 1)^3 = 2$

$G''_X(t) = \frac{3}{4}(1 + t)^2$

$G''_X(1) = \frac{3}{4}(1 + 1)^2 = 3$

$\text{Var}(X) = G''_X(1) + G'_X(1) - (G'_X(1))^2$

$\text{Var}(X) = 3 + 2 - 2^2 = 1$

15 a $\frac{1}{36}(11t + 9t^2 + 7t^3 + 5t^4 + 3t^5 + t^6)$

b i $\frac{91}{36}$ **ii** $\frac{\sqrt{2555}}{36}$

Challenge

a 0 **b** $k + 1$ **c** $\frac{1}{2}$

Exercise 7D

1 a $\frac{5}{24}t + \frac{1}{4}t^2 + \frac{11}{24}t^3 + \frac{1}{12}t^4$

b $E(Z) = G'_Z(1) = \frac{5}{24} + \frac{1}{2}(1) + \frac{33}{24}(1)^2 + \frac{1}{3}(1)^3 = \frac{29}{12}$

$E(X) = G'_X(1) = \frac{9}{4}$

$E(Y) = G'_Y(1) = \frac{1}{6}$

$E(X) + E(Y) = \frac{9}{4} + \frac{1}{6} = \frac{29}{12} = E(Z)$

2 a $X \sim B(2, 0.5)$; $Y \sim B(2, 0.6)$

b $\frac{1}{25} + \frac{1}{5}t + \frac{37}{100}t^2 + \frac{3}{10}t^3 + \frac{9}{100}t^4$

c $E(Z) = G'_Z(1) = \frac{11}{5}$, $E(X) = G'_X(1) = 1$, $E(Y) = G'_Y(1) = \frac{6}{5}$

$E(X) + E(Y) = 1 + \frac{6}{5} = \frac{11}{5} = E(Z)$

3 a $e^{1.3(t-1)}$; $e^{2.4(t-1)}$

b $e^{3.7(t-1)}$

c $E(Z) = G'_Z(1) = \frac{37}{10}$, $E(X) = G'_X(1) = \frac{13}{10}$, $E(Y) = G'_Y(1) = \frac{12}{5}$

$E(X) + E(Y) = \frac{13}{10} + \frac{12}{5} = \frac{37}{10} = E(Z)$

4 a $X \sim \text{Negative B}\left(1, \frac{1}{6}\right)$, $G_X(t)$

$$= \left(\frac{pt}{1 - 1 - pt}\right)^r = \left(\frac{\frac{1}{6}t}{1 - \frac{5}{6}t}\right)^1$$

Multiply fraction through by 6: $G_X(t) = \frac{t}{6 - 5t}$

b $\left(\frac{\frac{1}{10}t}{1 - \frac{9}{10}t}\right)^2 = \left(\frac{t}{10 - 9t}\right)^2$ **c** $\left(\frac{t}{6 - 5t}\right)\left(\frac{t}{10 - 9t}\right)^2$

d $E(Z) = E(X) + E(Y) = G'_X(1) + G'_Y(1) = 6 + 20 = 26$

5 a $\frac{1}{27}$ **b** $\frac{4}{9}$

c $G'_X(t) = \frac{6}{27} + \frac{24}{27}t + \frac{24}{27}t^2$

$E(X) = G'_X(1) = \frac{6}{27} + \frac{24}{27} + \frac{24}{27} = 2$

$G''_X(t) = \frac{24}{27} + \frac{48}{27}t$

$\text{Var}(X) = G''_X(1) + G'_X(1) - (G'_X(1)^2)\ \frac{24}{27} + \frac{48}{27} + 2 - (2)^2 = \frac{2}{3}$

d $\frac{3}{4}; \frac{11}{4}$

6 a $\frac{4t}{(3 - t)^2(3 - 2t)^3}$ **b** 8

c $\text{Var}(Z) = G''_z(1) + G'_z(1) - (G'_z(1)^2) = \frac{151}{2} + 8 - (8)^2 = \frac{39}{2}$

7 a $(0.7 + 0.3t)^5(0.6 + 0.4t)^5$

b $G'_X(t) = 2\left(\frac{2t}{5} + \frac{3}{5}\right)^4 \times \left(\frac{3t}{10} + \frac{7}{10}\right)^5$

$$+ \frac{3\left(\frac{2t}{5} + \frac{3}{5}\right)^5 \times \left(\frac{3t}{10} + \frac{7}{10}\right)^4}{2}$$

$E(X) = G'_X(1) = \frac{7}{2} = 3.5$

8 a $\frac{1}{2}t^3 + \frac{1}{2}t^9$ **b** $\left(\frac{1}{2}t^5 + \frac{1}{2}t^9\right)$ **c** $\left(\frac{1}{2}t^{-1} + \frac{1}{2}t^7\right)$

9 a $\frac{2}{3}$ **b** 2.5 **c** $\left(\frac{1}{6}t + \frac{1}{6}t^3 + \frac{2}{3}t^5\right)$

d $G'_Y(t) = \frac{10t^4}{3} + \frac{t^2}{2} + \frac{1}{6}$

$E(Y) = G'_Y(1) = 4$

$G'_X(t) = \frac{1}{6} + \frac{1}{3}t + 2t^2$

$E(X) = G'_X(1) = \frac{5}{2}$

$2E(X) - 1 = 4 = E(Y)$

Challenge

1 Let $G_X(t) = i_0 + i_1 t + i_2 t^2 + i_3 t^3 + \ldots$

So, $E(X) = G'_X(1) = i_1 + 2i_2 + 3i_3 + \ldots$

From the question, we have:

$G_Y(t) = t^b G_X(t^a) = t^b i_0 + t^b i_1 t^a + t^b i_2 t^{2a} + t^b i_3 t^{3a} + \ldots$

$= i_0 t^b + i_1 t^{a+b} + i_2 t^{2a+b} + i_3 t^{3a+b} + \ldots$

$E(Y) = G'_Y(1)$

$G'_Y(t) = bi_0 t^{b-1} + (a + b)i_1 t^{a+b-1} + (2a + b)i_2 t^{2a+b-1} + (3a + b)$
$i_3 t^{3a+b-1} + \ldots$

$= (bi_0 t^{b-1} + bi_1 t^{a+b-1} + bi_2 t^{2a+b-1} + bi_3 t^{3a+b-1} + \ldots)$
$+ (ai_1 t^{a+b-1} + 2ai_2 t^{2a+b-1} + 3ai_3 t^{3a+b-1} + \ldots)$

$G'_Y(1) = (bi_0 + bi_1 + bi_2 + bi_3 + \ldots) + (ai_1 + 2ai_2 + 3ai_3 + \ldots)$

$= b(i_0 + i_1 + i_2 + i_3 + \ldots) + a(i_1 + 2i_2 + 3i_3 + \ldots)$

$= b(1) + aE(X) = aE(X) + b$ as required

2 a $\frac{0.6t}{1 - 0.4t}$

b Let the random variable X represent the number of shots required to hit the bullseye twice,

$X \sim \text{Negative B}(2, 0.6)$

We know that the probability generating function for a negative binomial is $\left(\frac{pt}{1 - (1 - p)t}\right)^r$

Let $H(t)$ be the pgf of X

So $H(t) = \left(\frac{0.6t}{1 - 0.4t}\right)^2 = (G(t))^2$ as required.

c $(G(t))^4$

Mixed exercise 7

1 a $\frac{1}{7}$

b $\frac{2}{7}$

c $\frac{1}{112}(2 + 2t + 3t^2)(1 + t + 2t^2)^2$

d $\frac{17}{112}$ or 0.1518 (4 d.p.)

2 $X \sim \text{Geo}(p)$

We know that the pgf of a geometric distribution is

$$G_X(t) = \frac{pt}{1 - (1 - p)t}$$

$$G'_X(t) = \frac{p}{t(p - 1) + 1} - \left(\frac{pt(p - 1)}{t(p - 1) + 1}\right)^2$$

$$E(X) = G'_X(1) = 1 - \frac{p - 1}{p} = \frac{1}{p}$$

$$G''_X(t) = \frac{2pt(p - 1)^2}{(t(p - 1) + 1)^3} - \frac{2p(p - 1)}{(t(p - 1) + 1)^2}$$

$$G''_X(1) = \frac{2(p - 1)^2}{p^2} - \frac{2(p - 1)}{p}$$

$$\text{Var}(X) = G''_X(1) + G'_X(1) - (G'_X(1))^2$$

$$= \frac{2(p - 1)^2}{p^2} - \frac{2(p - 1)}{p} + \frac{1}{p} - \frac{1}{p^2}$$

$$= \frac{2 - 2p}{p^2} + \frac{p - 1}{p^2} = \frac{p^2 - p^3}{p^4} = \frac{1 - p}{p^2}$$

3 $X \sim B(5, 0.4)$

$P(X = x) = \binom{5}{x} \times p^x \times (1-p)^{5-x}$

$$G_X(t) = \sum_0^5 t^x P(X = x)$$
$$= \sum_0^5 t^x \times \binom{5}{x} \times 0.4^x \times (1-0.4)^{5-x}$$
$$= \sum_0^5 \binom{5}{x} \times (0.4t)^x \times (1-0.4)^{5-x}$$
$$= \left(\binom{5}{0} \times (0.4t)^0 \times (1-0.4)^5\right) + \left(\binom{5}{1} \times (0.4t)^1 \times (1-0.4)^{5-1}\right) + \left(\binom{5}{2} \times (0.4t)^2 \times (1-0.4)^{5-2}\right) + \dots$$

Binomial expansion of $(a+b)^n$ with $a = 0.6$ and $b = 0.4t$ and $n = 5$

Hence $G_X(t) = (0.6 + 0.4t)^5$

4 a $X \sim \text{Geo}\left(\frac{4}{15}\right)$

b $G_Y(t) = \frac{4}{15}t\left(\frac{1}{1-\left(1-\frac{4}{15}\right)t}\right) = \frac{4t}{15-11t}$

c i $\frac{15}{4}$ ii $\frac{165}{16}$

d $\frac{20t^2}{(15-11t)(12-7t)}$

e $\frac{123}{20}$; 3.6976 (4 d.p.)

5 a i $e^{-0.5}$ ii $1 - (e^{-0.5} + \frac{1}{2}e^{-0.5} + \frac{1}{8}e^{-0.5}) = 1 - \frac{13}{8}e^{-0.5}$

b From the previous part of the question we have:

$P(X = 0) = e^{-0.5}$; $P(X = 1) = \frac{1}{2}e^{-0.5}$;

$P(X = 2) = \frac{1}{8}e^{-0.5}$; $P(X = 3) = 1 - \frac{13}{8}e^{-0.5}$

$G_X(t) = e^{-0.5} + \frac{1}{2}e^{-0.5}t + \frac{1}{8}e^{-0.5}t^2 + \left(1 - \frac{13}{8}e^{-0.5}\right)t^3$

$= t^3 + e^{-0.5}\left(1 + \frac{1}{2}t + \frac{1}{8}t^2 - \frac{13}{8}t^3\right)$

c 0.498; 0.488

6 a 7; $\sqrt{10}$

b i 0 ii $\frac{1}{32}$

c i $\frac{t}{(4-3t^2)^2}$ ii $\frac{t^3}{(2-t)^5(4-3t)^2}$

d $E(Z) = G'_Z(1)$

$$G'_Z(t) = \frac{5t^3}{(3t-4)^2(t-2)^6} - \frac{3t^2}{(3t-4)^2(t-2)^5} + \frac{6t^3}{(3t-4)^3(t-2)^5}$$

$$G'_Z(1) = \frac{5}{(3-4)^2(1-2)^6} - \frac{3}{(3-4)^2(1-2)^5} + \frac{6}{(3-4)^3(1-2)^5} = 14$$

7 a $G_X(t) = k(9t^6 + 12t^5 + 4t^4 + 6t^3 + 4t^2 + 1)$

$1 = 9k + 12k + 4k + 6k + 4k + k$

$1 = 36k$ hence, $k = \frac{1}{36}$

b $\frac{1}{9}$

c $G'_X(t) = \frac{3t^5}{2} + \frac{5t^4}{3} + \frac{4t^3}{9} + \frac{t^2}{2} + \frac{2t}{9}$,

$E(X) = G'_X(1) = \frac{3}{2} + \frac{5}{3} + \frac{4}{9} + \frac{1}{2} + \frac{2}{9} = \frac{13}{3}$; $\frac{41}{18}$

d $\frac{1}{36}t^{-2}(1 + 2t^6 + 3t^9)^2$

Challenge

a i $P(X = 0) = G_X(0) = \tan(0) = 0$

ii $G_X(t) = \tan\left(\frac{\pi t}{4}\right)$

Use the Maclaurin Expansion of $\tan\left(\frac{\pi t}{4}\right) = \frac{\pi t}{4} + \frac{\left(\frac{\pi t}{4}\right)^3}{3} + \dots$

Which gives $P(X = 1) = \frac{\pi}{4}$

iii Use the Maclaurin expansion from part **ii**, which gives $P(X = 2) = 0$

b $E(X) = G'_X(1) = \frac{\pi\left(\tan\left(\frac{\pi(1)}{4}\right)^2 + 1\right)}{4} = \frac{\pi}{2}$

$\text{Var}(X) = G''_X(1) + G'_X(1) - (G'_X(1))^2$

$G''_X(1) = \frac{\pi^2\tan\left(\frac{\pi}{4}\right)\left(\tan\left(\frac{\pi}{4}\right)^2 + 1\right)}{8} = \frac{\pi^2}{4}$

Hence, $\text{Var}(X) = \frac{\pi^2}{4} + \frac{\pi}{2} - \left(\frac{\pi}{2}\right)^2 = \frac{\pi}{2} = E(X)$

c $\frac{\pi^3}{192}$

Prior knowledge 8

1 H_0: $\mu = 10°C$; H_1: $> 10°C$. Accept H_0.

2 a $X \geqslant 10$; $X \leqslant 1$

b $X \geqslant 20$

Exercise 8A

1 a $X \geqslant 6$ b 0.0197, 0.9527

2 a $X \leqslant 1$ b 0.0076, 0.9757

3 a $\{X \leqslant 1\} \cup \{X \geqslant 9\}$ b 0.0278, 0.9519

4 a $X \geqslant 11$ b 0.0426, 0.9015

5 a $X = 0$ b 0.0111, 0.9698

6 a $\{X \leqslant 3\} \cup \{X \geqslant 16\}$ b 0.0433, 0.9494

7 a $X \geqslant 15$ b 0.0440, 0.5123

8 a $X \geqslant 229$ b 0.0100, 0.8989

9 a $\{X \leqslant 2\} \cup \{X \geqslant 369\}$ b 0.0447, 0.8100

10 a i Type 1 error: reject H_0 when H_0 true

ii Type 2 error: accept H_0 when H_0 false

b $\{X \leqslant 13\} \cup \{X \geqslant 748\}$ c 0.1007

11 a $X \geqslant 5$ b 0.048 c $X = 1$

d 0.05 e 0.9412 f 0.9162

Exercise 8B

1 a $\bar{x} > 51.5605...$

b 0.01

c 0.0162, awrt 0.016

2 a $\bar{x} < 29.178$

b 0.05

c 0.0869, awrt 0.087 ~ 0.088

3 a $\{\bar{X} < 37.939\} \cup \{\bar{X} > 42.061\}$

b 0.01

c 0.5319, awrt 0.53

4 a $\bar{x} < 14.608$ or $\bar{x} > 15.392$

b 0.1492, awrt 0.1492

5 a $\bar{x} > 42.4025...$

b 0.6103, awrt 0.61

c Only way to reduce P(Type II) error without changing the significance level is to increase the sample size. Altering the significance level can increase the chances of a Type 1 error occurring.

Exercise 8C

1 a $\bar{x} > 20.9869...$ b 0.3757, awrt 0.378

2 a 0.0196 b 0.0247

3 **a** 0.0111 **b** 0.0166 (3 s.f.)
4 0.3522, awrt 0.352
5 **a** 0.0548
b 0.8791
c The test is more powerful for values of p further away from $p = 0.3$
6 **a** A Type I error is when H_0 is rejected when H_0 is in fact true.
b The size of a significance test is the probability of a Type I error occurring.
c 0.0569
7 **a** $X \geqslant 5$ **b** 0.5904
8 **a** $2 \leqslant X \leqslant 368$ **b** 0.0402
9 **a** 0.0490 **b** 0.6723
10 **a** 0.8714 **b** 0.3127

Challenge
a No more than 5 boxes can be inspected
b 0.4583

Exercise 8D

1 **a** 0.0430
b $P(x \leqslant 2|\lambda) = e^{-\lambda} + (e^{-\lambda} \times \lambda) + \frac{e^{-\lambda} \times \lambda^2}{2}$ take out a factor of $e^{-\lambda}$ to get desired answer
c $\lambda = 2 \Rightarrow s = 0.6767 = 0.68$ (2 d.p.)
$\lambda = 5 \Rightarrow t = 0.1247 = 0.12$ (2 d.p.)
d

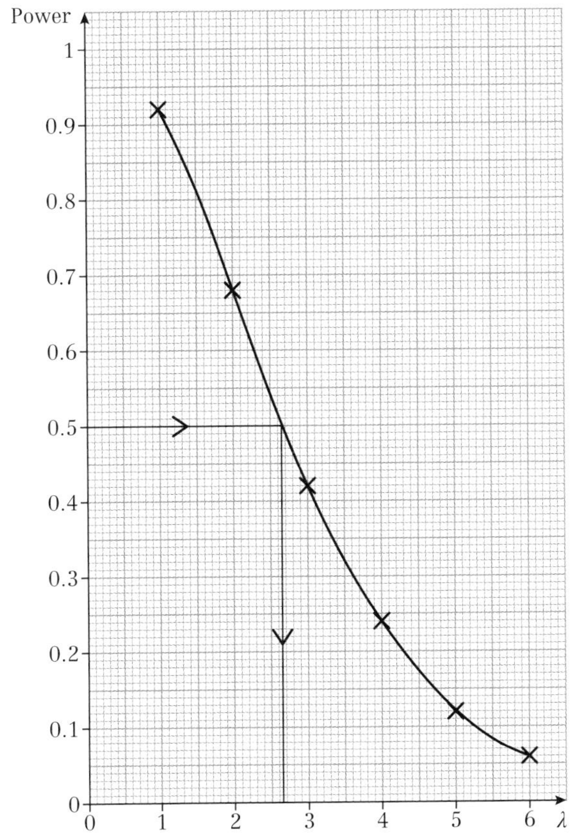

e Correct conclusion is arrived at when $\lambda = 6.5$, H_0 is accepted. So since size is 0.0430 probability of accepting $\lambda = 6.5$ is 0.957 $\therefore \lambda = 6.5$ or for $\lambda < 6.5$, correct conclusion is to reject H_0.
So require where power > 0.5 i.e. $\lambda < 2.65$ (from graph).

2 **a** 0.0421
b $P(X \leqslant 2) = \binom{12}{0} \times p^0 \times (1-p)^{12} + \binom{12}{1} \times p^1 \times (1-p)^{11} + \binom{12}{2} \times p^2 \times (1-p)^{10}$
$= (1-p)^{12} + 12p(1-p)^{11} + 66p^2(1-p)^{10}$
c 0.2528
3 **a** 0.0547 **b** 0.6778
c The test is more powerful for values of p further away from 0.4
4 **a** 0.0547
b Test A uses the binomial expansion and the power function is part of the binomial expansion for $x = 8$, 9, 10
c 0.0615
d $(1-p)^5 (2 - (1-p)^5)$
e $p = 0.25 \Rightarrow \text{power}_A = 0.5256$
$p = 0.35 \Rightarrow \text{power}_A = 0.2616$
f Use test A as this is always more powerful.
5 **a** 0.009 **b** $(1-p)^{29}$
6 **a** 0.01 **b** p^{10} **c** p^{12}
d The test with 10 trials has larger power.

Mixed exercise 8

1 **a** $X \geqslant 9$ **b** 0.0422 **c** 0.3036
2 **a** $X = 0$ **b** 0.0302 **c** 0.0498
3 **a** $\bar{X} < 6.614...$ or $\bar{X} > 9.3859...$ **b** 0.05
c awrt 0.707 ~ 0.708
d 0.293 ~ 0.292
4 **a** 0.0001318 **b** 0.3566545
5 **a** 0.0866
b **i** $r = 1 - 0.9489 = 0.0511$
$s = 1 - 0.7440 = 0.2560$
$t = 1 - 0.3239 = 0.6761$
ii

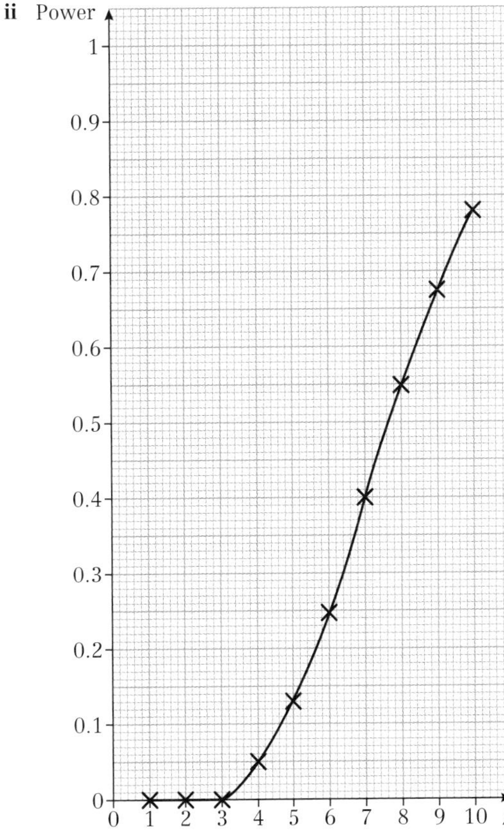

6 **a** 0.0424

b Power = $P(X \leqslant 3 \mid X \sim B(15, p))$

$p = 0.2 \Rightarrow s = 0.6482$

$p = 0.4 \Rightarrow t = 0.0905$

c

7 **a** $H_0: \lambda = 2$ (Quality the same) $H_1: \lambda > 2$ (Quality is poorer)

b Critical region $X \geqslant 5$

c 0.3712

8 **a** $H_0: \lambda = 2$ $H_1: \lambda > 2$

b Critical region $X \geqslant 5$

c 0.1847

d Critical region $X \geqslant 11$

e 0.294

f Second test is more powerful as it uses more days.

9 **a** 0.0620

b $P(x \leqslant 2 \mid \mu) = e^{-\mu} + (e^{-\mu} \times \mu) + \dfrac{e^{-\mu} \times \mu^2}{2}$ take out a factor of $\frac{1}{2}e^{-\lambda}$ to get desired answer

c $s = 0.6767$

$t = 0.1247$

d

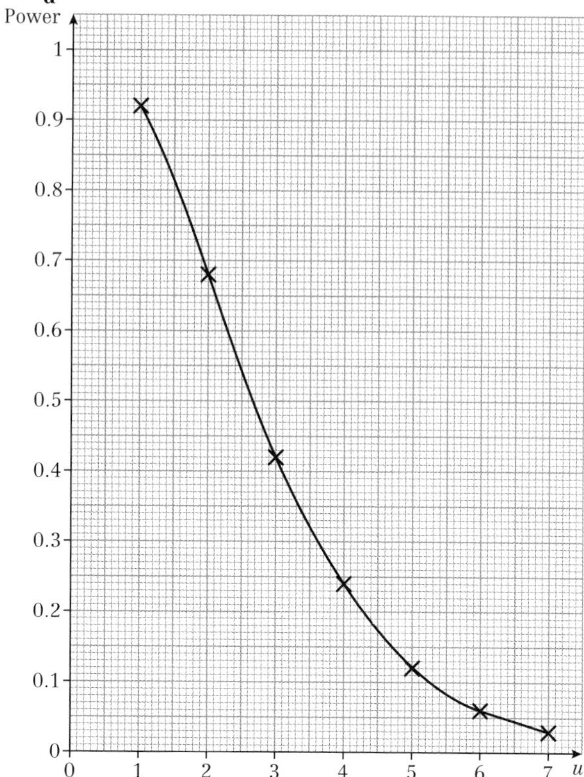

e $\mu < 1.55$

10 **a** 0.0016

b 0.08

c 0.0086

d 0.24

e

f **i** 0.325

ii With p greater than this value, the technician's test is stronger than the supervisor's.

g Test is more powerful for probabilities closer to zero, quicker to test 5 than to test 10

11 **a** 0.0839

b 0.38 (2 d.p.)

c $X \geqslant 9$

d 0.0403

e 0.91 (2 d.p.)

f

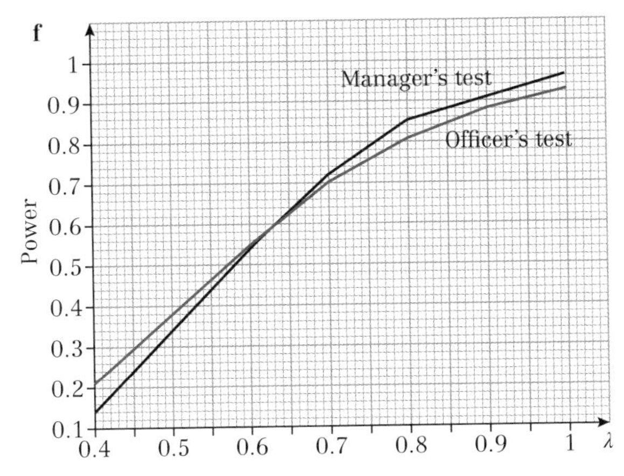

g i 0.63

ii With λ greater than this value, the manager's test is more powerful.

Challenge

a 0.0423

b $1 - (1 - p)^{12} - 12p(1 - p)^{11}$

c 0.0173

d $1 - (1 - p)^{12} - 12p(1 - p)^{11}$ from part **b**, with $q = 6p$, substitute $p = \frac{q}{6}$ to get thee answer

e $0.0087 + 14.8695q^4 - 23.7912q^5 + 9.913q^6$

f

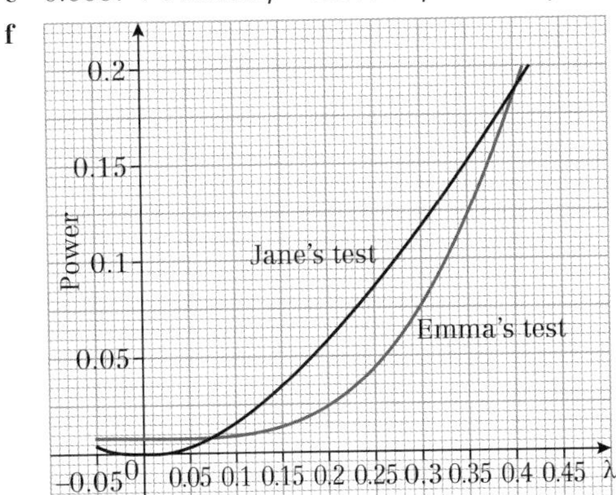

g The power of Jane's test is greater than that of Emma's when $0.1 < q < 0.4$

Review exercise 2

1 a $p = 0.1$

b $r = 28.5$ (1 d.p.), $s = 100 - 91 = 9.0$ (1 d.p.)

c t.s. > c.v. so reject H_0.
(significant result) binomial distribution is not a suitable model

d Defective items do not occur independently *or* not with constant probability.

2 a B(5, 0.5)

b Insufficient evidence to reject H_0.
B(5, 0.5) is a suitable model.
No evidence that coins are biased.

3 a $p = \frac{223}{1000} = 0.223$

b $r = 10.74$, $s = 30.20$, $t = 3.28$

c H_0: B(10, 0.2) is a suitable model for these data.
H_1: B(10, 0.2) is *not* a suitable model for these data.

d Since $t < 5$, the last two groups are combined and $v = 5 - 1 = 4$. Since there are then 5 cells and the parameter p is given.

e $4.17 < 9.488$ so not significant or do not reject null hypothesis.
The binomial distribution with $p = 0.2$ is a suitable model for the number of cuttings that do not grow.

4 Critical value $\chi_3^2(5\%) = 5.991$
from Poisson 5.47 is not in the critical region so accept H_0. Number of computer failures per day can be modelled by a Poisson distribution.

5 a Reject H_0.
Conclude there is evidence of an association between Mathematics and English grades.

b May have some expected frequencies <5 (and hence need to pool rows/columns).

6 $3.841 > 2.59$. There is insufficient evidence to reject H_0.
There is no association between a person's gender and their acceptance of the offer of a flu jab.

7 $14.19 > 9.210$ so significant result or reject null hypothesis.
There is evidence of an association between course taken and gender.

8 $3.47619 < 9.488$
There is no evidence of association between treatment and length of survival.

9 $2.4446 < 5.991$
so insufficient evidence to reject H_0.
No association between age and colour preference.

10 $3.2 < 15.1$ therefore no evidence to suggest it is not uniformly distributed

11 a 24.8, 14.88, 8.928, 5.3568, 8.0352

b H_0: $X \sim$ Geo(0.4) is a suitable model;
H_1: $X \sim$ Geo(0.4) is not a suitable model.

c $12.5047 < 13.3$ therefore no evidence to suggest the model is not suitable.

d Reject null hypothesis since critical value < test statistic.

12 a $\frac{1}{36}$ b $\frac{1}{3}$

13 a Geo(0.24) b 0.0462 c $\frac{0.24t}{1 - 0.76t}$

d $(0.76 + 0.24t)^{15}$ e $\left(\frac{0.24t}{1 - 0.76t}\right)^4$

14 $\mathrm{P}(X = x) = \frac{e^{-1.7} \times 1.7^x}{x!}$

$$G_x(t) = \sum \mathrm{P}(X = x)t^x$$
$$= \sum \frac{e^{-1.7} \times 1.7^x}{x!} t^x$$
$$= e^{-1.7} \sum \frac{(1.7t)^x}{x!}$$
$$= e^{-1.7}\left(1 + 1.7t + \frac{(1.7t)^2}{2!} + \frac{(1.7t)^3}{3!} + \ldots\right)$$
$$= e^{-1.7} e^{1.7t}$$
$$= e^{1.7(t-1)}$$

15 a 1; 1.225 (4 s.f.)

b i $\frac{4}{9}$ ii $\frac{8}{27}$

16 a $\frac{1}{16}$ b $\frac{3}{8}$

c $E(X) = G_x'(1) = \frac{1}{8}(27 + 12 + 1) = 5$
$\mathrm{Var}(X) = G_x''(1) + G_x'(1) - (G_x'(1))^2 = \frac{3}{8}$

d $\frac{3}{2}$; 6.5

17 a $G_x(1) = 1 \Rightarrow k(1 + 4 + 2)^2 = 49k = 1 \Rightarrow k = \frac{1}{49}$

b $\frac{8}{49}$

c $E(X) = G_x'(1) = \frac{2}{49} \times 1(2 + 4 + 1)(6 + 8 + 1) = \frac{30}{7}$
$\mathrm{Var}(X) = \frac{40}{49}$

d $t^{-2}(t^3 + 4t^6 + 2t^9)^2$

18 a i Type I – H_0 rejected when it is true
ii Type II – H_0 is accepted when it is false
b $H_0 : \lambda = 5 \quad H_1 : \lambda > 5$
$P(X \geqslant 7 \mid \lambda = 5) = 1 - 0.7622 = 0.2378 > 0.05$
No evidence of an increase in the number of chicks reared per year.
c $P(X \geqslant c \mid \lambda = 5) < 0.05$
$P(X \geqslant 9) = 0.0681$, $P(X \geqslant 10) = 0.0318$, $c = 10$
P(Type I error) = 0.0318
d $\lambda = 8$
$P(X \leqslant 9 \mid \lambda = 8) = 0.7166$

19 a $\dfrac{\bar{X} - 250}{\frac{4}{\sqrt{15}}} > 2.3263$ or $\dfrac{\bar{X} - 250}{\frac{4}{\sqrt{15}}} < -2.3263$
$\bar{X} > 252.40...$ or $\bar{X} < 247.6...$

b $(PX < 252.4 \mid \mu = 254) - P(X < 247.6 \mid \mu = 254)$
$= P\left(Z < \dfrac{252.4 - 254}{\frac{4}{\sqrt{15}}}\right) - P\left(Z < \dfrac{247.6 - 254}{\frac{4}{\sqrt{15}}}\right)$
$= P(Z < -1.5492) - P(Z < -6.20)$
$= (1 - 0.9394) - (1 - 1)$
$= 0.0606$

20 a $P(X \leqslant c_1) \leqslant 0.05$; $P(X \leqslant 3 \mid \lambda = 8) = 0.0424 \Rightarrow X \leqslant 3$
$P(X \geqslant c_2) \leqslant 0.05$; $P(X \geqslant 14 \mid \lambda = 8) = 0.0342$
$P(X \geqslant 13 \mid \lambda = 8) = 0.0638 \Rightarrow X \geqslant 13$
$\therefore$ critical region is $\{X \leqslant 3\} \cup \{X \geqslant 13\}$
b i $P(4 \leqslant X \leqslant 12 \mid \lambda = 10) = P(X \leqslant 12) - P(X \leqslant 3)$
$= 0.7916 - 0.0103$
$= 0.7813$
ii Power $= 1 - 0.7813 = 0.2187$

21 a $1 - 0.8891 = 0.1109$
b Power of test $= 1 - P(X \leqslant 2 \mid X \sim B(12, p))$
$= 1 - P(X = 0) + P(X = 1) + P(X = 2)$
$= 1 - (1 - p)^{12} + 12p(1 - p)^{11} + 66p^2(1 - p)^{10}$
$= 1 - (1 - p)^{10}(1 + 10p + 55p^2)$
c $1 - 0.5583 = 0.442$
$1 - 0.00281 = 0.997$
d The test is more discriminating (powerful) for the larger value of p.

22 a i The size of a test is the probability of rejecting the null hypothesis when it is in fact true and this is equal to the probability of a Type I error.
ii The power of a test is the probability of rejecting the null hypothesis when it is not true.
Power = 1 – P(Type II error) = P(being in the critical region when H_0 is false)
b $X \sim B(8, 0.25)$
Size $P(X > 6) = 1 - P(X \leqslant 6 \mid n = 8, p = 0.25)$
$= 1 - 0.9996 = 0.0004$
c Power of test $= P(X > 6 \mid X \sim B(8, p))$
$= P(X = 8) + P(X = 7)$
$= p^8 + 8p^7(1 - p)$
$= 8p^7 - 7p^8$
d Power $= 8(0.3^7) - 7(0.3^8) = 0.00129$
e Power = 1 – P(Type II)
P(Type II) = 1 – power
$= 1 - 0.00129$
$= 0.9087...$
f Increase the probability of a Type I error, e.g. increase the significance level of the test. Increase the value of p.

23 a 0.3311
b Power of test $= P(X \leqslant 2 \mid X \sim Po(\lambda))$
$(1 + P(3 \leqslant X \leqslant 4 \mid X \sim Po(\lambda))$
Leading to a power function of
$e^{-2\lambda}\left(1 + \lambda + \dfrac{\lambda^2}{2}\right)\left(e^{\lambda} + \dfrac{\lambda^3}{6} + \dfrac{\lambda^4}{24}\right)$
c 0.5891

24 a $H_0 : p = 0.35 \quad H_1 : p \neq 0.35$
b Let X = Number cured then $X \sim B(20, 0.35)$
α = P(Type I error) $= P(x \leqslant 3) + P(x \geqslant 11)$ given $p = 0.35$
$= 0.0444 + 0.0532$
$= 0.0976$
c β = P(Type II error) $= P(4 \leqslant x \leqslant 10)$

p	0.2	0.3	0.4	0.5
β	0.5880	0.8758	0.8565	0.5868

d Power $= 1 - \beta$
0.4120 0.1435
e Not a good procedure.
Better further away from 0.35 or this is not a very powerful test (power $= 1 - \beta$)

25 a 0.0226
b Power $= 1 - P(0) - P(1)$
$= 1 - (1 - p)^5 - 5p(1 - p)^4$
$= 1 - (1 - p)^4(1 + 4p)$
c 0.0115
d $a = 0.47$
e

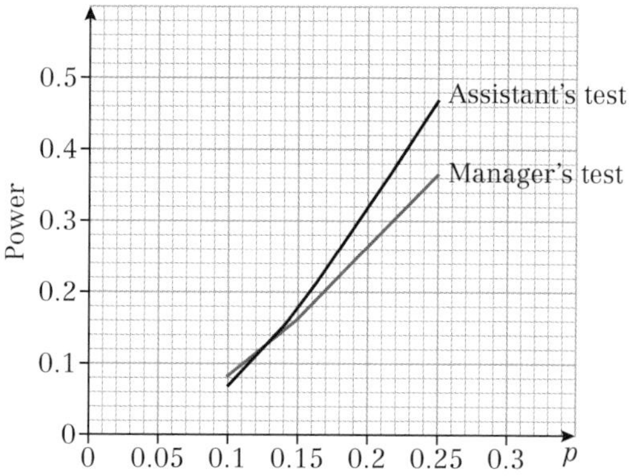

Power
0.5
0.4
0.3
0.2
0.1
0
0 0.05 0.1 0.15 0.2 0.25 0.3 p
Assistant's test
Manager's test

f The assistant's test as at $p > 0.2$ that test is more powerful.
g e.g. The manager believes the actual probability is close to 0.05, or that it would be more time or resource consuming to take larger samples.

Challenge

1 a $r = 19.15$, $s = 19.15$
b $12.12 < 15.086$ so accept H_0.
The distribution can be modelled by a $N \sim (360, 20)$.

2 a i $E(S) = G_S'(1)$
$G_S'(t) = \dfrac{d}{dt}(G_N(G_X(t))) = G_N'(G_X(t))G_X'(t)$
$G_S'(1) = G_N'(G_X(1))G_X'(1) = G_N'(1)\, G_X'(1) = E(N)\, E(X)$
ii $\text{Var}(S) = G_S''(1) + G_S'(1) - (G_S'(1))^2$
$= G_S''(1) + E(N)E(X) - E(N)^2 E(X)^2$
$G_S''(t) = \dfrac{d}{dt}(G_X'(t)\, G_N'(G_X(t)))$
$= G_X'(t)\, G_X'(t)\, G_N''(G_X(t)) + G_X''(t)\, G_N'(G_X(t))$
$G_S''(1) = E(X)^2 G_N''(1) + G_X''(1)\, E(N)$

$$\begin{aligned}\text{Var}(S) &= \text{E}(X)^2\text{G}_N''(1) + \text{G}_X''(1)\text{E}(N) + \text{E}(N)\text{E}(X) \\ &\quad - \text{E}(N)^2\text{E}(X)^2 \\ &= \text{E}(N)(\text{G}_X''(1) + \text{E}(X)) + \text{E}(X)^2(\text{G}_N''(1) - \text{E}(N)^2) \\ &= \text{E}(N)(\text{G}_X''(1) + \text{E}(X)) + \text{E}(X)^2(\text{G}_N''(1) - \text{E}(N)^2) \\ &\quad - \text{E}(N)\text{E}(X)^2 + \text{E}(N)\text{E}(X)^2 \\ &= \text{E}(N)(\text{G}_X''(1) + \text{E}(X) - \text{E}(X)^2) \\ &\quad + \text{E}(X)^2(\text{G}_N''(1) + \text{E}(N) - \text{E}(N)^2) \\ &= \text{E}(N)\text{Var}(X) + \text{E}(X)^2\text{Var}(N)\end{aligned}$$

b $N \sim \text{Po}(\lambda)$, so $\text{E}(N) = \text{Var}(N) = \lambda$,
$X \sim \text{B}(1, p)$, $\text{E}(X) = p$, $\text{Var}(X) = p(1 - p)$.
So $\text{E}(S) = \text{E}(N) \times \text{E}(X) = \lambda p$
$\text{Var}(S) = \text{E}(N)\text{Var}(X) + \text{E}(X)^2\text{Var}(N)$
$= \lambda p(1 - p) + p^2\lambda = \lambda p - \lambda p^2 + \lambda p^2 = \lambda p$
Trials are independent and occur randomly over a fixed interval.
$\text{E}(S) = \text{Var}(S)$, so $S \sim \text{Po}(\lambda p)$

c $N \sim B(n, q)$ $\text{E}(N) = nq$ $\text{Var}(N) = nq(1 - q)$
$\text{E}(X) = p$, $\text{Var}(X) = p(1 - p)$
$\text{E}(S) = npq$ $\text{Var}(S) = npq(1 - pq)$
Trails are independent, fixed number of trials, probability of success constant.
$S \sim \text{B}(n, pq)$

d **i** mean = £1575 σ = £227 (nearest £)
ii mean = £420 σ = £108 (nearest £)

Exam-style practice: AS level

1 **a** 0.12; 0.43 **b** 1.8924 **c** 0.78

2 **a** 0.0424 **b** 0.5940 **c** 0.5520

3 **a** H_0: There is no association between sport and gender.
H_1: There is an association between sport and gender.
b 6.150 **c** 2
d Fail to reject H_0 – the critical value for X^2 is 7.378 which is > test statistic, therefore not significant.
e Reject H_0 since new critical value is less than the test statistic (5.99).

4 **a** 3.75; 3.73125 **b** 0.5162
c n is large and p is small, thus mean ≈ variance

5 **a** 0.565
b H_0: Binomial is a suitable model; H_1: Binomial is not a suitable model
X^2 test statistic = 4.89 (2 d.p.), critical value = 6.25 (3 degrees of freedom)
Fail to reject H_0; there is evidence to suggest that the binomial model is suitable.

Exam-style practice: A level

1 H_0: Geo(0.4) is a good model; H_1: Geo(0.4) is not a good model
Test statistic = 14.87, critical value = 9.488 so reject null hypothesis; Geo(0.4) is not a good model

2 **a** $\text{G}_X(1) = 1 \Rightarrow k(1 + 2 + 3)3 = 1 \Rightarrow k = \frac{1}{216}$
b $\frac{7}{72}$
c $\text{G}_X'(1) = 4$, $\text{G}_X''(1) = \frac{41}{3}$
$\text{Var}(X) = \text{G}_X''(1) + \text{G}_X'(1) - (\text{G}_X'(1))^2$
$= \frac{41}{3} + 4 - 16 = \frac{5}{3}$
d $\text{G}_Y = \frac{t^3}{216}(1 + 2(t^2) + 3(t^2)^2)^3 = \frac{t^3}{216}(1 + 2t^2 + 3t^4)^3$

3 **a** $Y \sim$ Negative binomial(r, p); Throws are independent and probability is constant.
b 0.35 **c** 7

4 **a** 0.0255
b Probability (0.0780) is greater than significance level so not enough evidence at the 5% significance level to say number of flaws has been reduced.

5 **a** 0.3, 0.2, 0.05 **b** 0.3

6 **a** 15; 14.55
b Number of penalty kicks is large; probability of missing is small.
c 0.2511
d Probability using binomial = 0.2485 which agrees to 2 s.f. therefore accurate.

7 **a** $X \leqslant 1$; 0.0629
b Power of test $= \text{P}(X = 0) + \text{P}(X = 1)$
$= (1 - p)^{25} + 25p(1 - p)^{24}$
$= (1 - p)^{24}(1 + 24p)$
c 0.1122; $(1 - p)^{12}$
d Philip's test has a smaller size therefore better.
Philip's test has a greater power therefore better (Philip = 0.3286, Gemma = 0.3225).

Index

actual significance level 147
alternative hypotheses 59, 66, 92

binomial distribution
  degrees of freedom 105
  expected value/mean 32–3
  negative *see* negative binomial distribution
  parameter 67
  Poisson distribution as approximation to 34–5, 60
  probability generating function 132
  probability mass function 32
  testing as model 105–8
  variance 32–3

cells 96
central limit theorem 76–84
  applied to normal distribution 77–8
  applying to other distributions 80–1
  definition 77
chi-squared distribution 97–102
  critical regions 99
  critical value 98, 99
  degrees of freedom 97
  hypothesis testing 99–102
constant average rate 24
constraints 96
contingency tables 113–16
  degrees of freedom 114–15
  expected frequency 114
  selecting model 114
  setting hypotheses 114
continuous distributions 154
critical regions
  chi-squared distribution 99
  geometric distribution 69–71
  Poisson distribution 62–4
critical values 62, 98, 99
cumulative distribution functions 21–2
  geometric 44–5
  Poisson 59
cumulative distribution tables 21

degrees of freedom 96–7
  binomial distribution 105
  chi-squared distribution 97
  contingency tables 114–15
  discrete uniform distribution 104
  Poisson distribution 109
dice rolling 44, 49–50, 92–3
discrete data, testing goodness of fit with 103–10
discrete random variables 1–18
  expected value 2–3
  probability generating function 129–30
  solving problems involving 11–13
  variance 5–6
discrete uniform distribution
  degrees of freedom 104
  testing as model 104–5
distributions
  binomial *see* binomial distribution
  chi-squared *see* chi-squared distribution
  continuous 154
  geometric *see* geometric distribution
  normal *see* normal distribution
  Poisson *see* Poisson distribution

expected value (mean)
  binomial distribution 32–3
  discrete random variable 2–3
  finding by differentiating p.g.f. 135–6
  of function of X 7–10
  geometric distribution 47
  negative binomial distribution 52–3
  Poisson distribution 30, 59–60
  of X^2 3
exponential function, as infinite series expansion 20

geometric distribution 43–7
  central limit theorem applied to 62
  critical regions 69–71
  cumulative 44–5
  expected value/mean 47
  goodness of fit tests 119–20
  hypothesis testing 66–8, 69–71
  parameter 66–8
  probability function 44
  probability generating function 133
  variance 47
goodness of fit 92–4
  geometric distributions 119–20
  testing with discrete data 103–10

hypothesis formation 92–4
hypothesis testing 58–75
  alternative hypotheses 59, 66, 92
  chi-squared distribution 99–102
  comparing tests 164
  geometric distribution 66–8, 69–71
  null hypotheses 59, 66, 92
  one-tailed tests 59, 62, 99
  Poisson distribution 59–60, 62–4
  power function 162–5
  power of test 157–60
  quality of tests 146–72
  size of test 157–60
  two-tailed tests 60, 62, 64
  types of error *see* Type I errors; Type II errors

Maclaurin expansion 20
mean *see* expected value modelling, with Poisson distribution 23–4

negative binomial distribution 49–53
  central limit theorem applied to 80, 81
  expected value/mean 52–3
  and number of trials needed 49
  probability function 50
  probability generating function 133–4
  variance 52–3
normal distribution
  central limit theorem applied to 59–60
  finding Type I and Type II errors using 153–6
  sample mean approximately follows 77–78
null hypotheses 59, 66, 92

observed frequencies 96
one-tailed tests 59, 62, 99

parameters
  binomial distribution 67
  geometric distribution 66–8
  Poisson distribution 20, 23, 67
p.g.f. *see* probability generating functions
Poisson distribution 19–42
  adding 27–8
  as approximation to binomial distribution 34–5, 60
  central limit theorem applied to 80
  critical regions 62–4
  cumulative 59
  degrees of freedom 109
  expected value/mean 30, 59–60
  hypothesis testing 59–60, 62–4
  modelling with 23–4
  parameter 20, 23, 67
  probability generating function 132–3

testing as model 108–10
variance 30
power function 162–5
power of test 157–60
probability distributions *see* distributions
probability generating functions (p.g.f.) 128–45
binomial distribution 132
differentiating to find mean and variance 135–6
discrete random variables 129–30
as expectation of function of random variable 129
geometric distribution 133
$G_X(1) = 1$ property 129
negative binomial distribution 133–4
Poisson distribution 132–3
sums of independent random variables 139–40

random variables 2
discrete *see* discrete random variables
restrictions 96

sample mean 76–7
as normally distributed 77–8
sample variance 153
significance levels 60, 63, 99
actual 147
target 154
size of test 157–60
standard deviation 5, 47

target significance level 154
tests
one-tailed 59, 62, 99
two-tailed 60, 62, 64
see also goodness of fit; hypothesis testing
trials, number needed 49
two-tailed tests 60, 62, 64

Type I errors 147–51
finding using normal distribution 153–6
probability of 157–60
relationship with Type II errors 155–6
Type II errors 147–51
finding using normal distribution 153–6
probability of 157–60
relationship with Type I errors 155–6

variance
binomial distribution 32–3
discrete random variable 5–6
finding by differentiating p.g.f. 136
of function of X 7–9
geometric distribution 47
negative binomial distribution 52–3
Poisson distribution 30
sample 153